CARBON-13 NMR
SPECTRAL PROBLEMS

ORGANIC CHEMISTRY

CARBON-13 NMR SPECTRAL PROBLEMS · **Robert B. Bates** and **William A. Beavers** · *1981*

CARBON-13 NMR SPECTRAL PROBLEMS

Robert B. Bates

Department of Chemistry, University of Arizona, Tucson, Arizona

and

William A. Beavers

Texas Eastman Company, Longview, Texas

The HUMANA Press • Clifton, New Jersey

Library of Congress Cataloging in Publication Data

Main entry under title:
Bates, Robert Brown, 1933–
Carbon-13 NMR spectral problems.

(Organic chemistry)
Includes index.
1. Nuclear magnetic resonance spectroscopy.
2. Carbon—Isotopes. I. Title. II. Series.
QD96.N8B37 583′.0877 79-92216
ISBN 0-89063-010-5

Crescent Manor
P. O. Box 2148
Clifton, N. J. 07015

Printed in the United States of America

ISBN hard: 0-89603-010-5
soft: 0-89603-016-4

Preface

With the advent of Fourier transform spectrometers of great sensitivity, it has become practical to obtain carbon-13 nuclear magnetic resonance (C-13 NMR; ^{13}C NMR; CMR) spectra routinely on organic molecules, and this technique has become one of the highest utility in determining structures of organic unknowns. When the usual spectrometric techniques—proton magnetic resonance (H-1 NMR; ^{1}H NMR; PMR), infrared (IR), mass (MS), and ultraviolet (UV)—do not readily reveal a compound's structure, a C-13 NMR spectrum will often provide sufficient additional information to yield it unequivocally. With this in mind, the present work was designed to give advanced undergraduates, graduate students, and practicing chemists a working knowledge of and facility with the use of this valuable technique. Some familiarity with other spectrometric techniques is assumed (recommended book: Silverstein, Bassler, and Morrill, *Spectrometric Identification of Organic Compounds*), but no prior knowledge of C-13 NMR—which is treated very lightly, if at all, in the widely used elementary organic texts—is necessary.

A discussion of C-13 NMR spectroscopy is followed by 125 problems, each consisting of a molecular formula, two types of C-13 NMR spectra (partially and completely proton decoupled, with connecting lines to facilitate multiplicity assignments), an integrated H-1 NMR spectrum, and the most important IR, UV, and MS data. These problems have been very carefully prepared, thoroughly tested by students at the University of Arizona, and we believe that very few errors remain. Though the structures are, in general, more complex than those in problem books in which C-13 NMR is not used, with the combination of C-13 NMR and the other methods, the reader should be able in most cases to deduce the exact structure, and in those that prove refractory, to reduce the possible structures to just a few. Answers are given by reference to the literature to avoid inadvertant viewing of a structure. A reasonable amount of scratch space has been left on each page for the reader's use in solving the problems.

The problems are arranged roughly in order of increasing difficulty, with those in the first section having completely resolved C-13 NMR spectra in the sense that they contain no coincidental overlaps of carbon absorptions. If the molecular formula shows six carbons, and only five are visible in the C-13 NMR spectrum, then two carbons in the unknown absorb at the same location owing to symmetry in the molecule, rather than by coincidence.

The formula

$$C + 1 - \frac{1}{2}(H + X - N)$$

in which C = the number of carbons, X = the number of halogens, etc., can be used to obtain the sum of double bonds and rings from the molecular formula. For example, the first unknown, with molecular formula $C_6H_{12}O$, has 6 + 1 − ½(12) = 1 double bond or ring. The formula works for covalent compounds containing C, H, O, N, S, and X, but not for salts.

The compounds selected are of a variety of structural types, and include macrocycles, a polymer, transition metal complexes, and 16 elements. Be on the lookout for biologically important compounds: vitamins, amino acids, nucleic acid derivatives, hormones, and pharmaceuticals. Also, the "isoprene rule" may be helpful for those C_{10} compounds that are monoterpenes with carbon skeletons derived from

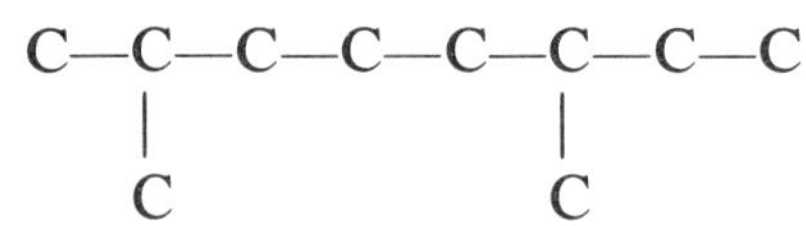

Finally, the tables of C-13 NMR shifts that appear inside the front and back covers merit special mention. In our view they are a considerable improvement on those currently available, and render the book valuable long after the problems have been worked through.

Acknowledgments

We thank Drs. S. K. Arora, R. S. Glass, V. J. Hruby, D. L. Lichtenberger, C. S. Marvel, and C. Steelink of the Department of Chemistry, Dr. M. A. Cusanovich of the Department of Biochemistry, and Dr. J. R. Cole of the College of Pharmacy of the University of Arizona for providing many of the compounds used, and C. A. Ogle and S. R. Walter for running the spectra of the last 25 compounds.

Table of Contents

Section 1

Section 2

Introduction to Carbon-13 NMR Spectral Problems

Because the natural abundance of the NMR-active carbon isotope ^{13}C is only 1%, and because, moreover, it gives an intrinsically weaker signal than does a proton, ^{1}H NMR spectroscopy developed before ^{13}C NMR spectroscopy. This does not mean that C-13 NMR spectra are less useful than ^{1}H NMR in the structure determination of organic compounds, for in fact C-13 NMR spectra provide certain important advantages:

1. C-13 NMR offers considerably better resolution, largely because the C-13 absorptions for most ordinary organic molecules are spread over 200 instead of 10 ppm.
2. Carbons bearing no protons are directly visible.
3. A count of the number of protons attached to each carbon results from comparison of the broad-band decoupled C-13 NMR spectrum with the off-resonance C-13 spectrum. Thus the number of methyl, methylene, methinyl, and quaternary carbons in a fairly complex molecule is far more readily determined by C-13 than by H-1 NMR.

There are certain disadvantages to C-13 NMR:

1. Larger sample size (up to 100 mg) or longer sampling time (up to several days) is sometimes necessary; however, if 100 mg of a sample is available and that sample has high solubility, the time requirement may be only a few minutes; a good spectrum may even be obtained on as little as a 1-mg sample when several days are available for scanning the sample.
2. Owing to variations in relaxation times and nuclear Overhauser effects (NOE), the areas of absorption for individual carbons vary considerably (up to a factor of about 10). Thus it is not as easy to tell relative numbers of carbons from C-13 as it is protons

from H-1 NMR; for this reason only the H-1 spectra are integrated in the problems below.

3. Protons attached to heteroatoms are not directly visible.

Thus, as with most new methods, C-13 NMR complements rather than replaces the earlier methods.

Types of C-13 NMR Spectra

Undecoupled Spectra

Figure 1A shows the undecoupled C-13 spectrum of vinyl acetate, with the methyl carbon absorbing approximately as a 1:3:3:1 quartet arising from splitting by three equivalent protons, the methylene as a 1:2:1 triplet, the methinyl as a 1:1 doublet, and the quaternary carbon as a singlet. Besides the large (~125 Hz) one-bond CH couplings, some much smaller (~5 Hz) two-bond CH couplings are observable. Nineteen hours of scanning were required to obtain this spectrum, and because undecoupled spectra take so long, they are rarely used.

Off-Resonance Spectra

Figure 1B is a corresponding "off-resonance" spectrum that required only 50 minutes to generate. This is a partially decoupled spectrum that still shows the one-bond couplings, but requires less time to take owing to nuclear Overhauser enhancement and also to sharpening of the peaks since the two-bond couplings are no longer resolved. In this type of spectrum the outer peaks in spin–spin multiplets are often weaker than expected, and if they do not occur above the noise level, it is impossible to distinguish singlets from triplets or doublets from quartets.

Broad-Band Decoupling

A "broad-band decoupled" spectrum, in which all coupling arising from protons is removed by irradiation throughout the proton region, is shown in Fig. 1C. Each carbon now absorbs as a sharp line, though because of its lack of attached protons to aid in relaxation, the

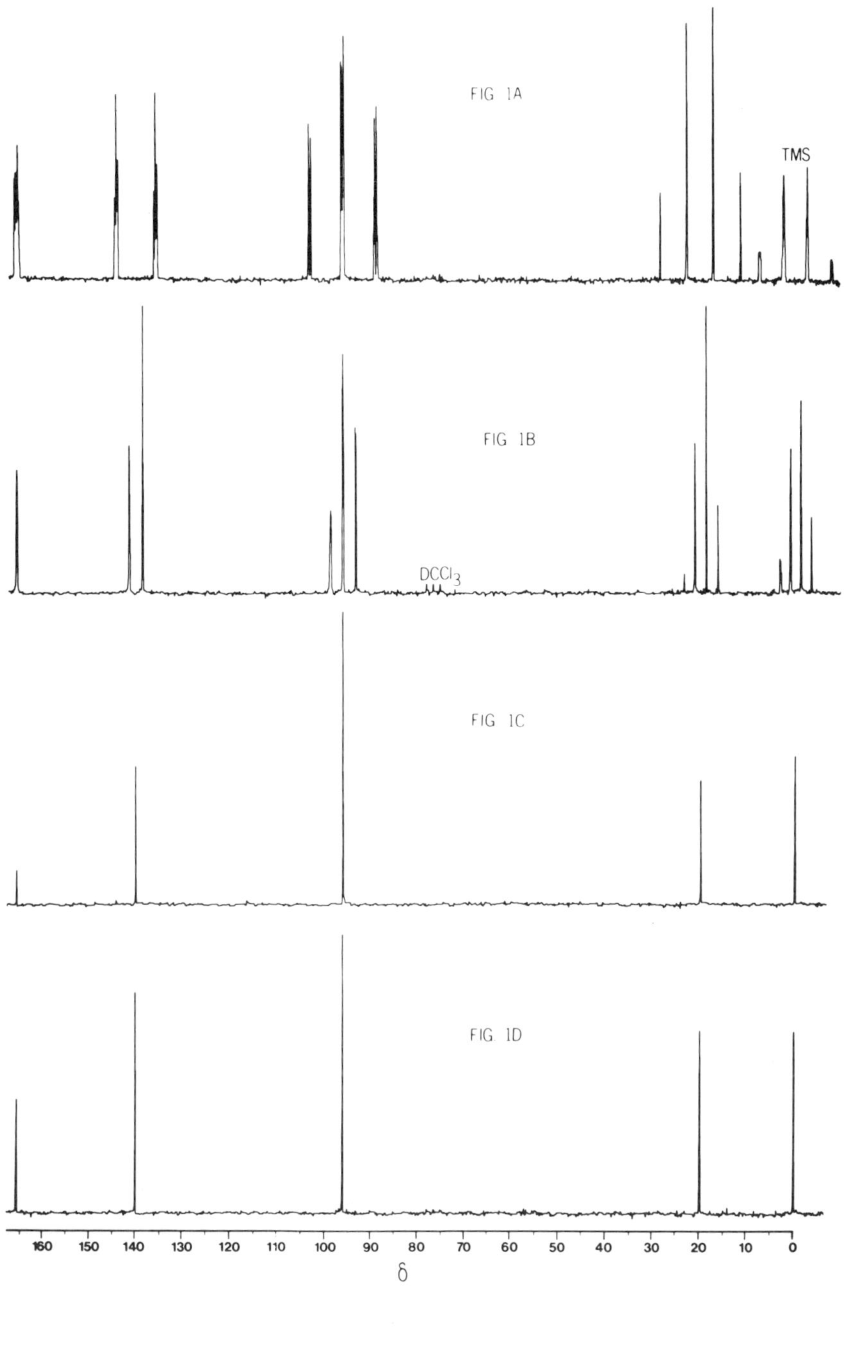
FIG 1A
TMS
FIG 1B
DCCl3
FIG 1C
FIG 1D
160
150
140
130
120
110
100
90
80
70
60
50
40
30
20
10
0
δ

quaternary carbon gives a much smaller peak than the other carbons. Although this weakness can help in distinguishing quaternary carbons, it hurts in cases where their absorption approximates the same magnitude as the noise, and a "time-delayed" broad-band decoupled spectrum (Fig. 1D) is often used to build quaternary absorptions. Each of these types of broad-band decoupled spectra required 10 minutes of scan time for this vinyl acetate sample.

In each of the problems below, a broad-band decoupled spectrum (with time delay when so indicated) and an off-resonance spectrum are given. The times required to obtain the spectra are not noted, but were generally less than those for vinyl acetate: the very rapid tumbling of such a small molecule lengthens the relaxation times (T_1 and T_2) for the various carbon atoms. It is possible, though time-consuming, to measure relaxation times for individual carbons in a molecule, thus gaining information about conformations and intramolecular rotations.

Carbon Chemical Shifts

Typical chemical shift ranges for many types of carbons are shown in the table inside the front cover, as are shifts of the most common solvents. Shifts for certain aliphatic and aromatic substances are given in the tables inside the back cover. In using these tables, it should be remembered that the values given are for *mono*functional compounds, and as in PMR, additivity effects should be taken into account if two or more functional groups are found close together in a molecule. For example, in the series CH_4—CH_3Cl—CH_2Cl_2—$CHCl_3$—CCl_4, the respective shifts are –3, 25, 54, 77, and 96; each chlorine therefore adds about 25 ppm. The additivity increment for a particular functional substituent is not always this close to being identical throughout a series, unfortunately, as is illustrated by the corresponding iodides, whose shifts are –3, –20, –54, –140, and –292! Still, approximate additivity parameters are better than none, and can be deduced for many common substituents from the values in the tables inside the back cover. For example, the shift of the methinyl carbon in lysine hydrochloride, $^+NH_3(CH_2)_4CH(NH_3^+)COOH \cdot 2\ Cl^-$, could be estimated by taking the shift value δ34 for hexanoic acid from the table, and adding the shift of 25.5 ppm caused by replacing a methyl hydrogen in pentane by NH_3^+ (39.5 – 14); the resulting estimate of δ59.5 is not far from the δ55.5 observed experimentally.

The values for aliphatic groupings in the two tables are, except for the cyclopropyl values, derived from acyclic models. They work somewhat less well with alicyclic than acyclic compounds, but are still useful with alicyclics. They were derived from relatively unhindered compounds, and in cases of highly hindered carbons, upfield "steric shifts" are observed. For example, the 1, 3, and 5 carbons in *axial* methylcyclohexane absorb 5 ppm upfield from the corresponding carbons in *equatorial* methylcyclohexane.

The carbon shifts in acyclic alkanes are well correlated (generally within 1 ppm) by the empirical equations:*

$$\delta\text{Methyl} = 6.80 + 9.56\beta_1 + 17.83\beta_2 + 25.48\beta_3 - 2.99\gamma + 0.49\Delta$$
$$\delta\text{Methylene} = 15.34 + 9.75\beta_1 + 16.70\beta_2 + 21.43\beta_3 - 2.69\gamma + 0.25\Delta$$
$$\delta\text{Methinyl} = 23.46 + 6.60\beta_1 + 11.14\beta_2 + 14.70\beta_3 - 2.07\gamma$$
$$\delta\text{Quaternary} = 27.77 + 2.26\beta_1 + 3.96\beta_2 + 7.33\beta_3 + 0.68\gamma$$

for carbons with β_1 α-methylene groups, β_2 α-methinyl groups, β_3 α-quaternary carbons, γ γ-carbon atoms, and Δ δ-carbon atoms. For example, for the methinyl carbon in $(CH_3)_3CCH_2CH(CH_3)_2$, δmethinyl = 23.46 + 6.60 – 3(2.07) = 23.9; δobserved = 24.9.

Thus α and β alkyl groups can have a considerable effect (~8–9 ppm for an α or β methyl group, and up to 25 ppm for an α *tert*-butyl group), and should be taken into account when using the aliphatic shift table to estimate the shift for a carbon such as that in the methylene group of $CH_3(CO)CH_2CH(CH_3)_2$, for which the table value of 43.5 for $R(CO)CH_2CH_2CH_2CH_2CH_3$ should be increased by 16.70 – 9.75 + 2.69 – 0.25 = 9.39 to give δ52.9 (δobserved = 52.7).

β-Groupings other than alkyls (even F and OH) generally have about the same effect as a methyl group on the carbon shift; e.g., the α-methylene carbon in $CH_3CH_2CH_2OCH_3$ (δestimated = 73.4) is a reasonable model for the methylene carbons in $CH_3OCH_2CH_2OCH_3$ (δobserved = 72.3).

A similar equation gives the sp^2 carbon shifts in simple alkenes:† $\delta = 123.3 + 10.6\alpha + 7.2\beta - 1.5\gamma - 7.9\alpha' - 1.8\beta' + 1.5\gamma' - 1.1$ (only if *cis* 1,2-disubstituted), in which 123.3 is the shift for ethylene, the Greek letters without primes are the totals of carbons attached to the carbon of interest, and the letters with primes are those attached to the other sp^2 carbon. For example, the sp^2 methylene in $CH_2{=}C(CH_3)CH_2C(CH_3)_3$ is calculated to absorb at 123.3 –

*L. P. Lindeman and J. Q. Adams, *Anal. Chem.* **43,** 1245 (1971).
†D. E. Dorman, M. Jautelat, and J. D. Roberts, *J. Org. Chem.* **36,** 2757 (1971).

7.9(2) – 1.8 + 1.5(3) = 110.2, and observed at 114.4. To illustrate the effects of polar substituents on alkene carbon shifts, some shifts for sp^2 carbons in CH_2=CHQ are:

Alkene Carbon sp^2 Shifts

Q	$\delta_{methylene}$	$\delta_{methinyl}$
H	123.3	123.3
CH_3	115.4	135.7
COOEt	130.4	129.7
CN	137.8	107.7
Cl	117.3	126.0
Br	122.0	115.5
I	130.4	85.3
OCH_3	84.1	153.2
O-*t*-Bu	90.2	146.8
OAc	96.3	141.6
$SO_2CH=CH_2$	131.3	137.7

Coupling Constants

^{13}C—H one-bond couplings depend on the hybridization of the carbon, with the usual ranges sp^3 125–149, sp^2 156–222, and *sp* 248–269 Hz. The highest values in the range are for cases with a polar group attached. Two-bond and higher ^{13}C—H couplings are usually < 20 Hz, except in C(C=O)H (~30 Hz) and C≡C—H (~50 Hz). A ^{13}C—D coupling can be approximated by dividing the corresponding ^{13}C—H coupling by 7. Couplings in Hertz between ^{13}C and other elements are indicated in the accompanying table.

	$—(CH_2)_n—Z$			$\phi—Z$			
	1 bond	2 bond	3 bond	1 bond	Ortho (=2 bond)	Meta (=3 bond)	Para (=4 bond)
$^{13}C—H$	+125	−5	0	+158	+1	+7	−1
$^{13}C—F$	−167	+20	+5	−245	+21	+8	+3
$^{13}C—P$	−11	+12	13	12	20	7	0
$^{13}C—P^+$	+ 48	−4	15	88	11	13	3
$^{13}C—Hg$	+656	−26	270	1186	88	102	18

Recording Conditions

CMR spectra were recorded at 22.6 MHz using a Bruker WH-90 spectrometer. When the solubility of the unknown permitted, $DCCl_3$ was used as a solvent and tetramethylsilane (TMS) as internal standard; when some other solvent and/or standard was used, this is indicated on the spectrum. Solvent and standard peaks are labeled "S," except for the TMS peak at $\delta 0$. For reading $^{13}C—F$ and $^{13}C—P$ coupling constants, use 22.6 Hz/ppm; $^{13}C—H$ coupling constants cannot be determined well since partial or total $^{13}C—H$ decoupling was employed for all spectra except in Fig. 1A.

H-1 NMR spectra were obtained at 60 MHz with a Varian T-60 instrument, generally on the same solution used for the C-13 NMR spectra. The δ scale is again used for chemical shifts. Solvent and standard pcaks are again labeled "S," except for the TMS or sodium 2,2-dimethyl-2-silapentane-5-sulfonate (DSS) peaks; the latter standard, used when D_2O is the solvent, has the slight disadvantage of having, in addition to its methyl proton peak at 0.0, weak absorptions (visible in the PMR spectrum in problem 2-58) for its methylenes at $\delta 0.5$, 1.8, and 3.0. Peaks for OH and NH protons, which readily exchange with D_2O, are labeled "E"; in D_2O, they have usually completely equilibrated with solvent and what is seen is mainly HOD. The number of exchangeable hydrogens, obtainable by integration in which allowance is made for the HOD present as an impurity in the D_2O used, is indicated on the spectra when it is not easily determined from the integration.

Mass spectra were not experimentally obtained, but the location of the molecular ion peak (preceded by "m") is given followed by "s" if expected to be unusually strong or "w" if it would be especially weak or absent. The locations of some of the expected fragment peaks are also given. For amine hydrochlorides (used for the NMR spectra because of their high water solubility), the mass spectrum of the free base is reported.

Only peaks in the 1600–3700 cm^{-1} region of the infrared spectrum are listed; the CH stretching absorptions present near 3000 cm^{-1} in virtually all of the unknowns are omitted, as are the many peaks in the 500–1600 cm^{-1} or "fingerprint" portion of the spectrum. Measured locations on the actual compound are given in most cases, but in a few, expected locations from closely related model compounds are used.

Many of the unknowns are complex aromatics, and although their ultraviolet spectra are characteristic of chromophoric systems, it is difficult to deduce the nature and substitution pattern of aromatic groups from the UV spectrum alone without a knowledge of the absorption patterns of closely related model compounds. Because the other spectral methods generally serve better to provide this structural information, these substances are simply listed in the problems as having "strong absorption" in the UV. When a number is given, it represents λ_{max} in nanometers. "None" is used to indicate no $\lambda_{max} > 210$ nm.

Solution to Problem 1-1

The broad-band decoupled C-13 NMR spectrum given here has 5 peaks (other than TMS), while the molecular formula indicates 6 carbons; one of the peaks must represent 2 carbons, and since in the first section there are no unknowns with coincidental shift overlaps, two carbons must absorb at the same location because of the molecular symmetry. Comparison of the broad-band decoupled spectrum with the off-resonance spectrum shows the peak at $\delta 79$ to result from a quaternary carbon; $\delta 66$, 38, and 26 to be for methylenes; and $\delta 27.5$ for methyl(s). The largest peak by far is the methyl peak, suggesting the presence of two methyl groups; this is confirmed by comparing the total number of hydrogens from the molecular formula (12) with the total for 3 methylenes and 2 methyls (noting also the absence of protons attached to atoms other than carbon by the absence of —OH absorption at 3000–3700 cm^{-1} in the IR and/or exchangeable protons—labeled "E"—in proton NMR).

Next, the methylene at $\delta 66$ and the quaternary carbon absorb in a region suggesting (table inside front cover) attachment to a heteroatom, which must be oxygen from the molecular formula. Since there is only one oxygen, the compound must be an ether with the partial formula

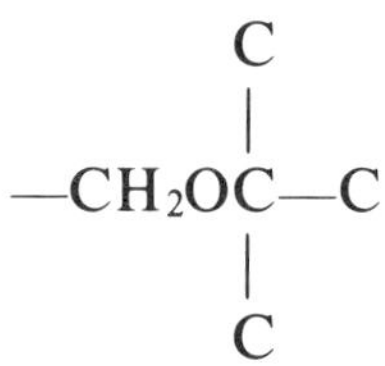

The only possible structure with two symmetry-equivalent methyl groups is

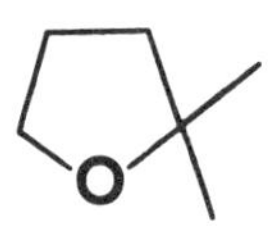

and there is supporting evidence. In the ^{1}H NMR spectrum, the peak for two equivalent methyl groups at $\delta 1.2$, the triplet at $\delta 3.7$ for the methylene next to oxygen, and the complex absorption at $\sim\delta 1.8$ for the other two methylenes are strongly confirmatory, as is the mass spectral peak at 85 (m − 15) for the loss of a methyl group.

In this relatively easy case, the structure might have been deduced from the H-1 NMR evidence and confirmed using the C-13 NMR.

Section 1

Resolved Carbon-13 NMR Spectra

Problem 1-1

1-1	MS:	m100, 85
$C_6H_{12}O$	IR:	-
	UV:	none

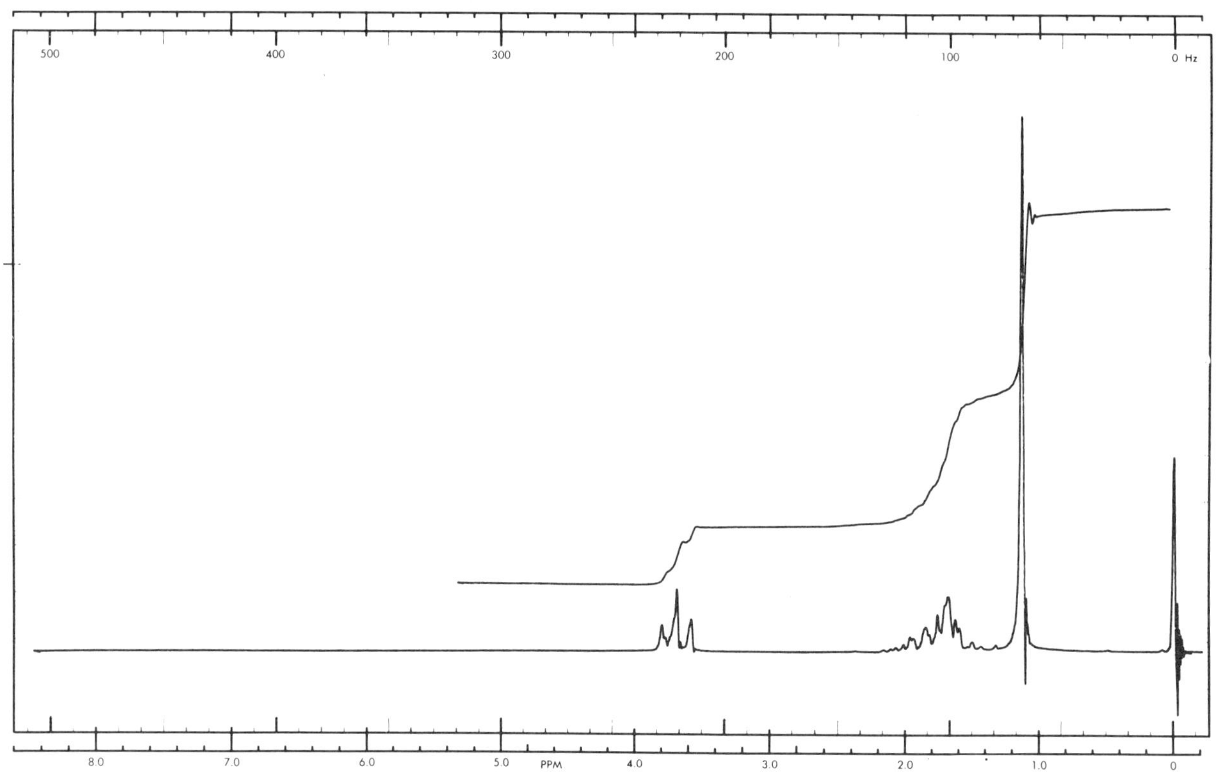

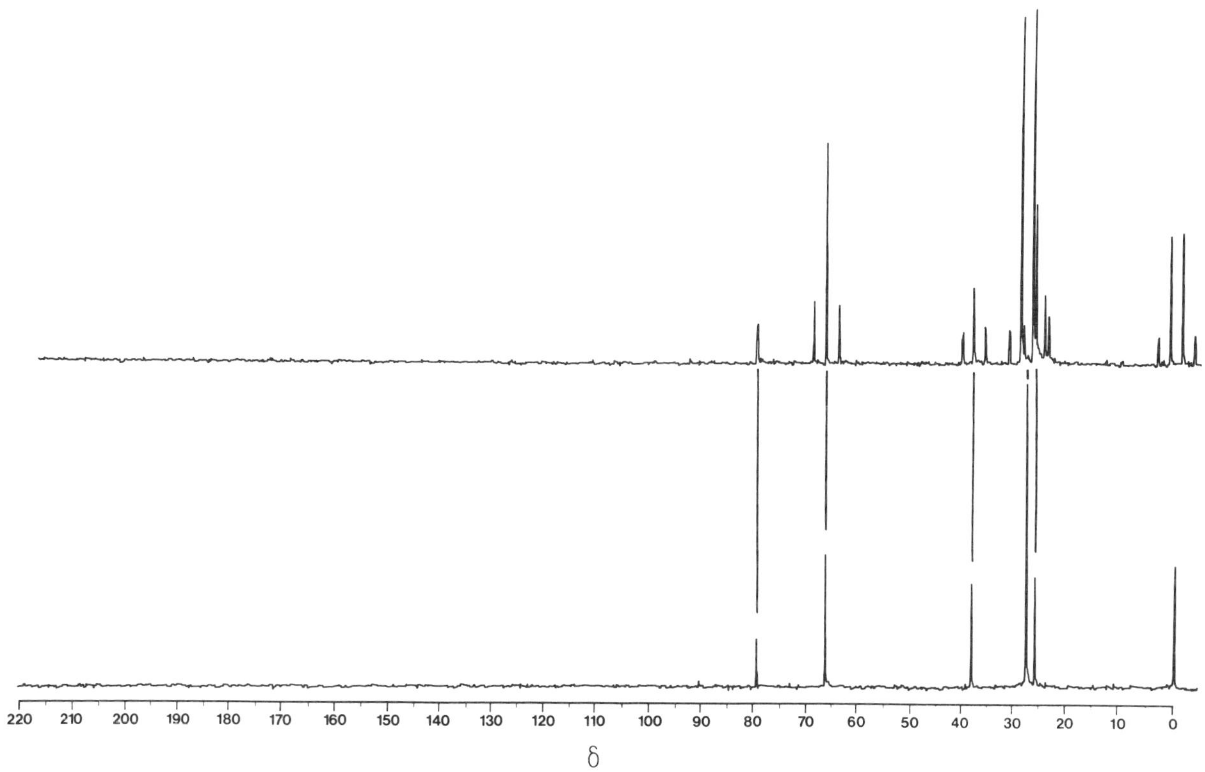
220
210
200
190
180
170
160
150
140
130
120
110
100
90
80
70
60
50
40
30
20
10
0
δ

Problem 1-2

1-2

$C_7H_8O_2 \cdot H_2O$

MS: m124(s), 96, 95

IR: 3300 cm^{-1}

UV: Strong absorption

PMR and CMR: d_6-Me_2CO + TMS

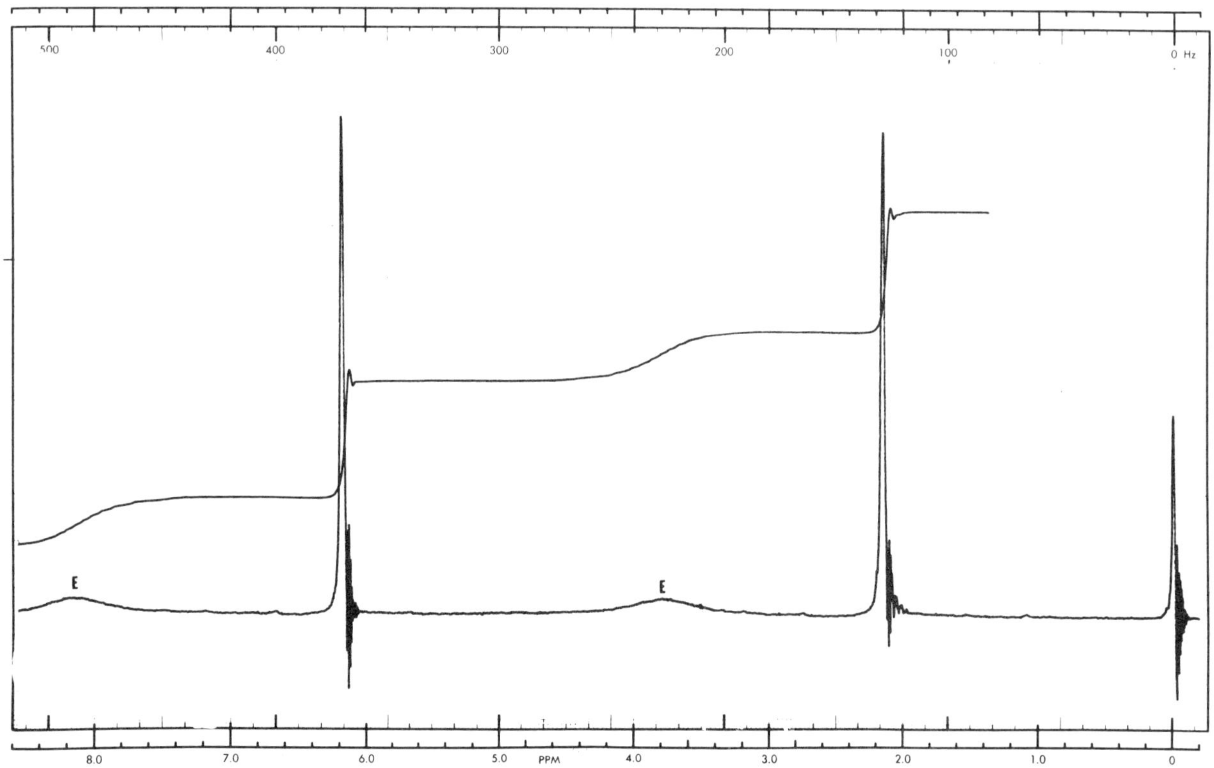

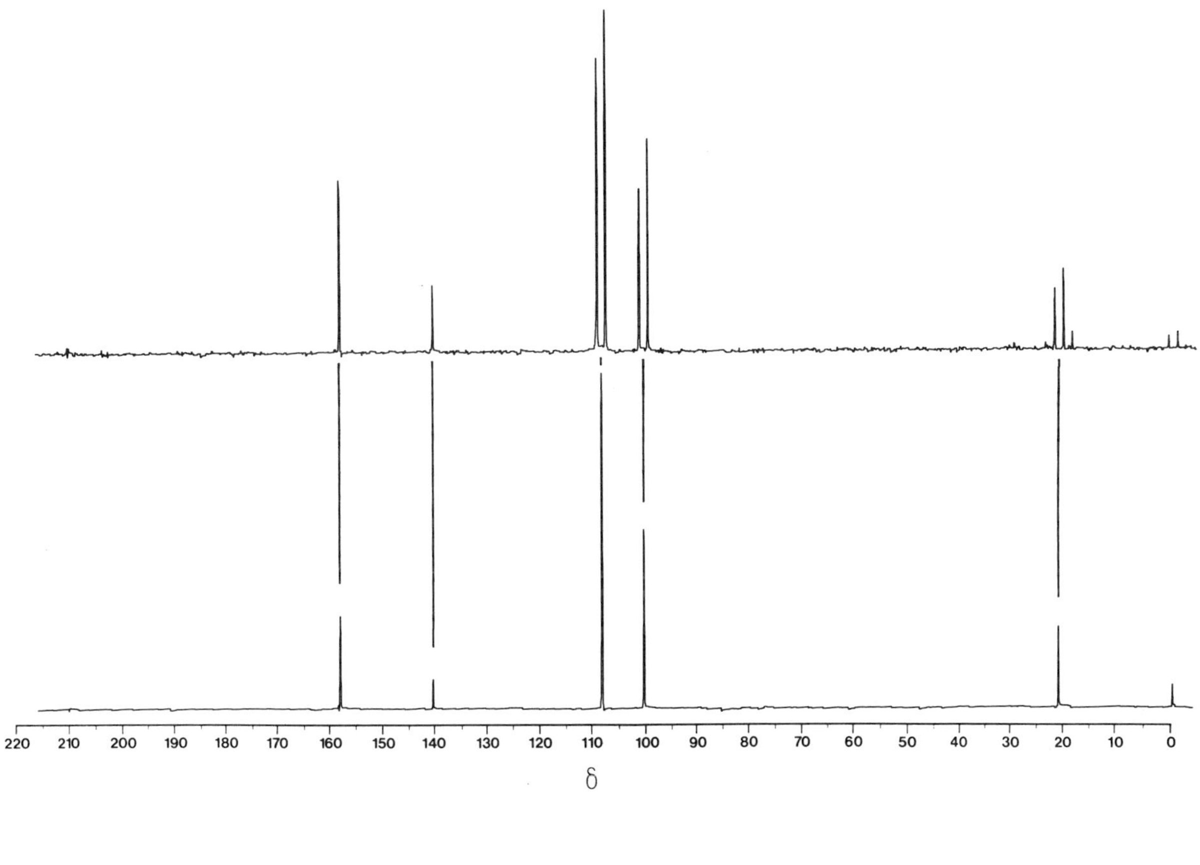
220 210 200 190 180 170 160 150 140 130 120 110 100 90 80 70 60 50 40 30 20 10 0
δ

1-3 MS: m104, 86, 31

$C_5H_{12}O_2$ IR: 3330 cm^{-1}

UV: none

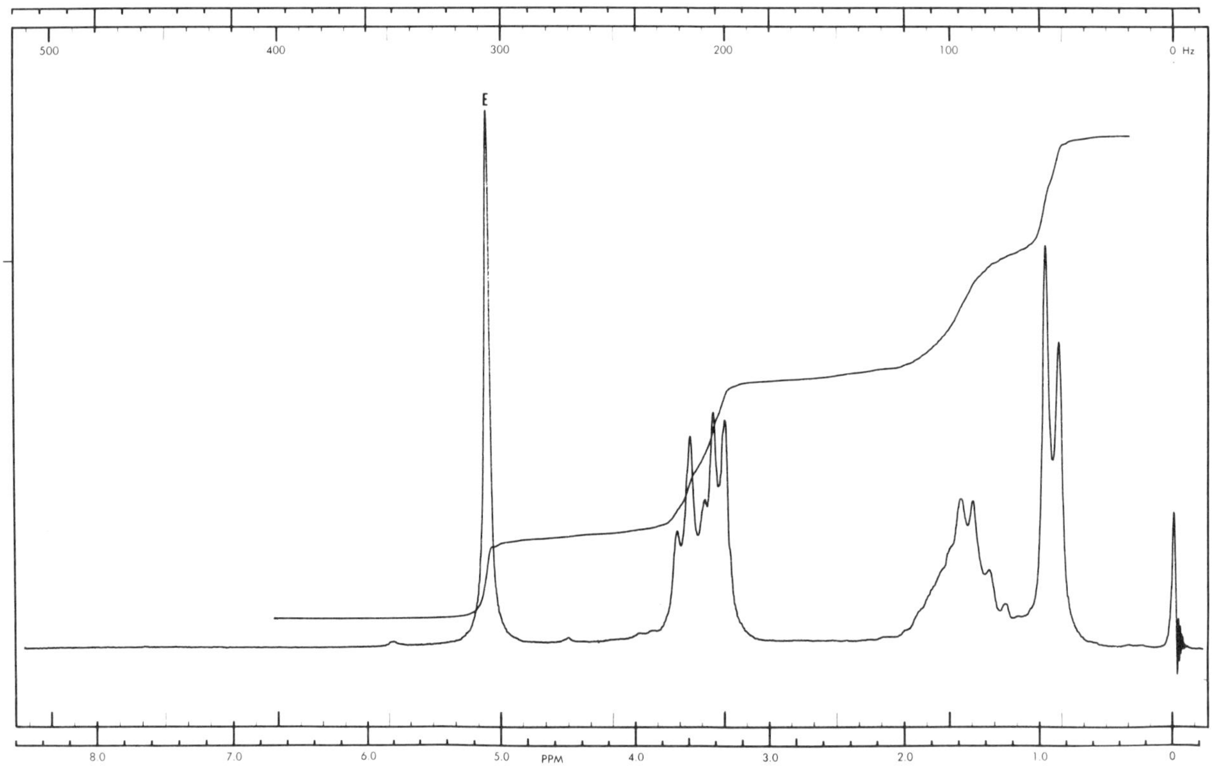

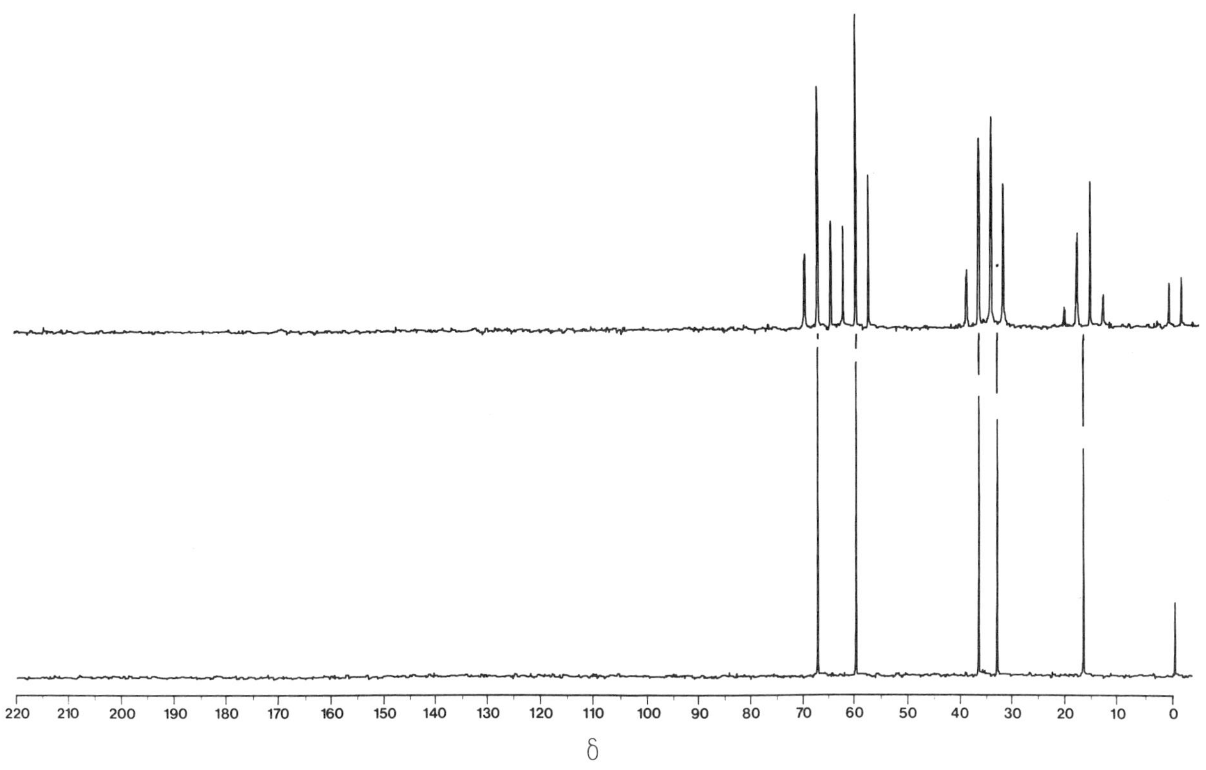
220
210
200
190
180
170
160
150
140
130
120
110
100
90
80
70
60
50
40
30
20
10
0
δ

Problem 1-4

1-4	MS:	m112(s), 84, 28
$C_6H_{12}N_2$	IR:	-
	UV:	none
	PMR:	D_2O + DSS
	CMR:	D_2O + dioxane

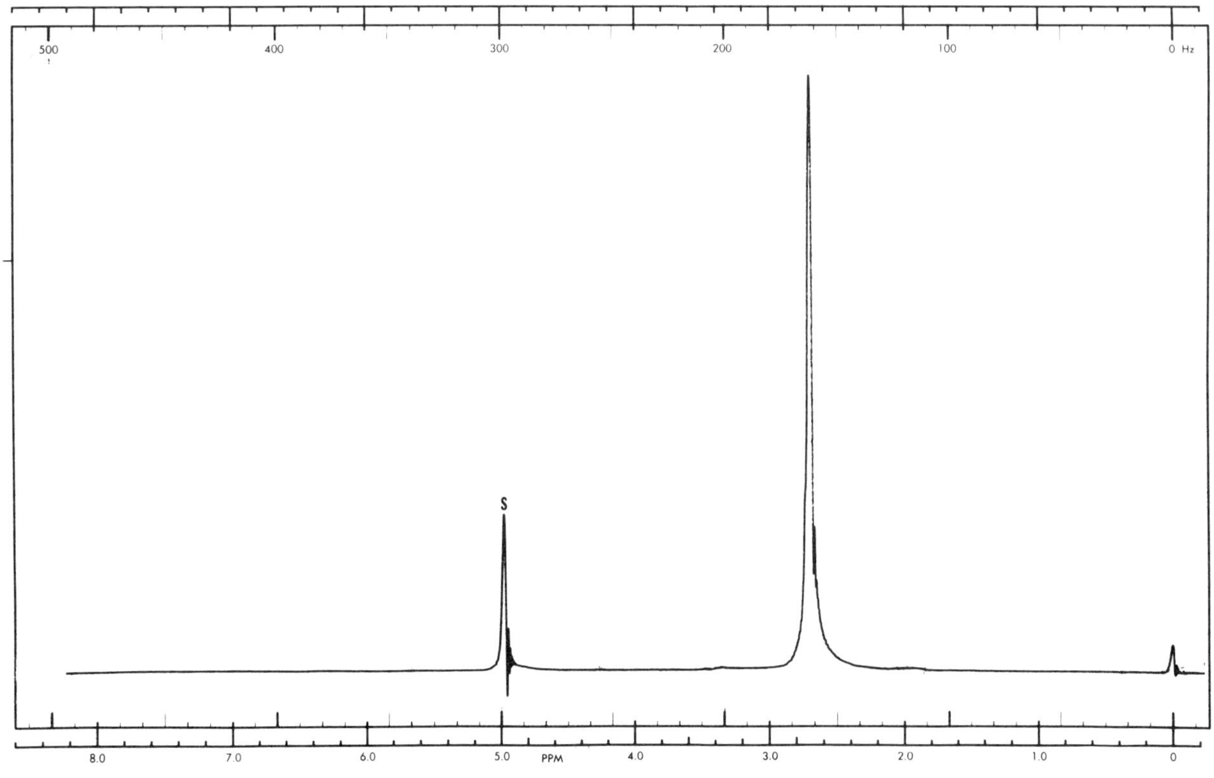

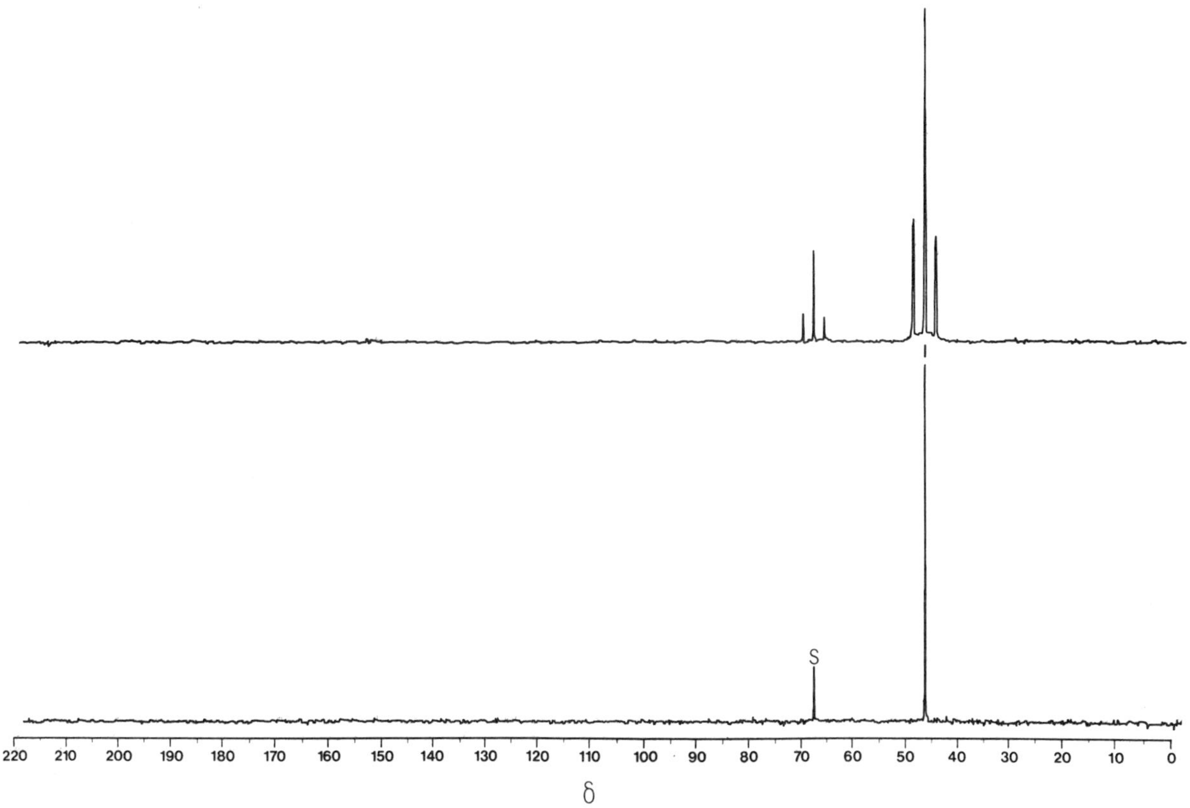
S
220 210 200 190 180 170 160 150 140 130 120 110 100 90 80 70 60 50 40 30 20 10 0
δ

Problem 1-5

1-5

$C_8H_{12}N_2$

MS: m136(s), 121, 95

IR: -

UV: Strong absorption

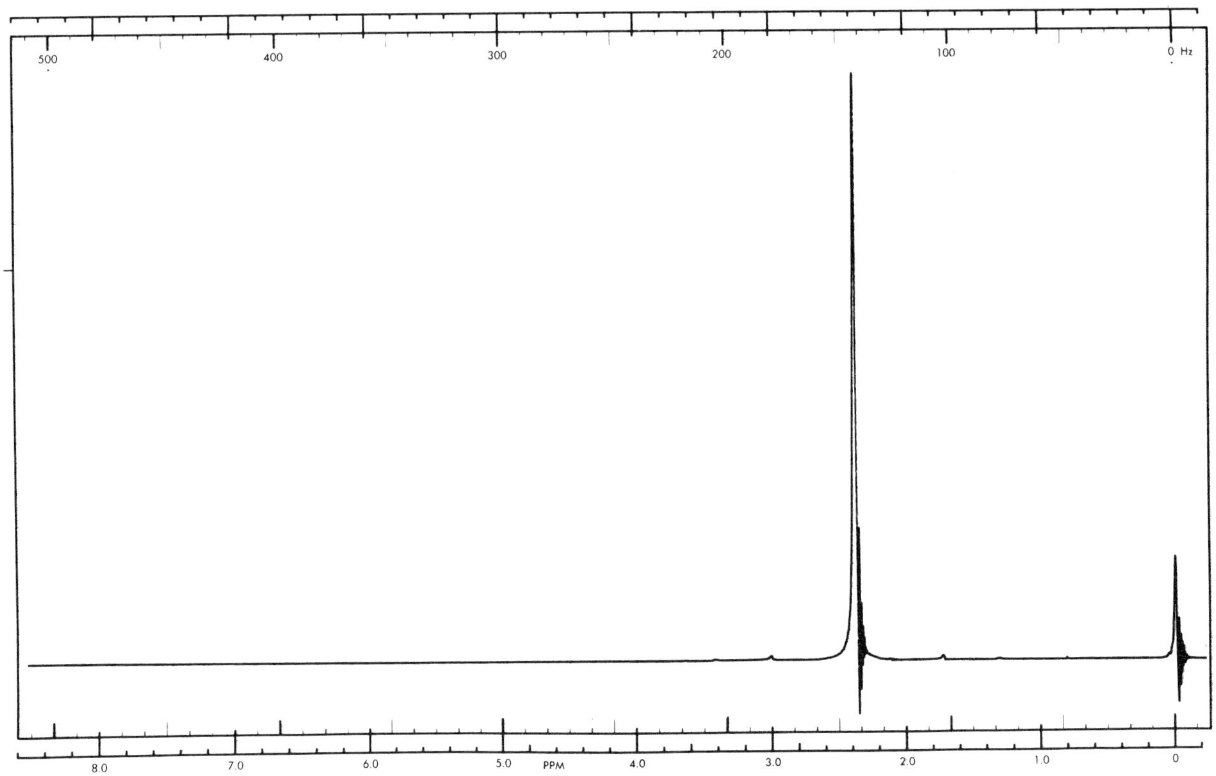

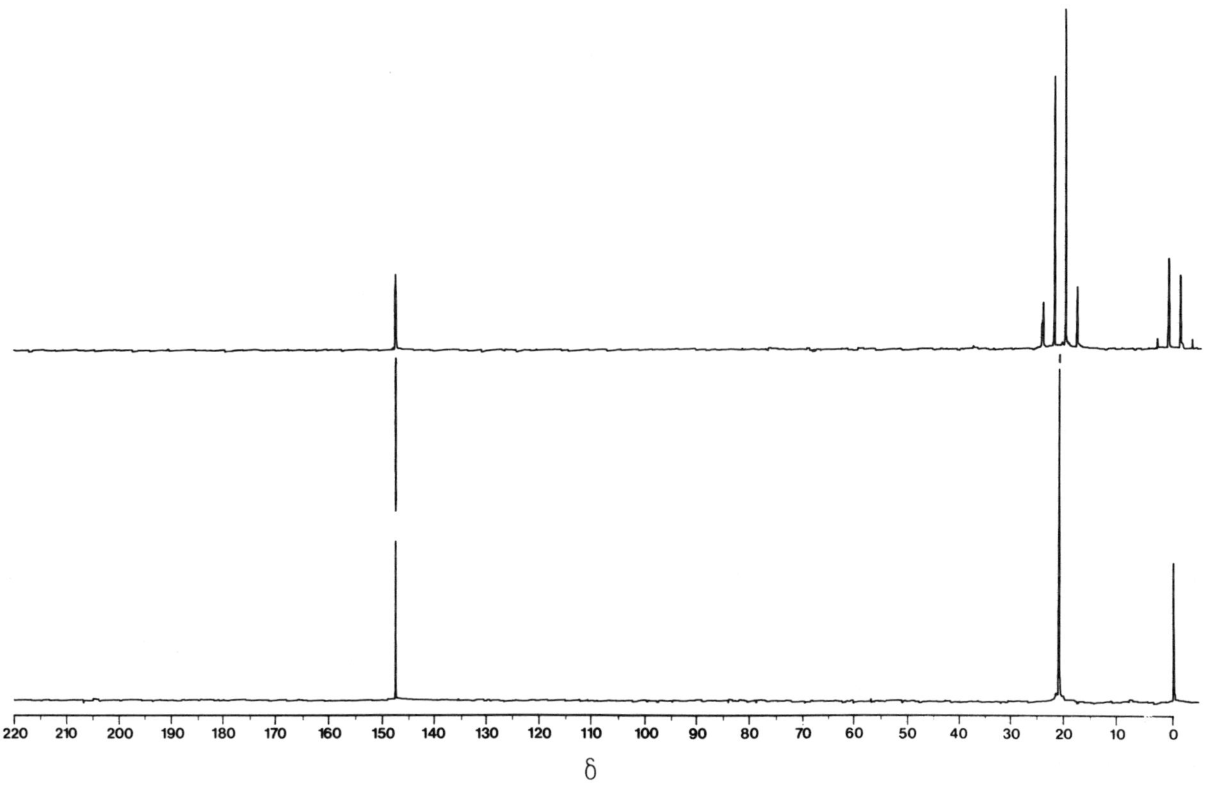
220 210 200 190 180 170 160 150 140 130 120 110 100 90 80 70 60 50 40 30 20 10 0
δ

Problem 1-6

1-6

$C_{10}H_{14}O_2Si$

MS: m194, 179, 177, 149, 76

IR: 2900 (broad), 1690 cm^{-1}

UV: Strong absorption

PMR: Upper scan offset 300 Hz

CMR: 5 sec delay

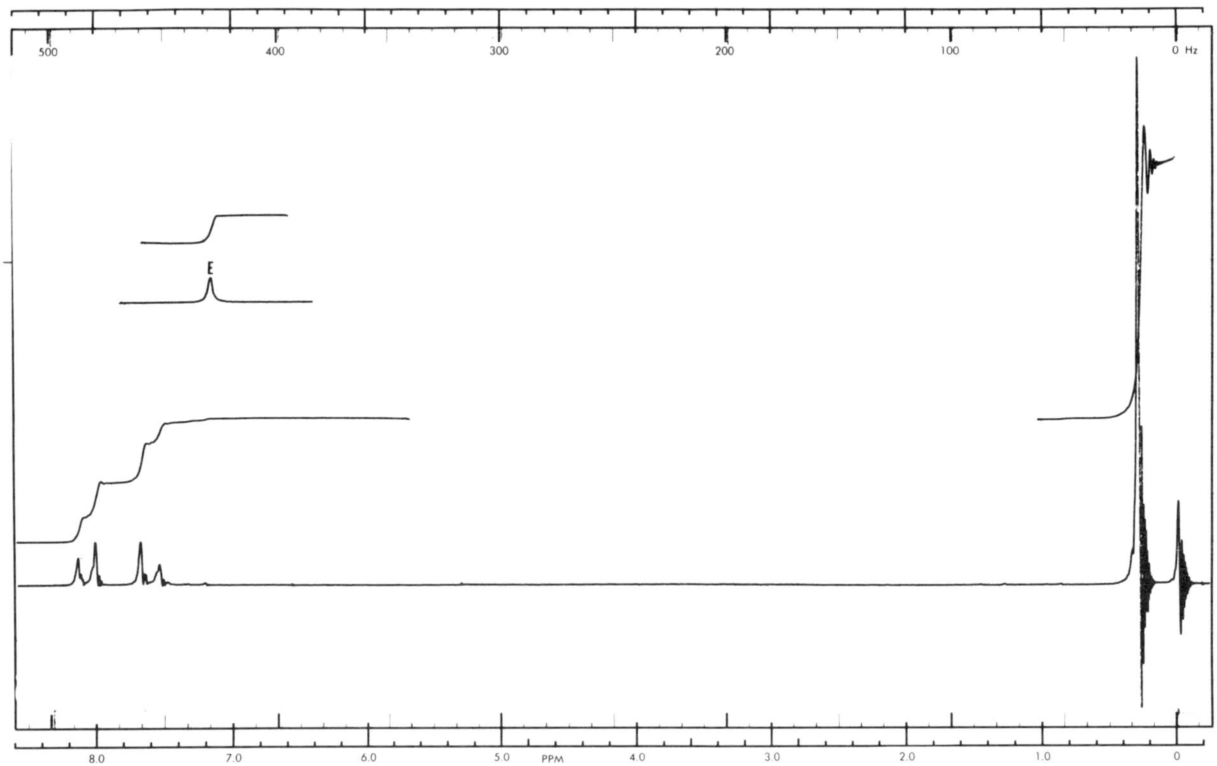

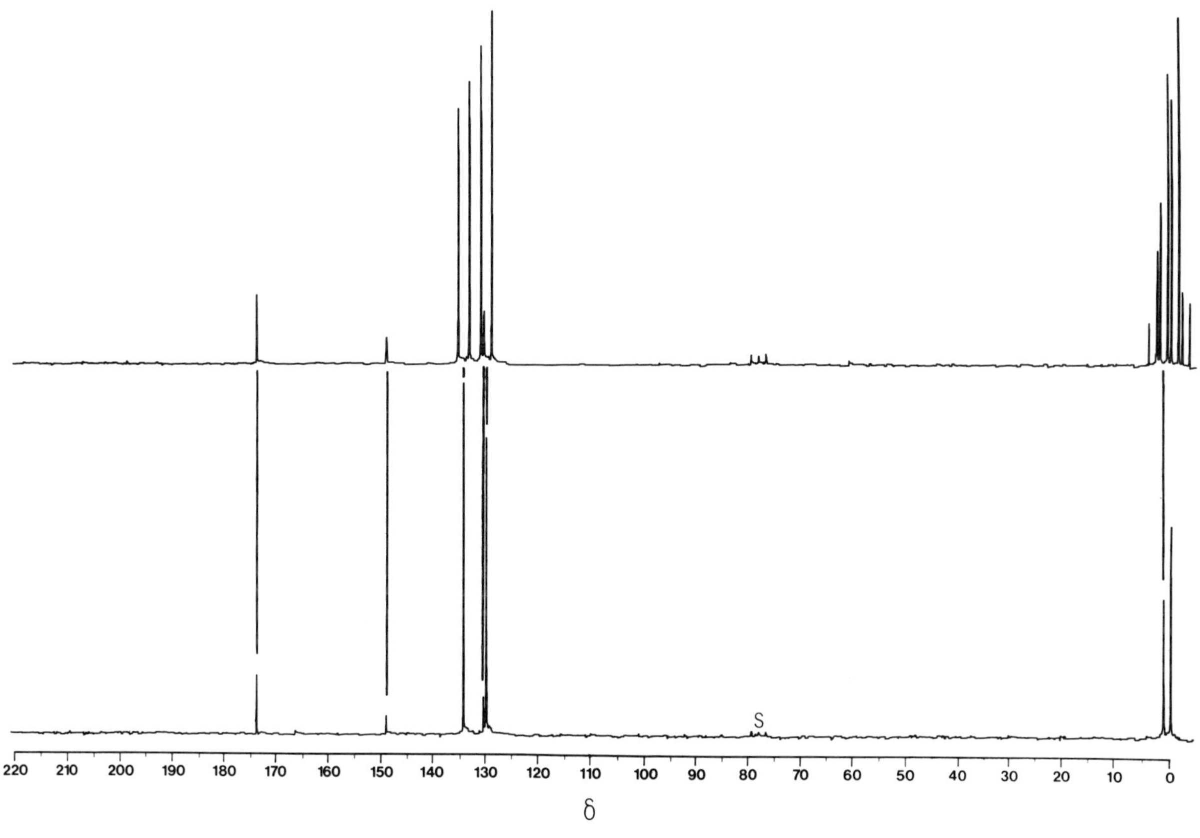
S
220
210
200
190
180
170
160
150
140
130
120
110
100
90
80
70
60
50
40
30
20
10
0
δ

Problem 1-7

1-7

$C_6H_6O_2$

MS: m110(s), 82

IR: 3410, 3280 cm^{-1}

UV: Strong absorption

PMR and CMR: d_6-Me_2CO + TMS

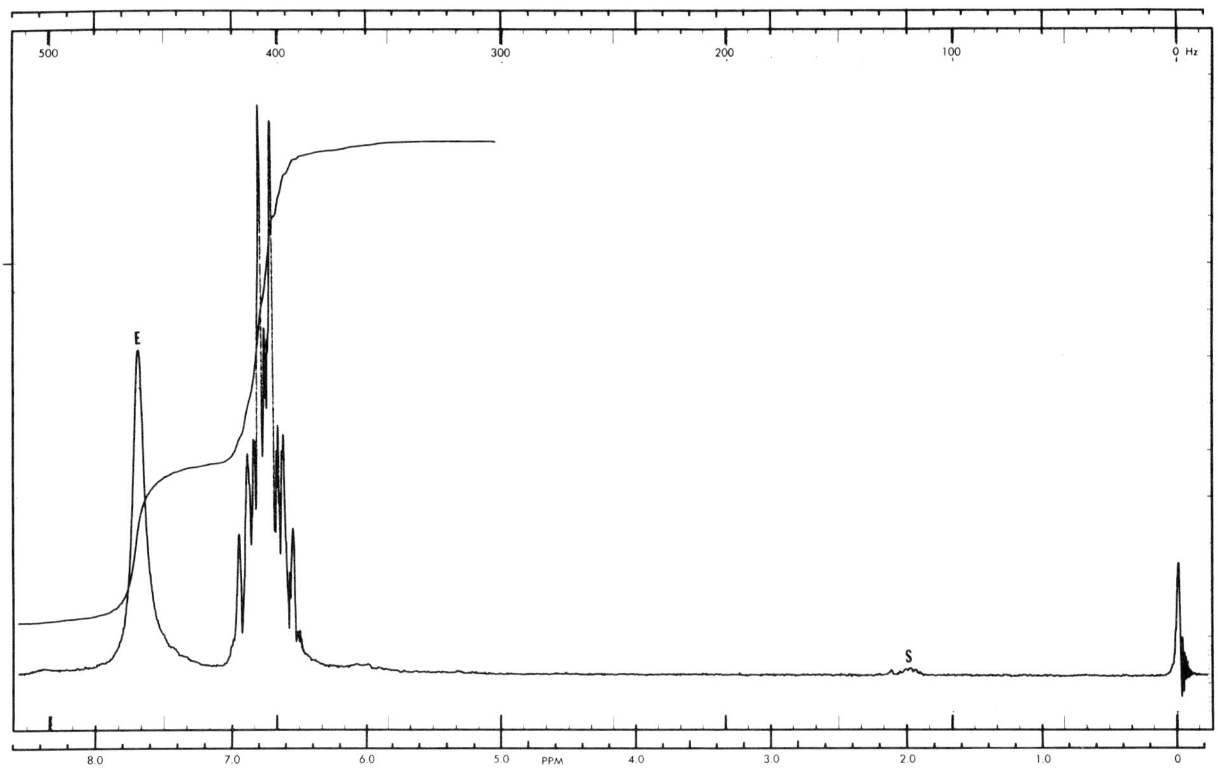

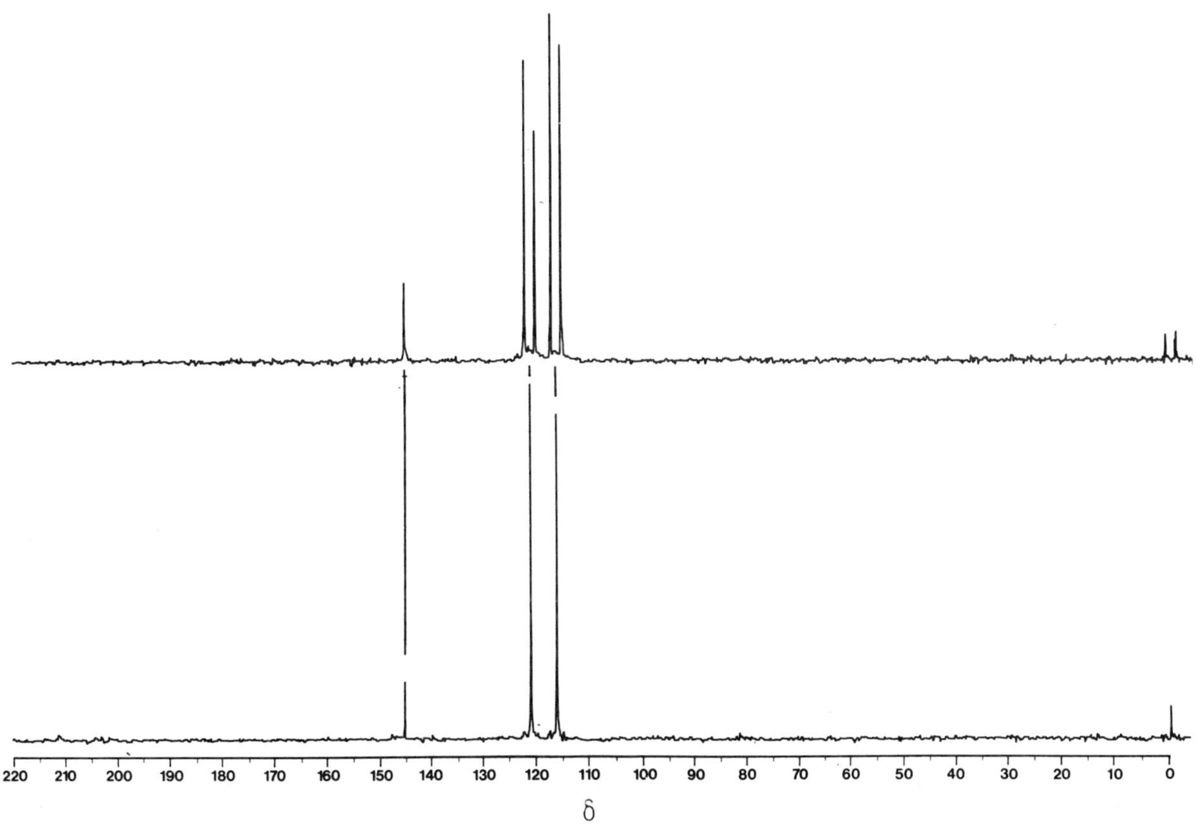

220 210 200 190 180 170 160 150 140 130 120 110 100 90 80 70 60 50 40 30 20 10 0
δ

Problem 1-8

1-8
$C_8H_{16}O_4$

MS: m176(s), 132, 88, 44
IR: -
UV: none

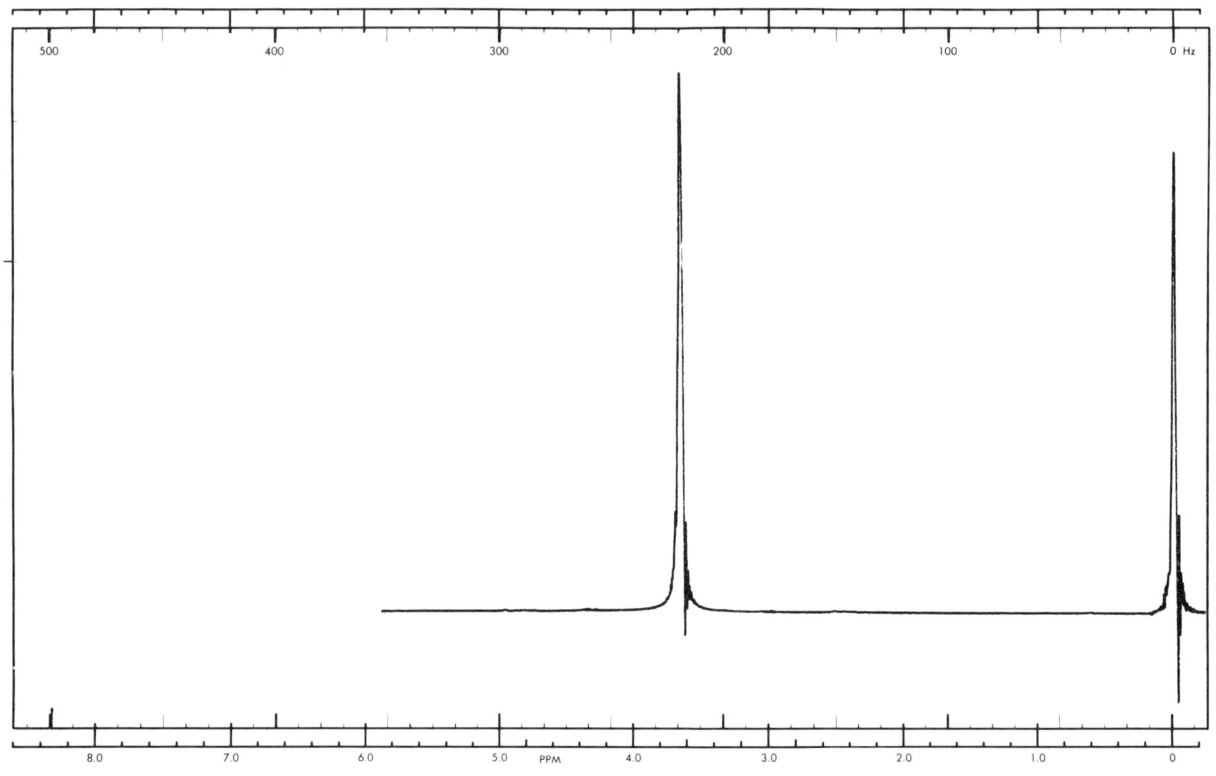

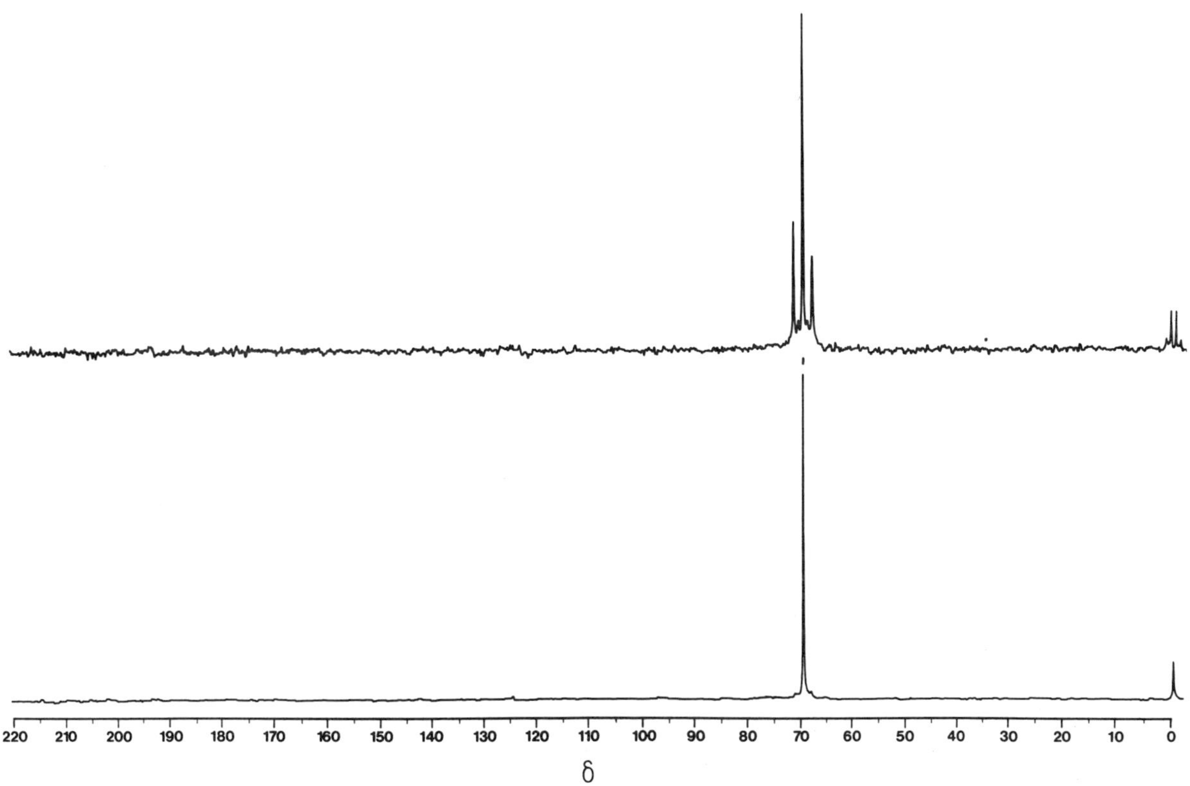
220 210 200 190 180 170 160 150 140 130 120 110 100 90 80 70 60 50 40 30 20 10 0
δ

Problem 1-9

1-9	MS:	m264(w), 242, 241, 240, 239, 238, 237, 236, 235, 233, 231, 229, 29
$C_7H_5Cl_5$	IR:	-
	UV:	308 (ε 1800)

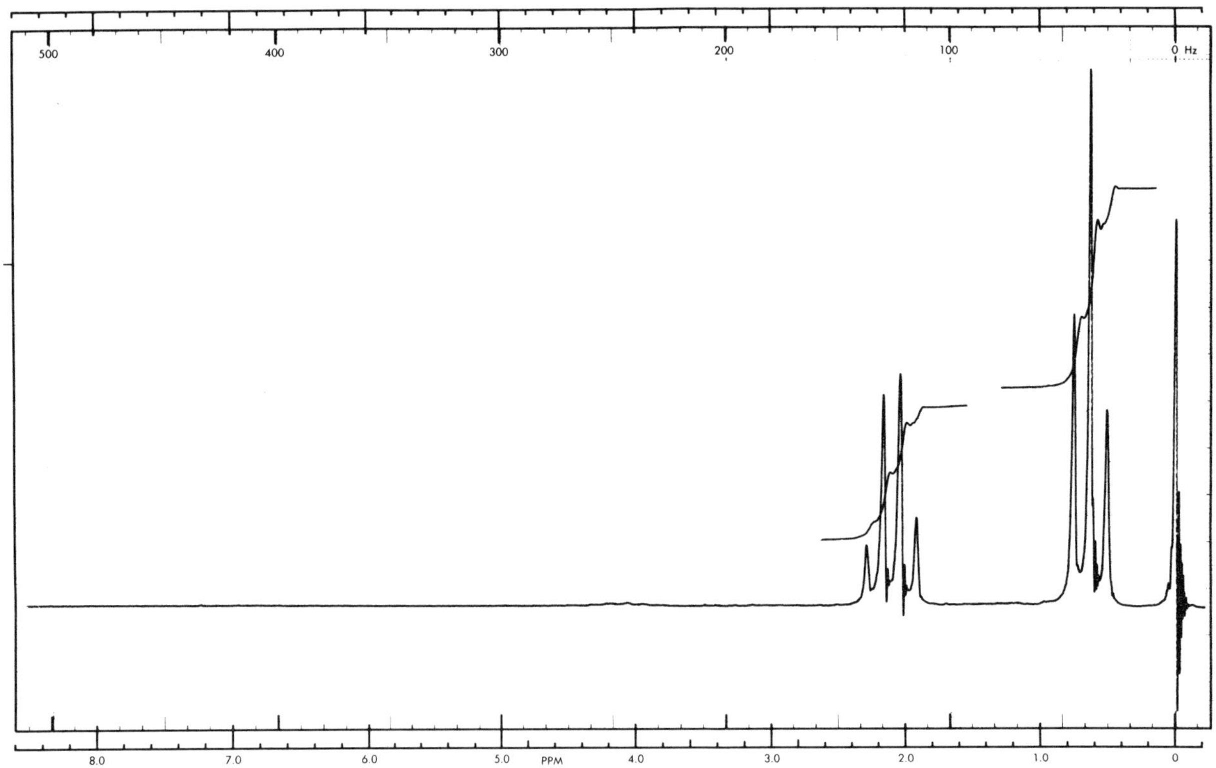

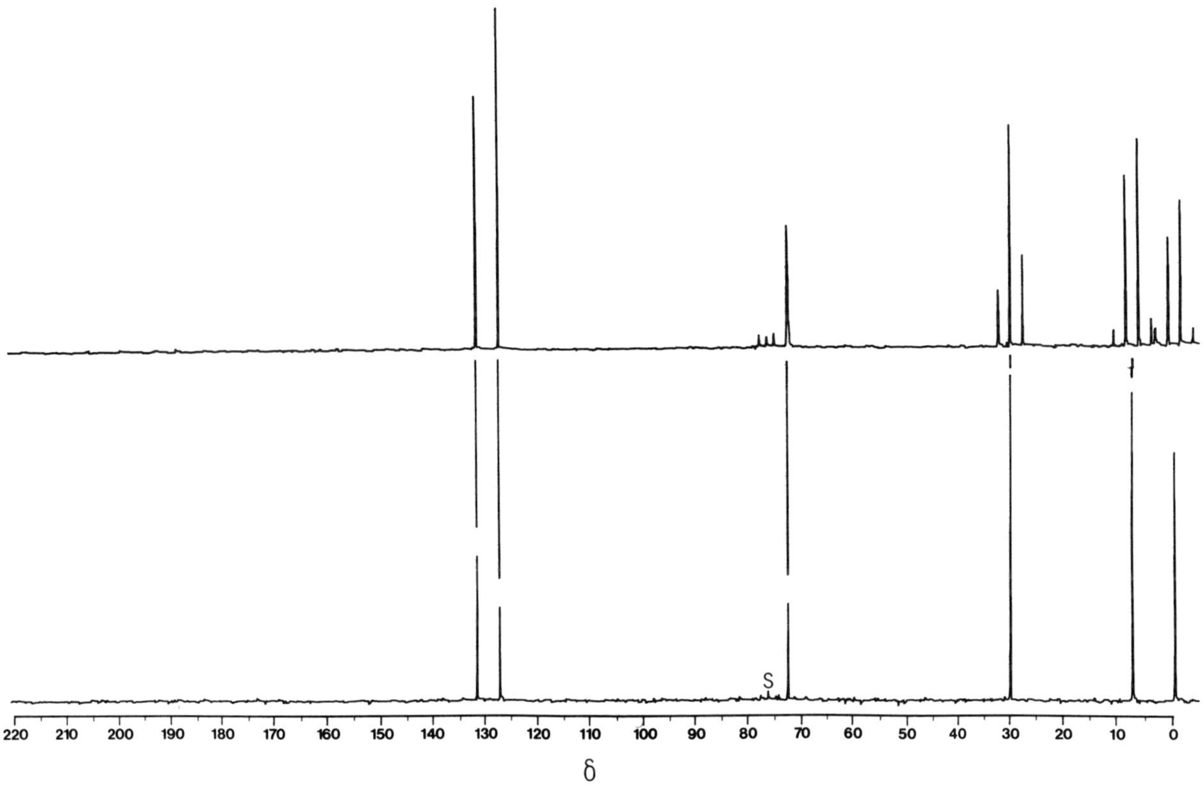
S
220
210
200
190
180
170
160
150
140
130
120
110
100
90
80
70
60
50
40
30
20
10
0
δ

Problem 1-10

1-10

$C_{13}H_8O$

MS: m180(s), 152, 76

IR: 1700 cm^{-1}

UV: Strong absorption

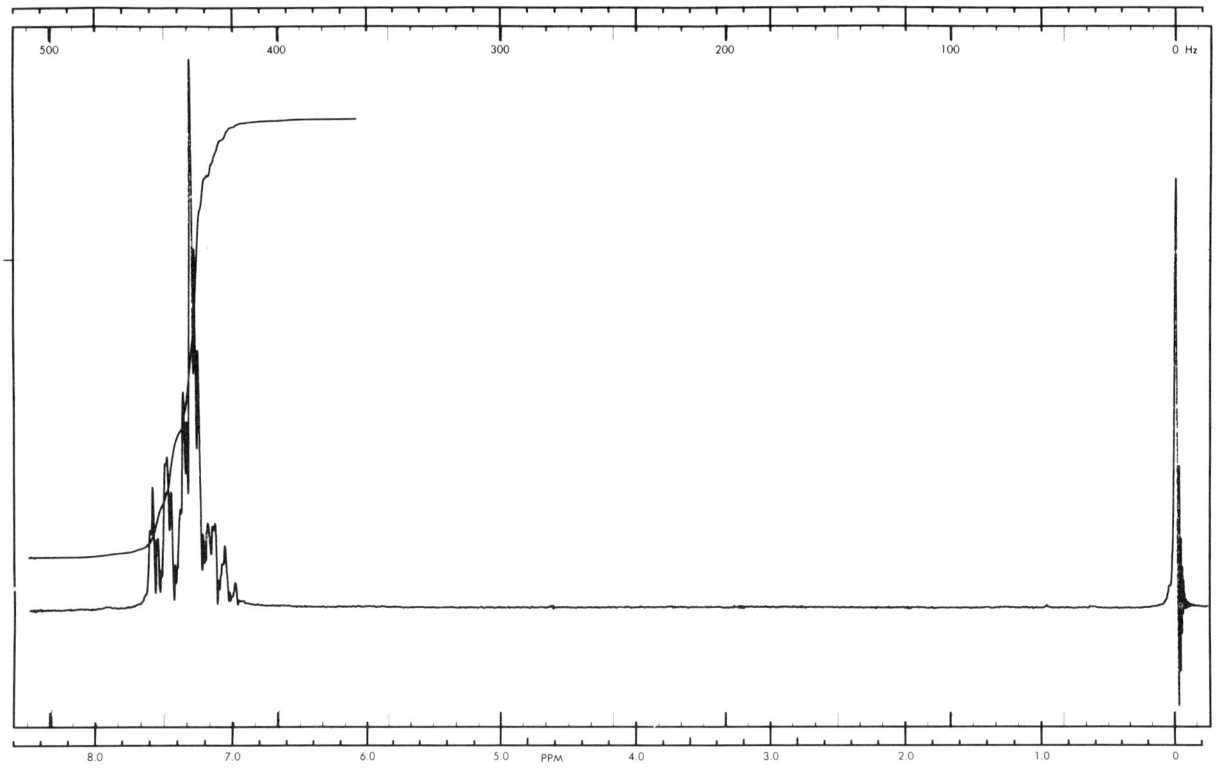

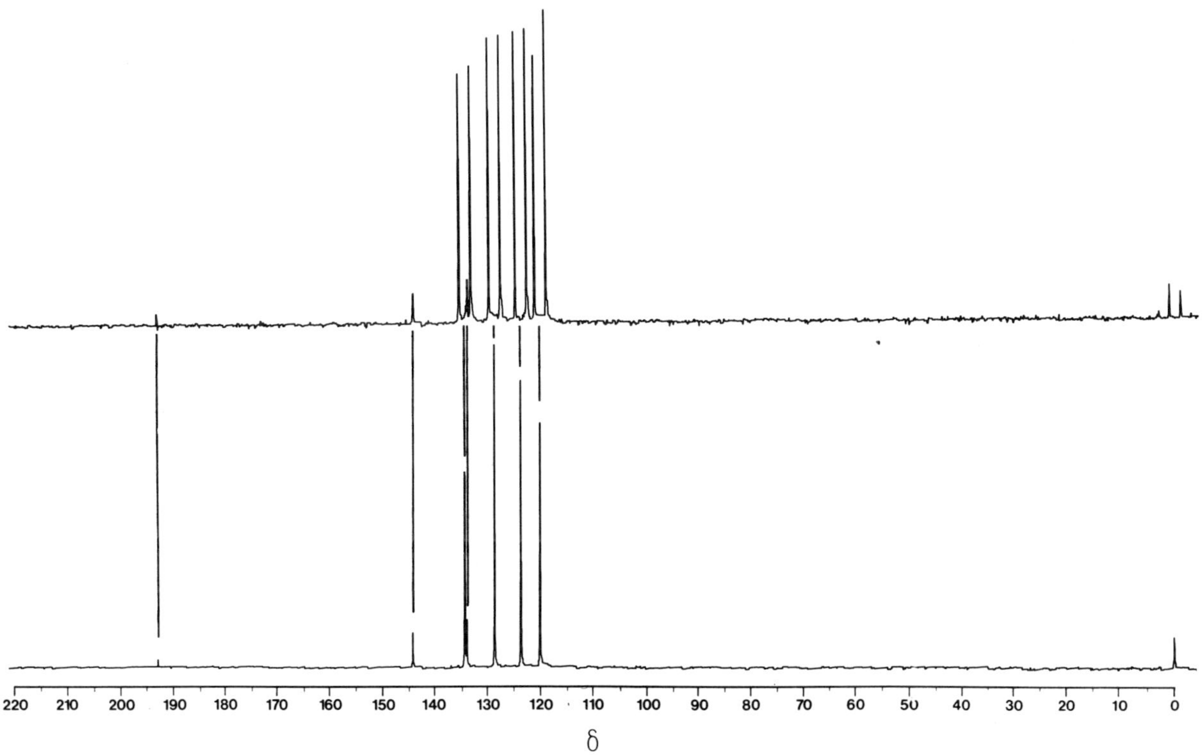
220
210
200
190
180
170
160
150
140
130
120
110
100
90
80
70
60
50
40
30
20
10
0
δ

Problem 1-11

1-11	MS:	m98(s)
$C_6H_{10}O$	IR:	-
	UV:	none

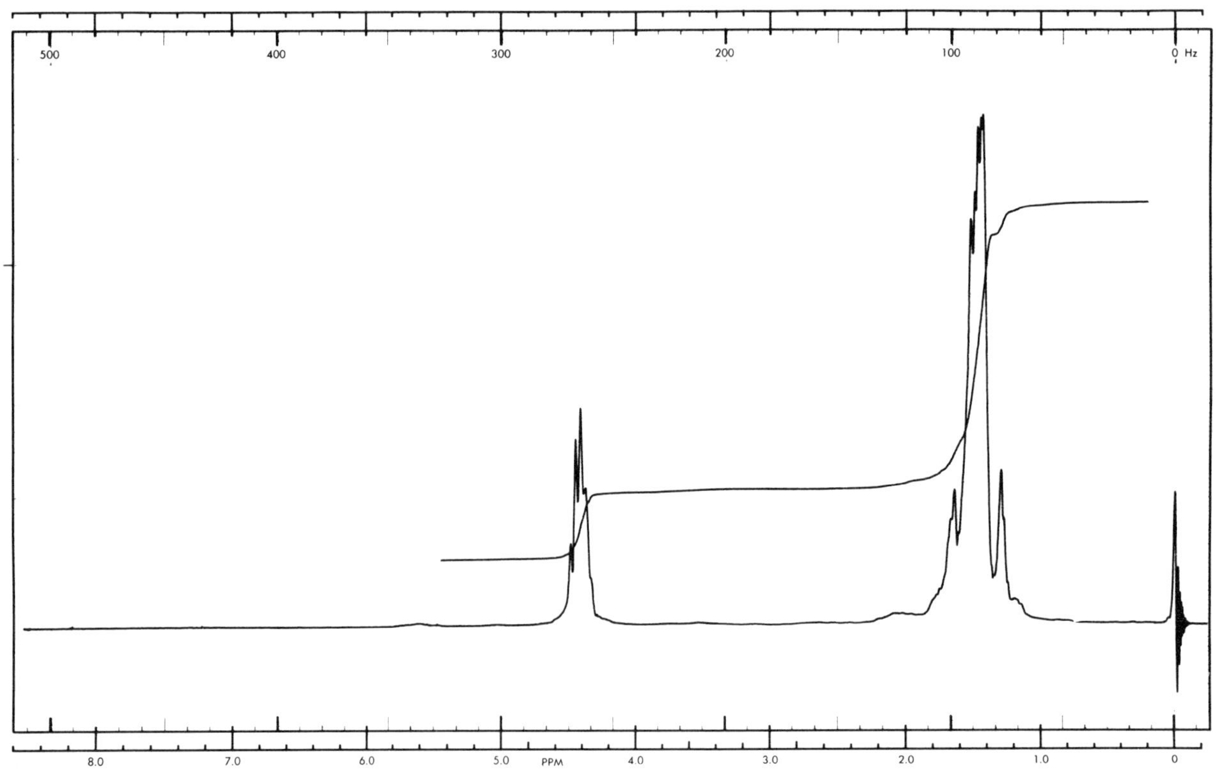

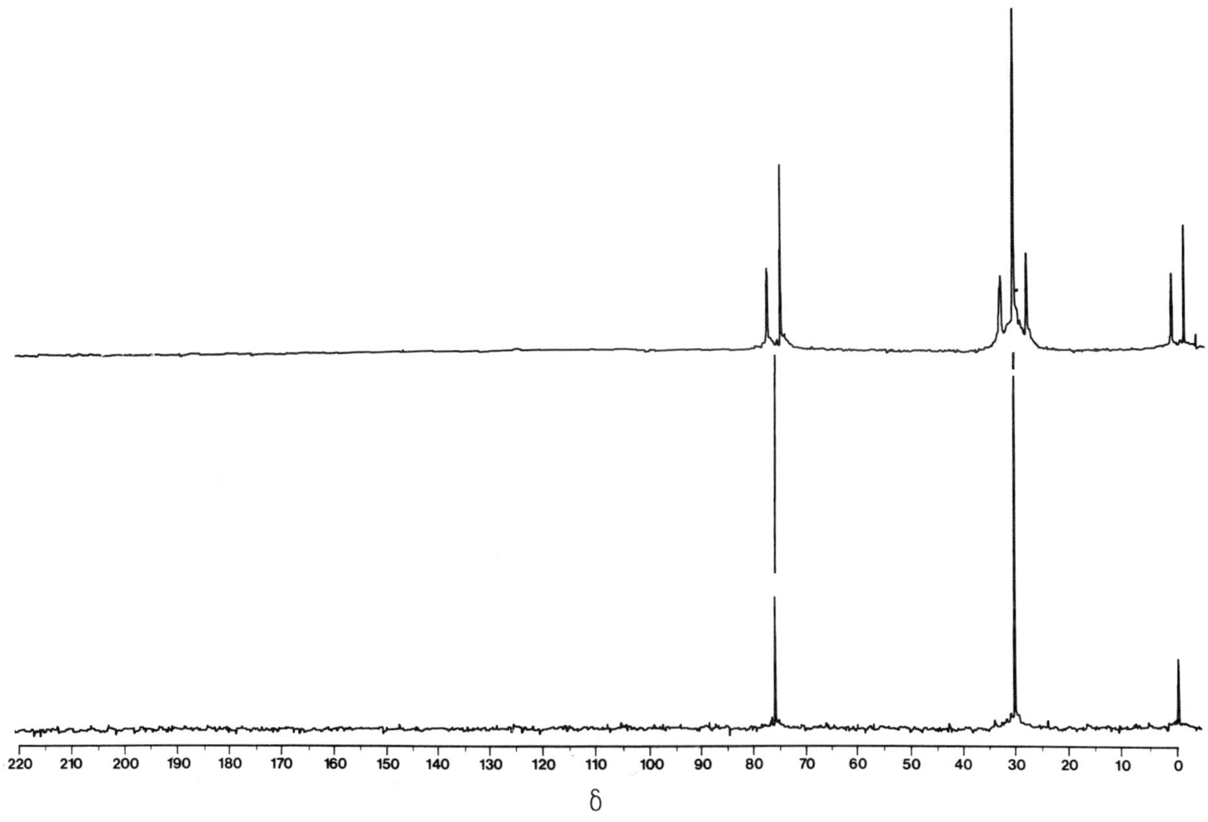
220
210
200
190
180
170
160
150
140
130
120
110
100
90
80
70
60
50
40
30
20
10
0
δ

Problem 1-12

1-12

$C_6H_{10}O$

MS: m98

IR: -

UV: none

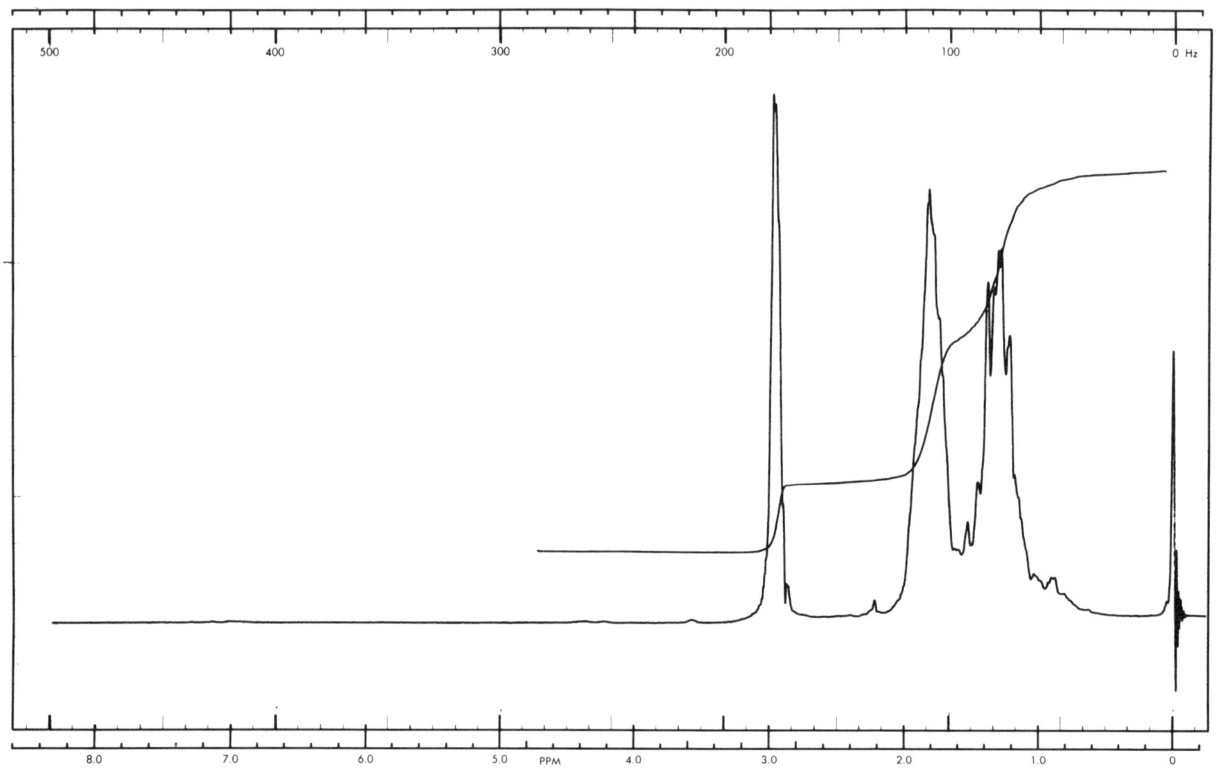

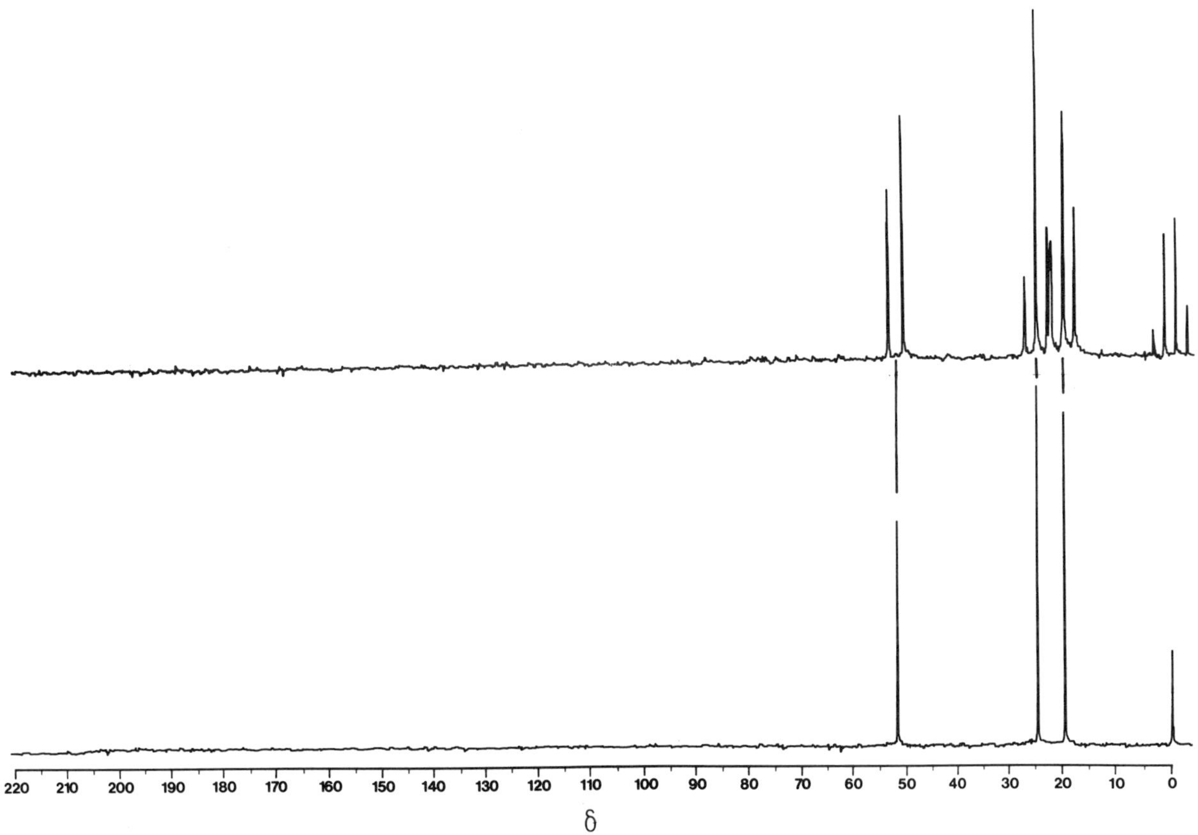
220
210
200
190
180
170
160
150
140
130
120
110
100
90
80
70
60
50
40
30
20
10
0
δ

Problem 1-13

1-13

$C_6H_{14}O_6$

MS: m182(w), 164, 146, 91, 61, 31

IR: 3200 cm^{-1} (broad)

UV: none

PMR: D_2O + DSS

CMR: D_2O + dioxane

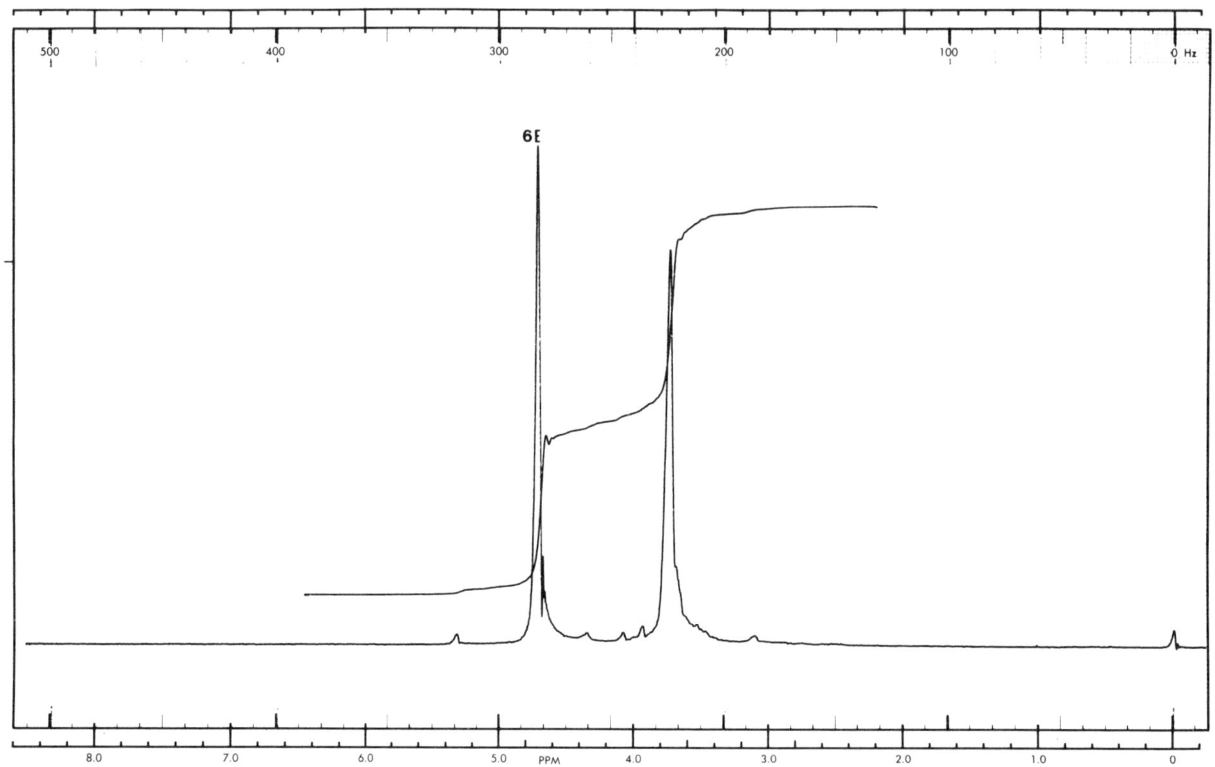

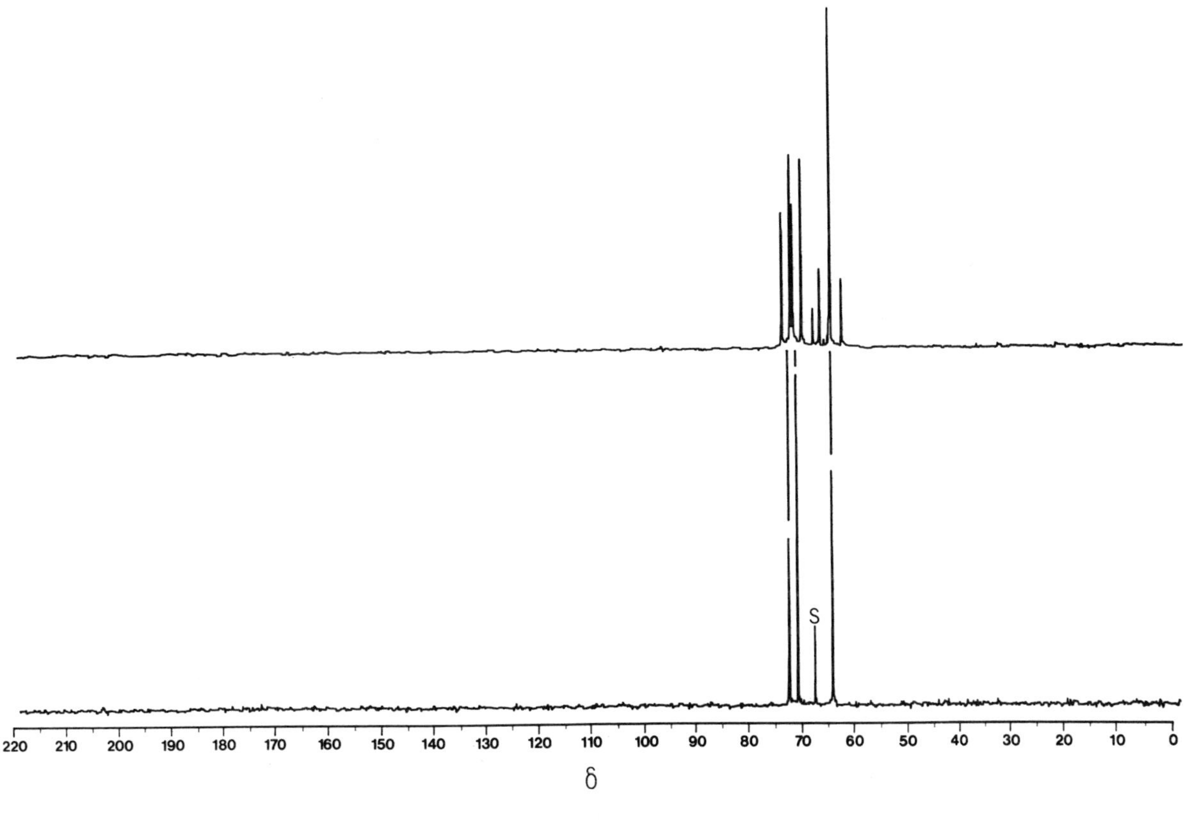
S
220 210 200 190 180 170 160 150 140 130 120 110 100 90 80 70 60 50 40 30 20 10 0
δ

Problem 1-14

1-14	MS:	m98, 83, 55, 41
C_7H_{14}	IR:	-
	UV:	none

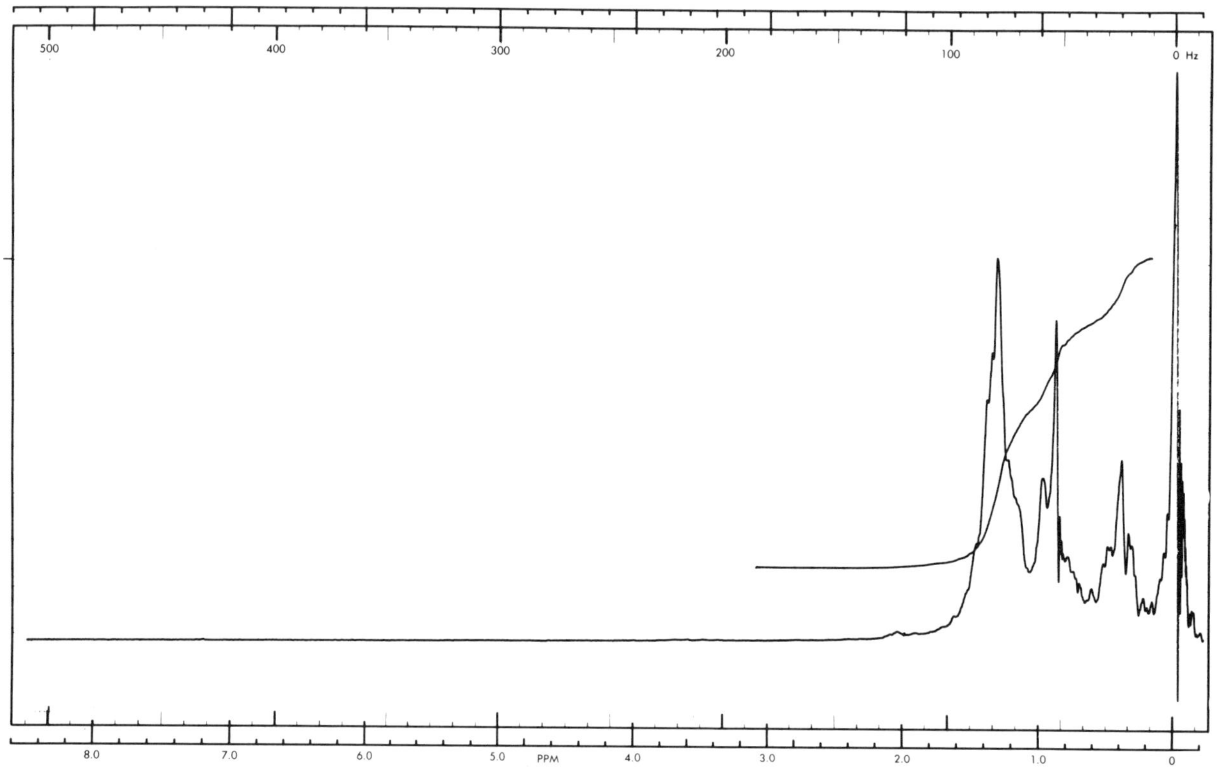

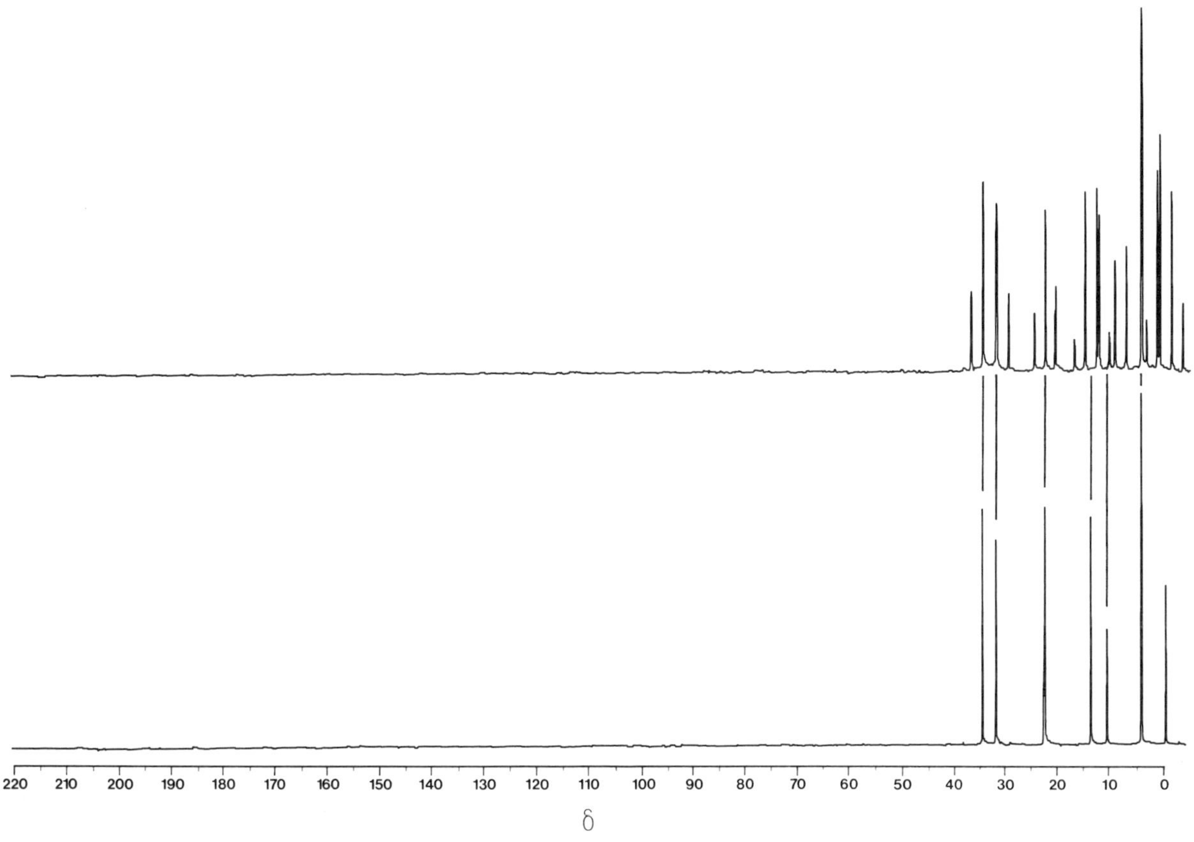
220
210
200
190
180
170
160
150
140
130
120
110
100
90
80
70
60
50
40
30
20
10
0
δ

Problem 1-15

1-15
$C_8H_4N_2O_2$

MS: m160(s), 134, 132
IR: 3550, 3450, 2220 cm^{-1}
UV: Strong absorption
PMR: d_6-Me_2CO + TMS; upper scan offset 200 Hz
CMR: d_6-Me_2CO + TMS

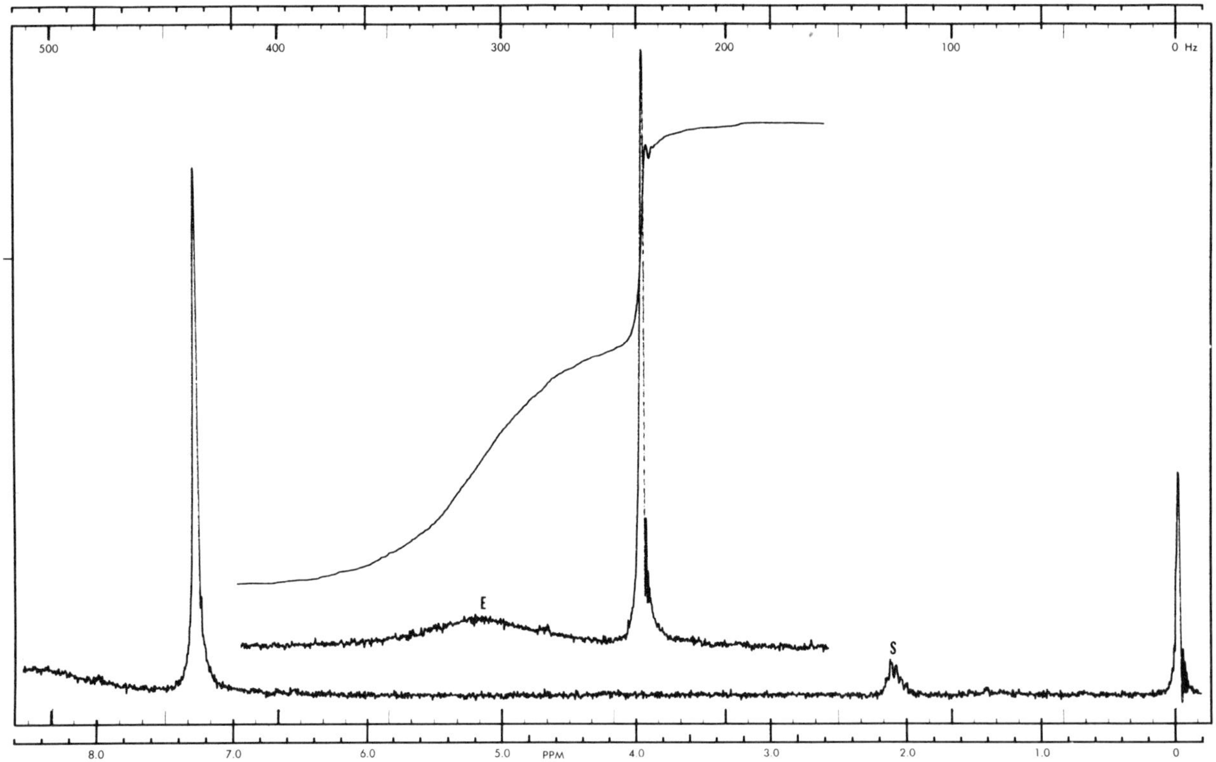

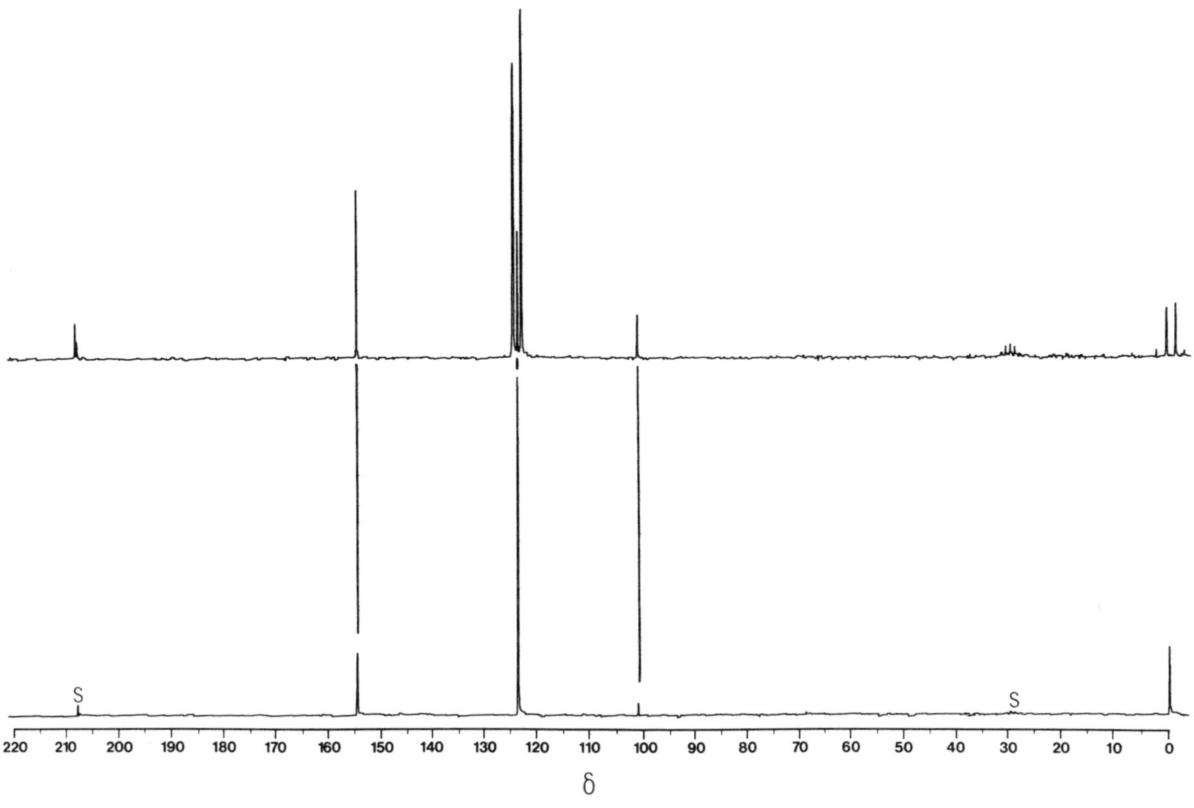
S
S
220
210
200
190
180
170
160
150
140
130
120
110
100
90
80
70
60
50
40
30
20
10
0
δ

Problem 1-16

1-16	MS:	m178(w), 133, 89, 45
$C_8H_{18}O_4$	IR:	-
	UV:	none

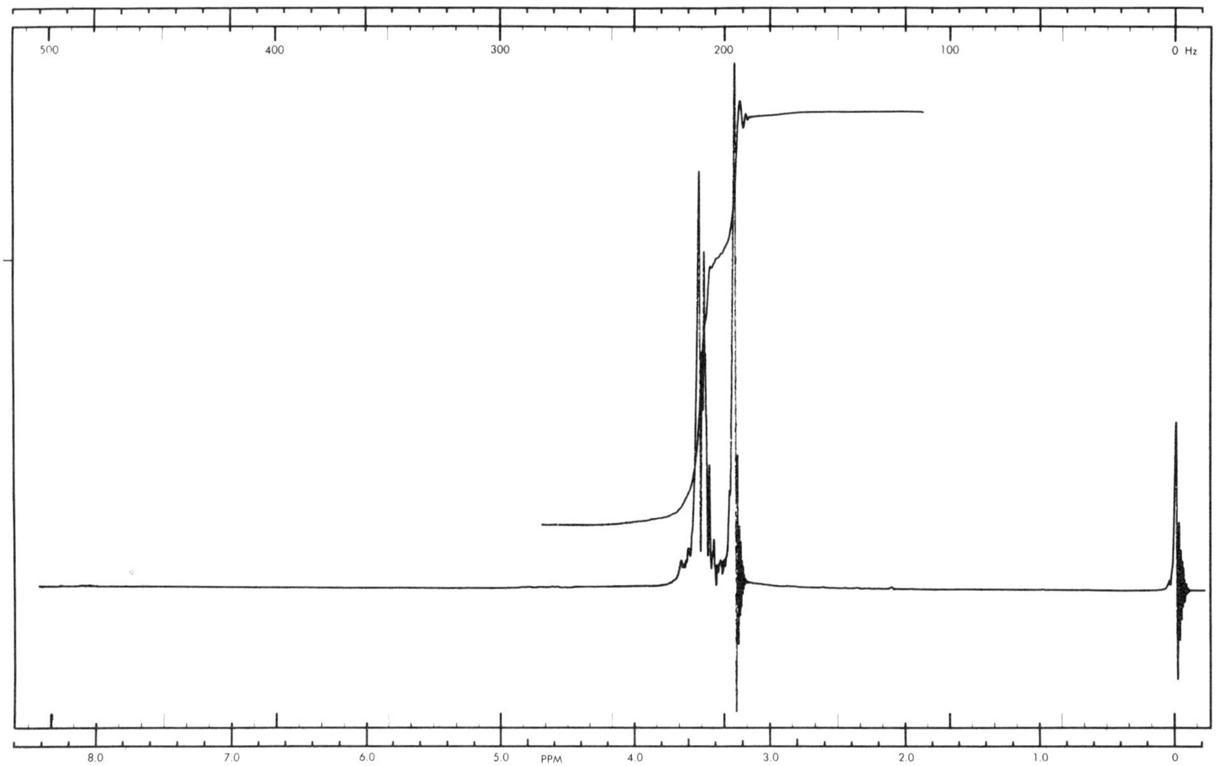

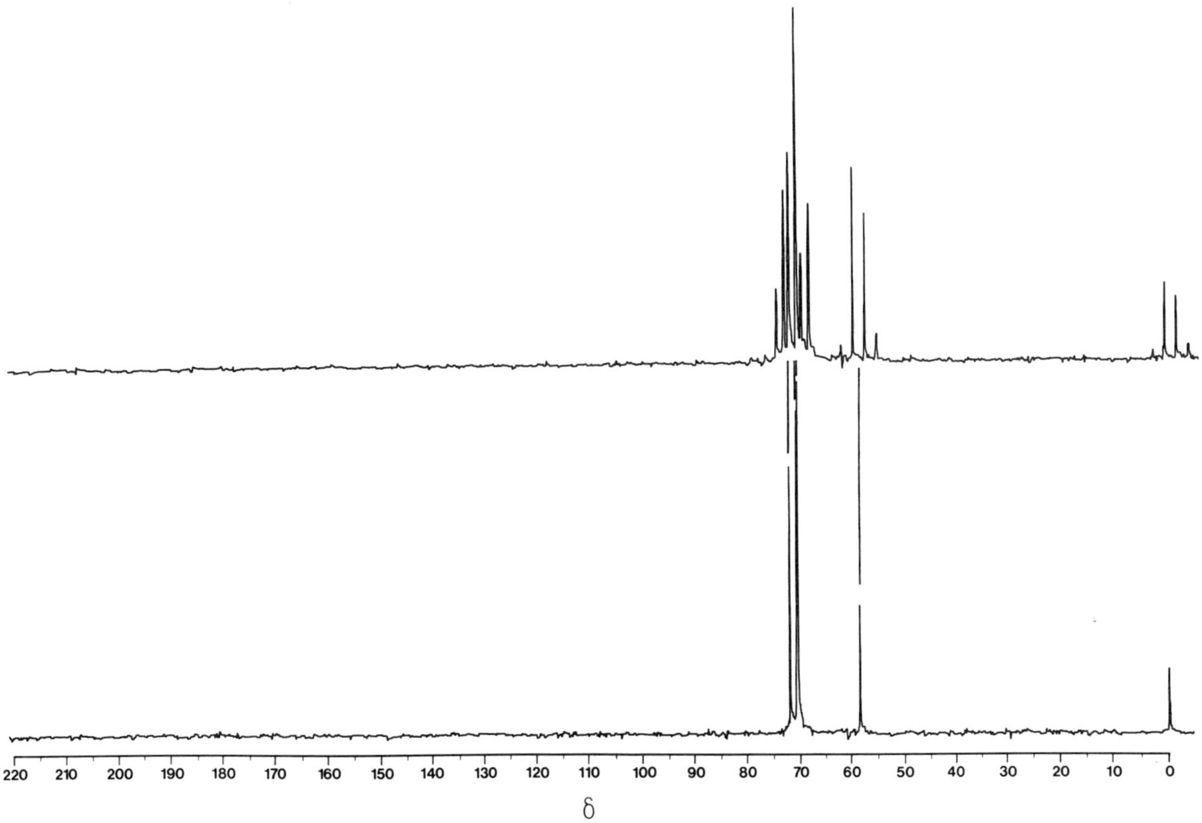
220
210
200
190
180
170
160
150
140
130
120
110
100
90
80
70
60
50
40
30
20
10
0
δ

Problem 1-17

1-17	MS:	m184(s), 108, 77, 76
$C_{12}H_8S$	IR:	-
	UV:	Strong absorption

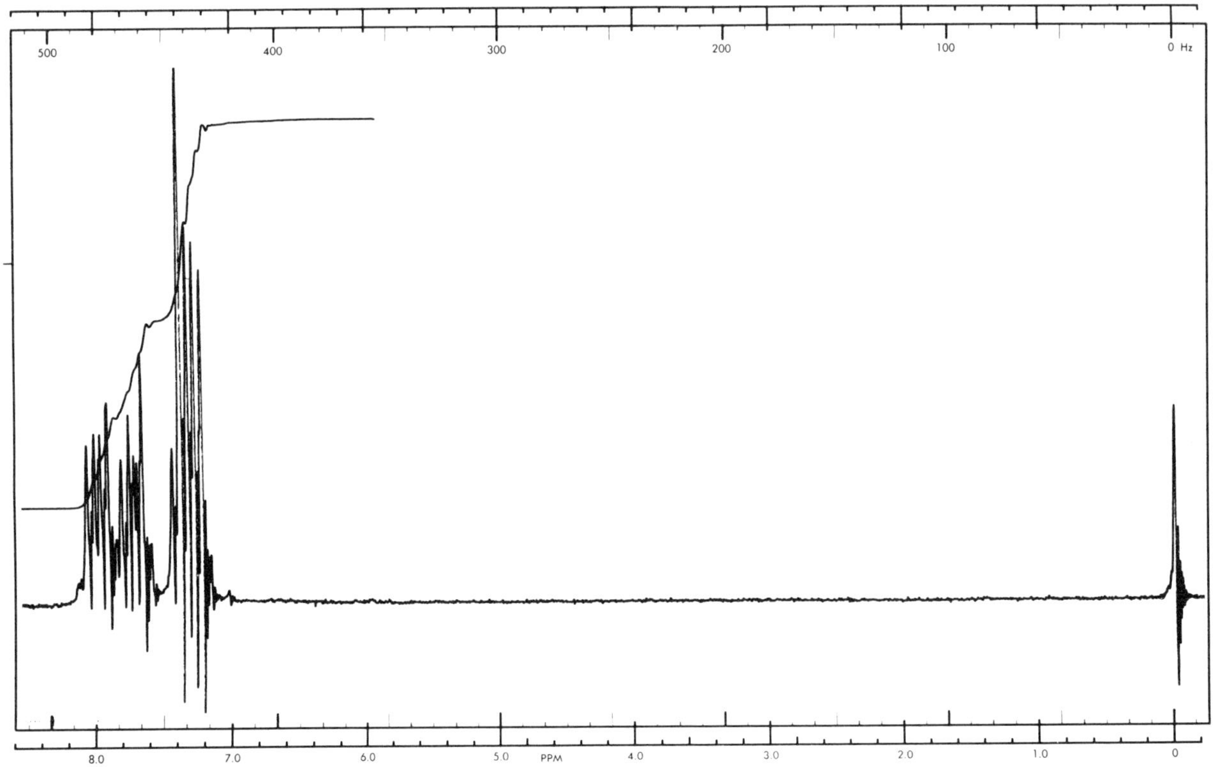

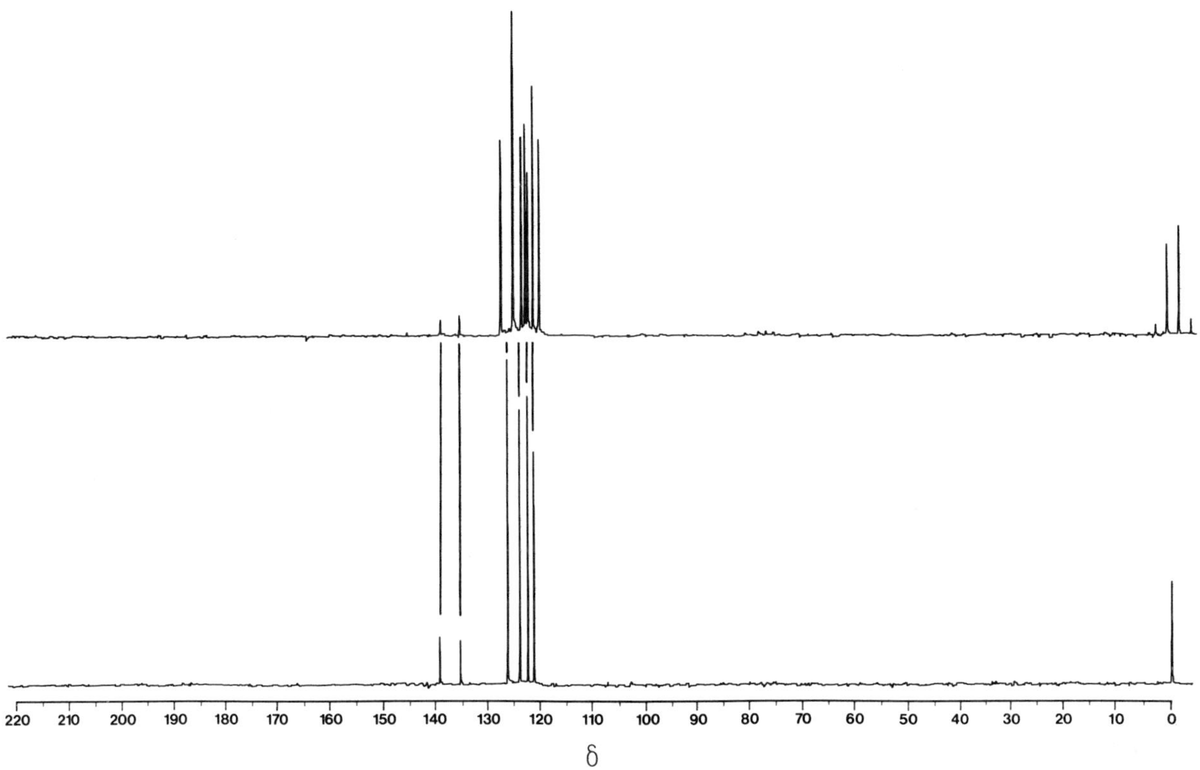

220 210 200 190 180 170 160 150 140 130 120 110 100 90 80 70 60 50 40 30 20 10 0
δ

Problem 1-18

1-18	MS:	Not volatile
$(C_5H_8)_n$	IR:	1670 cm^{-1}
n large	UV:	none

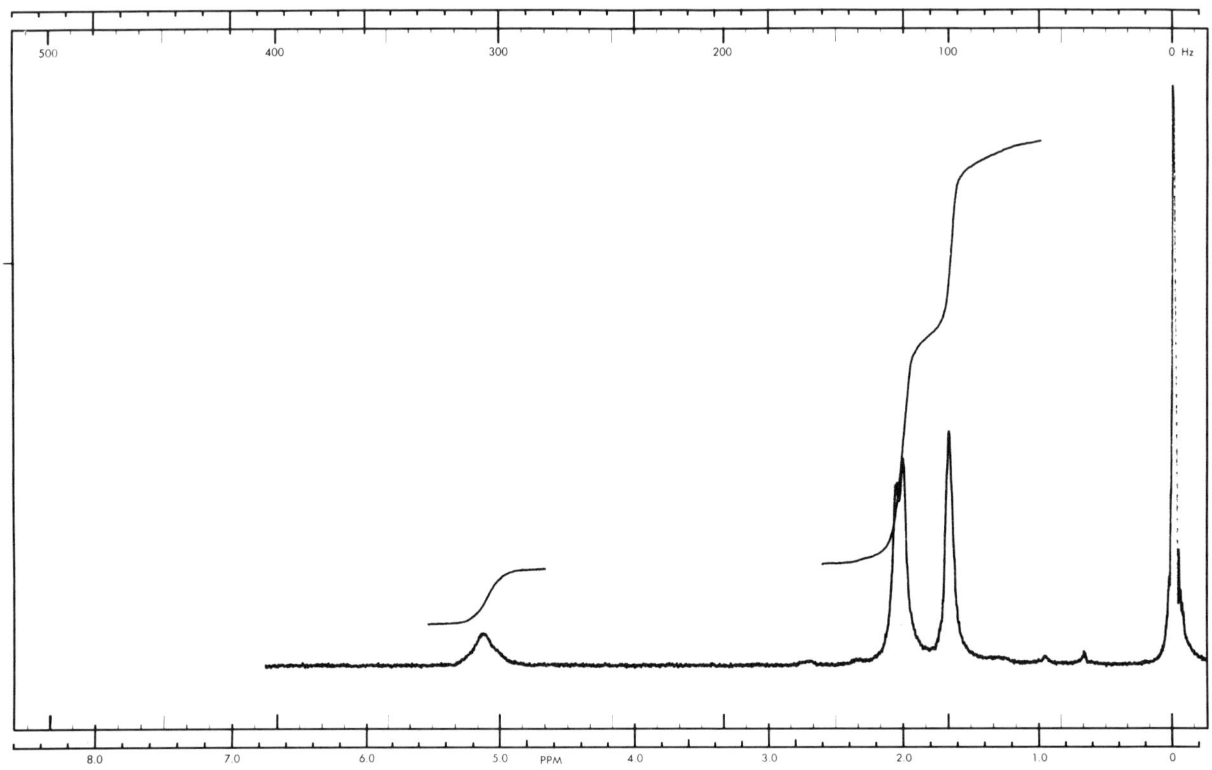

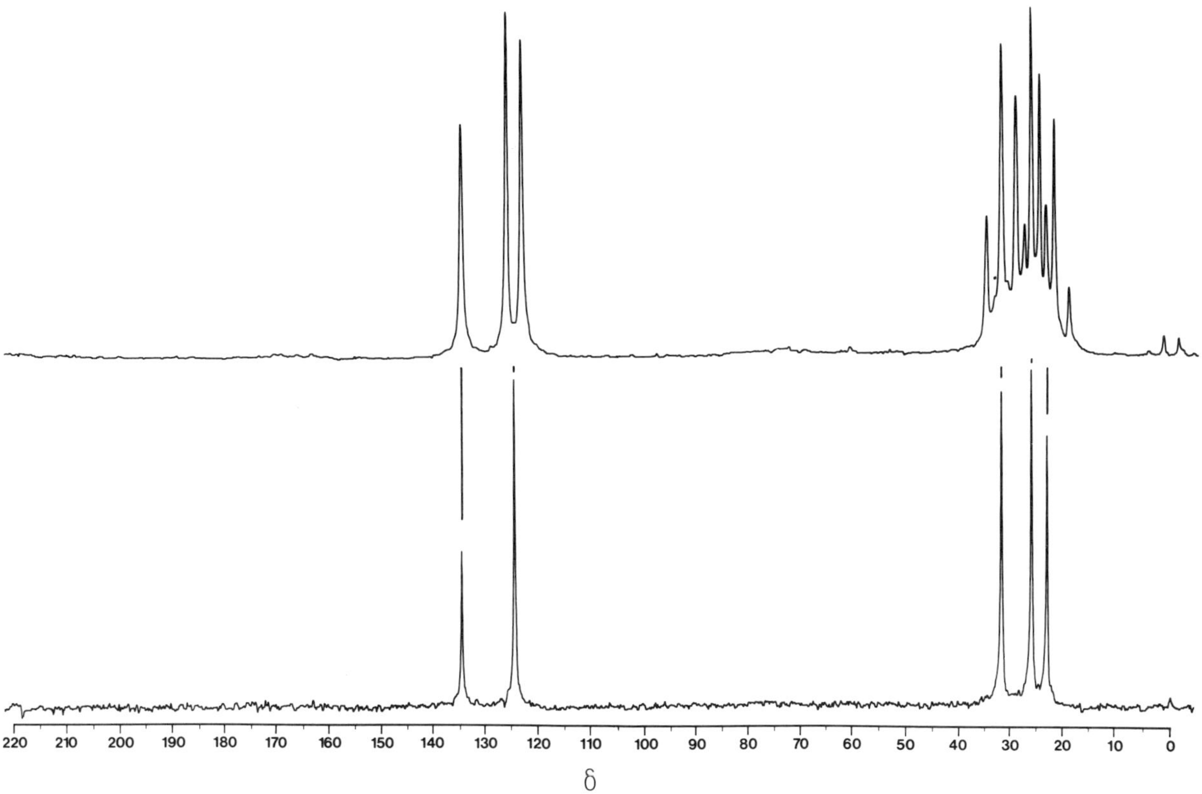
220
210
200
190
180
170
160
150
140
130
120
110
100
90
80
70
60
50
40
30
20
10
0
δ

Problem 1-19

1-19

$C_6H_6N_2O$

MS: m122(s), 106, 78

IR: 3350, 3200, 1700, 1630 cm^{-1}

UV: 262 (ε 2850)

PMR: D_2O + DSS; upper scan offset 300 Hz

CMR: D_2O + dioxane

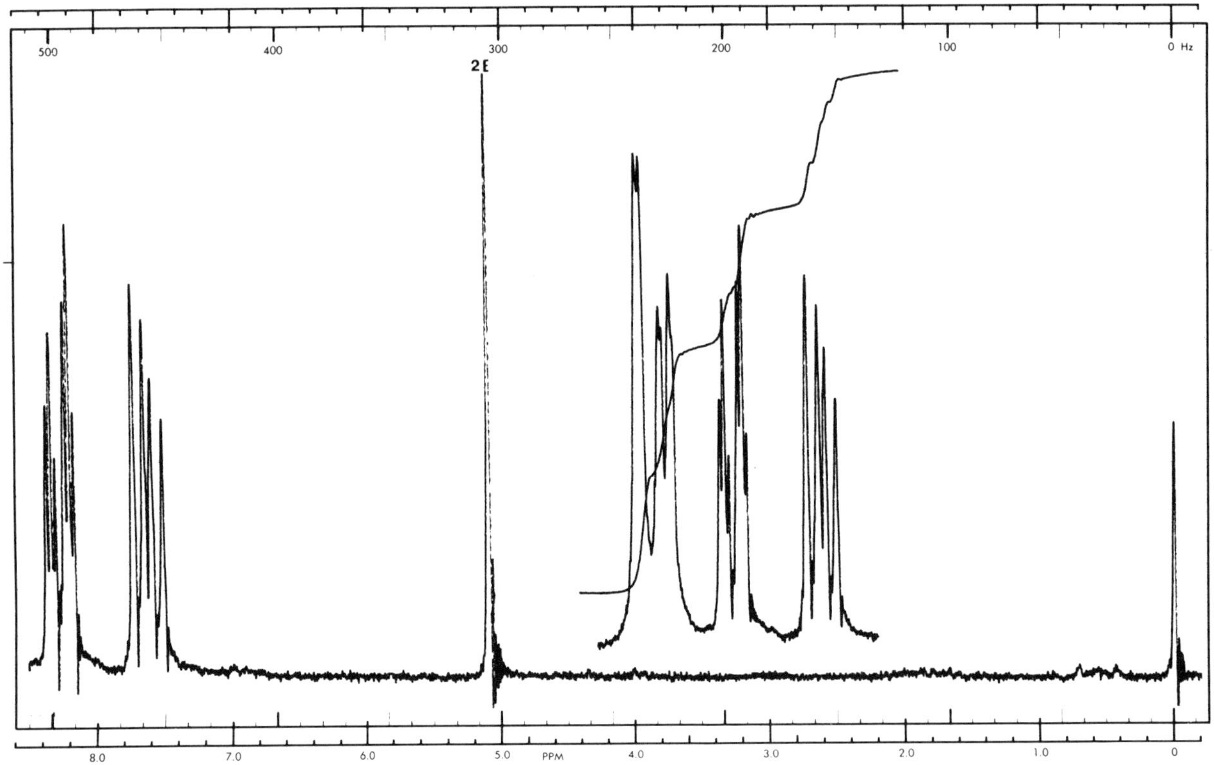

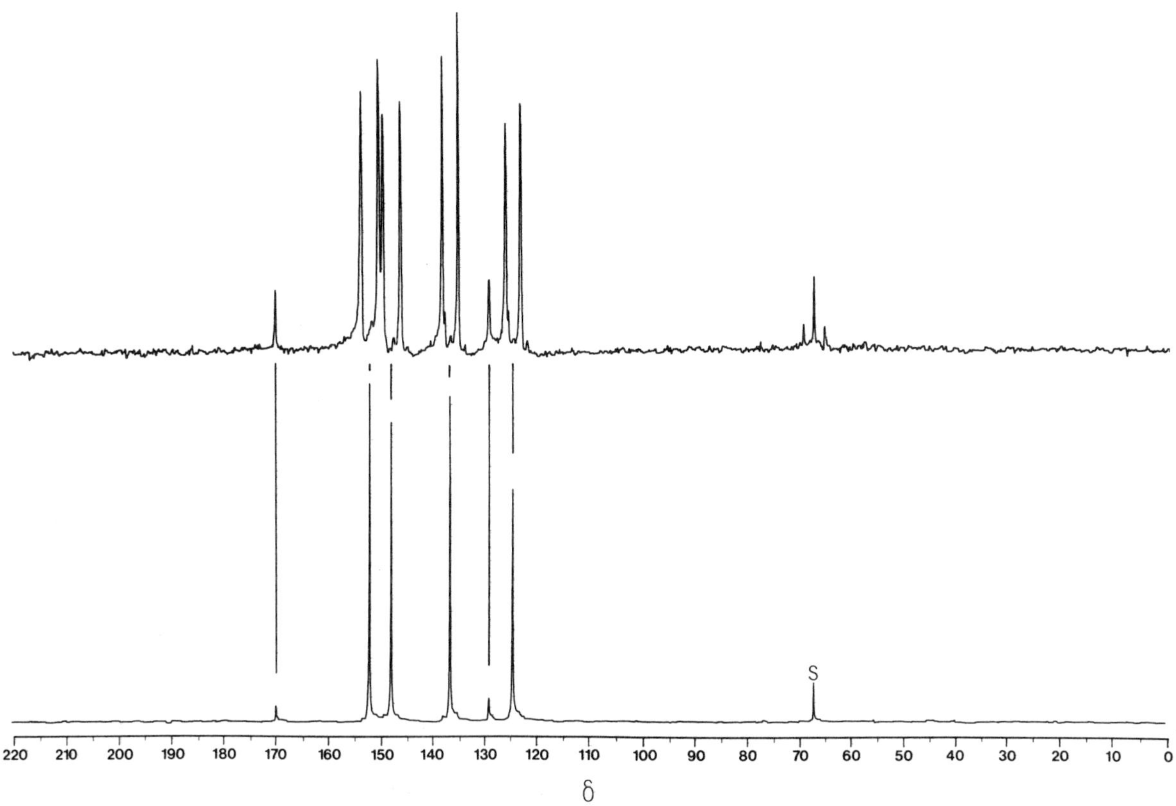
S
220
210
200
190
180
170
160
150
140
130
120
110
100
90
80
70
60
50
40
30
20
10
0
δ

Problem 1-20

1-20

$C_{18}H_{14}S_2$

MS: m294, 217, 185, 109, 77

IR: -

UV: Strong absorption

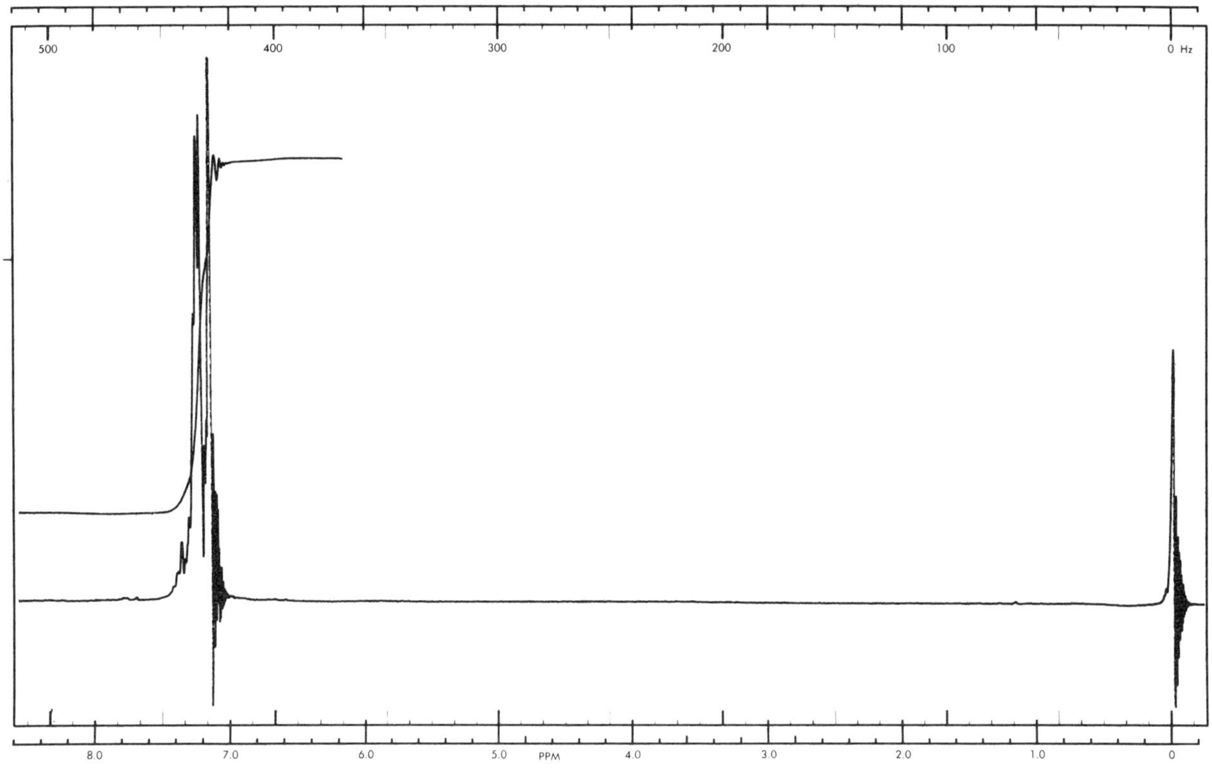

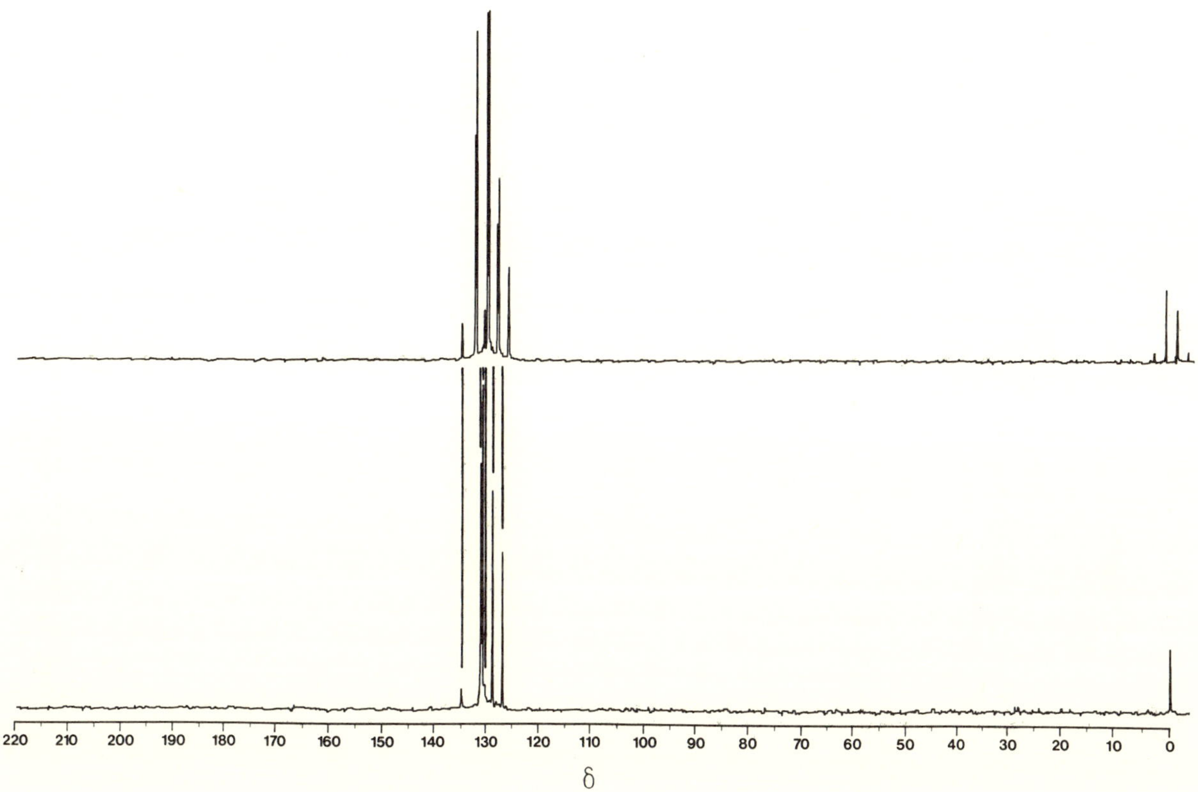
220
210
200
190
180
170
160
150
140
130
120
110
100
90
80
70
60
50
40
30
20
10
0
δ

Problem 1-21

1-21

$C_{12}H_{10}O_3S$

MS: m234, 141, 93, 77

IR: -

UV: Strong absorption

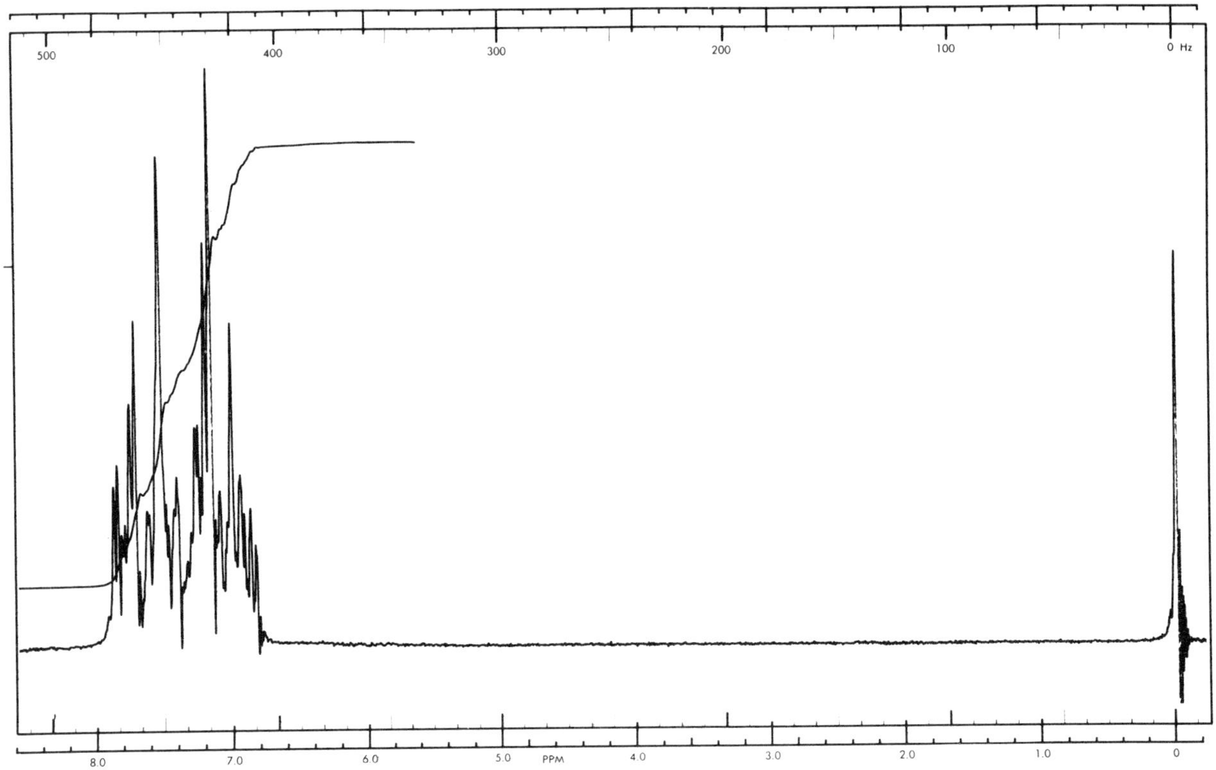

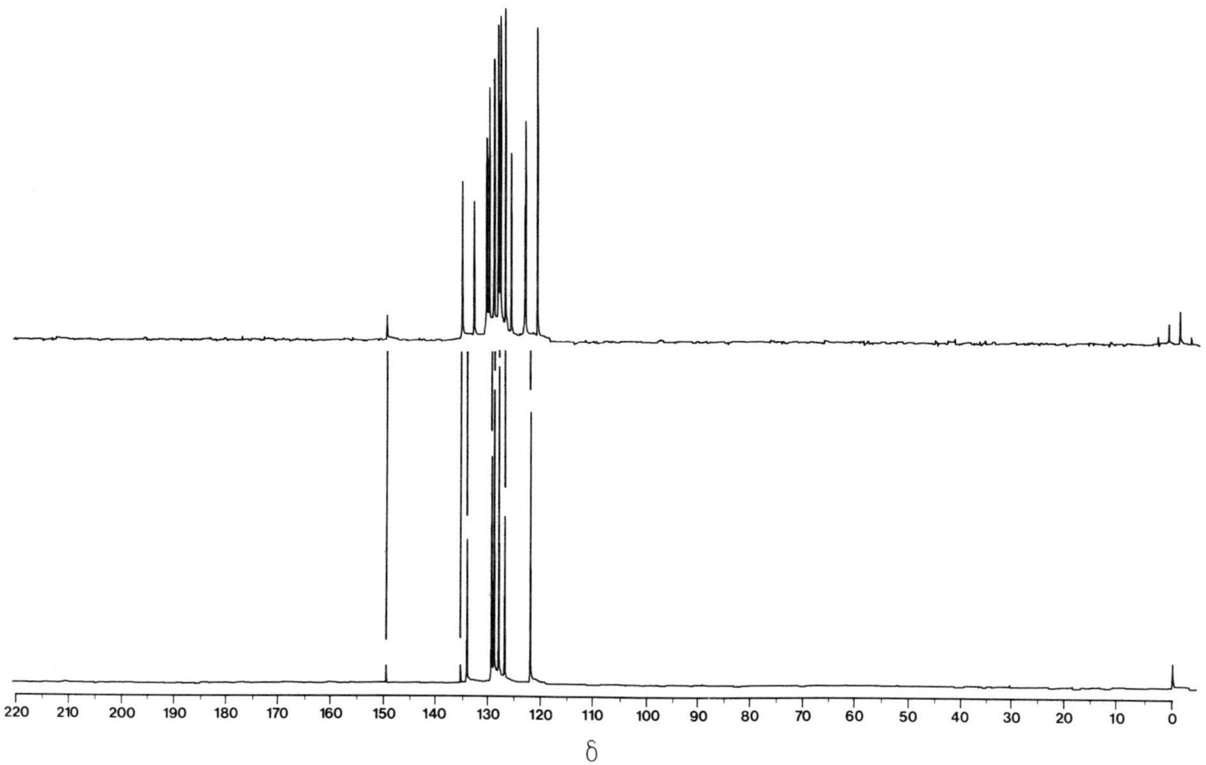
220
210
200
190
180
170
160
150
140
130
120
110
100
90
80
70
60
50
40
30
20
10
0
δ

Problem 1-22

1-22

$C_9H_{14}O_6$

MS: m218(w), 159, 158, 99, 43

IR: 1730 cm^{-1}

UV: none

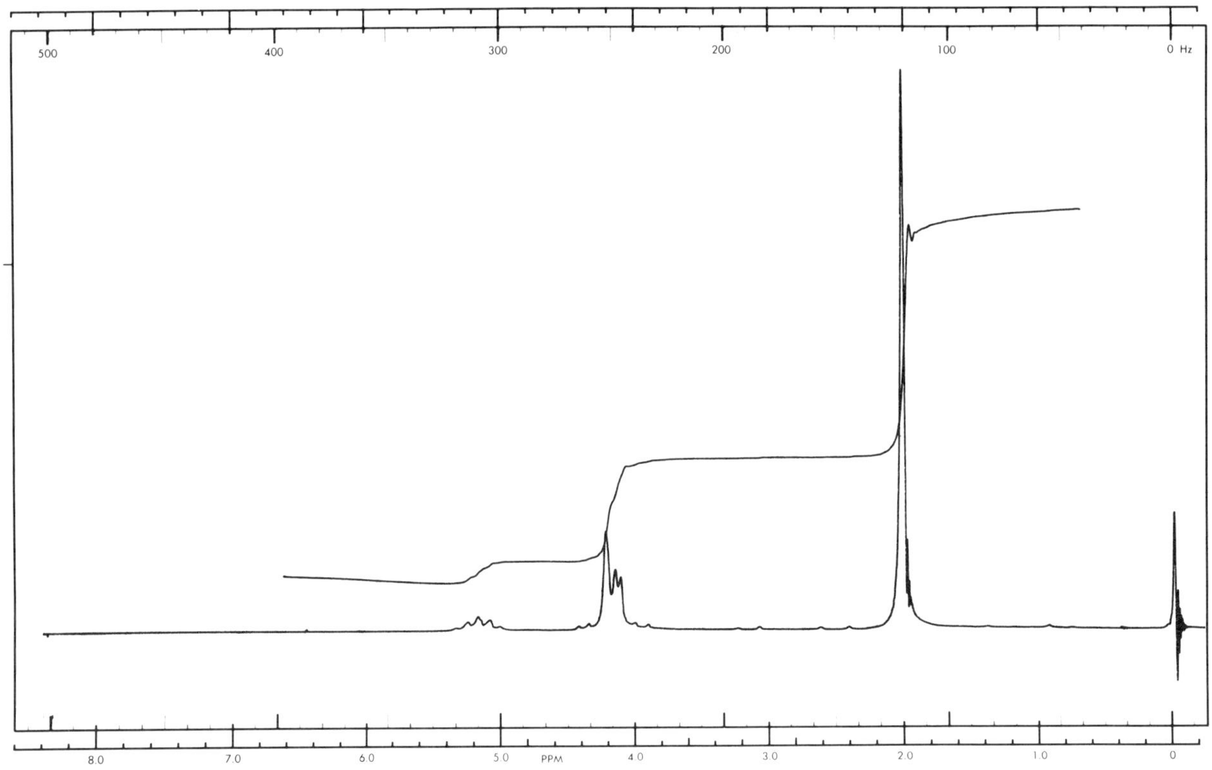

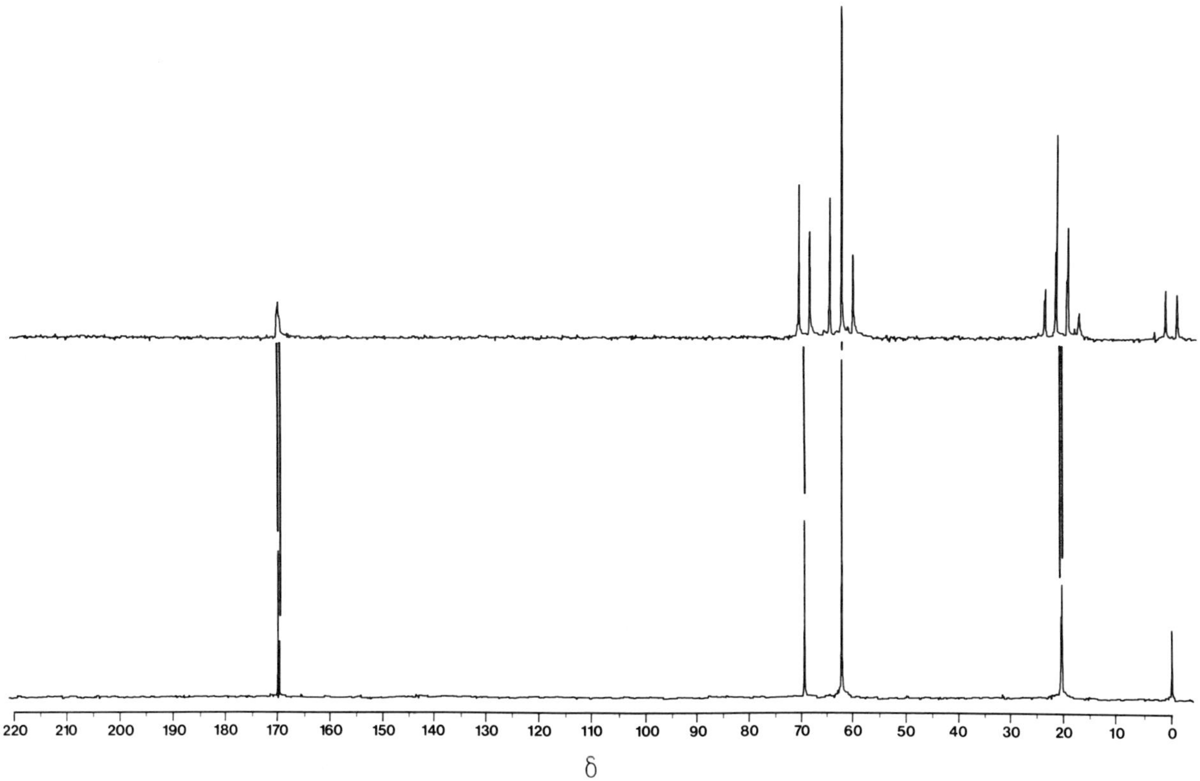
220 210 200 190 180 170 160 150 140 130 120 110 100 90 80 70 60 50 40 30 20 10 0
δ

Problem 1-23

1-23

$C_{12}H_{17}N$

MS: m175, 174, 148, 83

IR: 3420, 3350 cm^{-1}

UV: Strong absorption

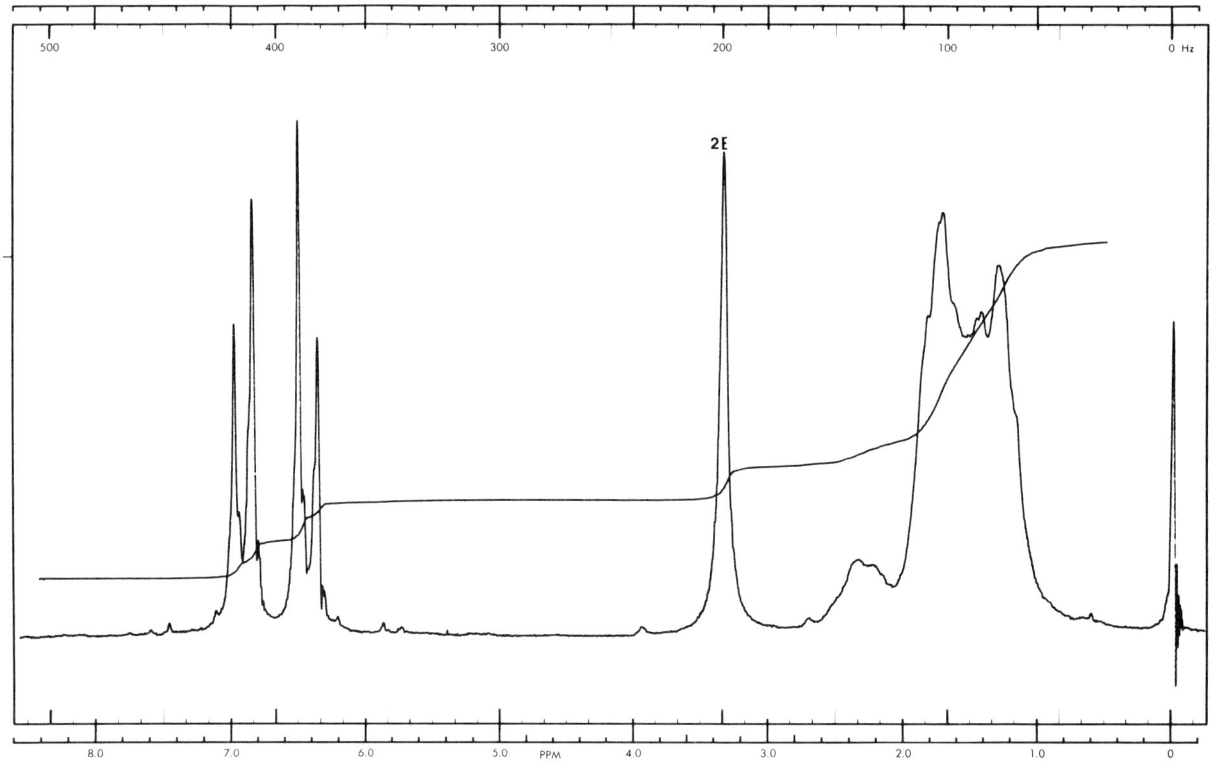

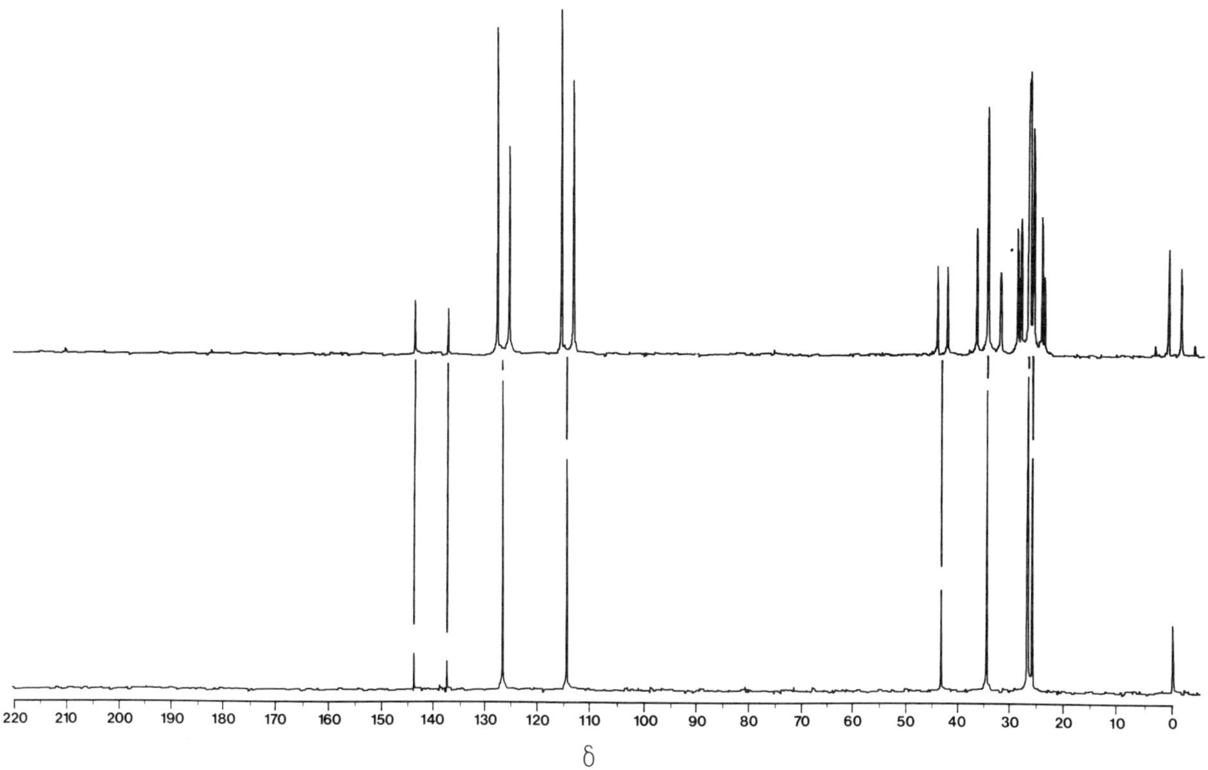
220 210 200 190 180 170 160 150 140 130 120 110 100 90 80 70 60 50 40 30 20 10 0
δ

Problem 1-24

1-24

$C_{15}H_{24}O$

MS: m220(w), 219, 205, 192, 191

IR: 3650 cm^{-1}

UV: Strong absorption

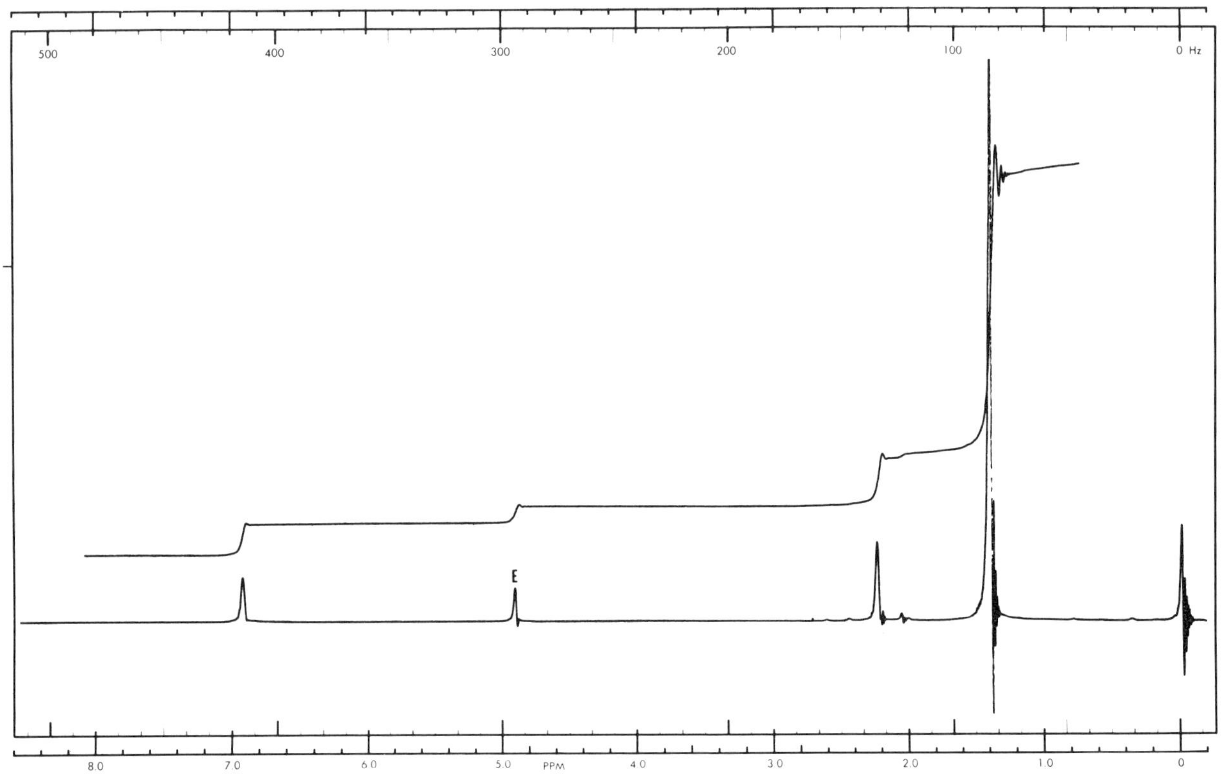

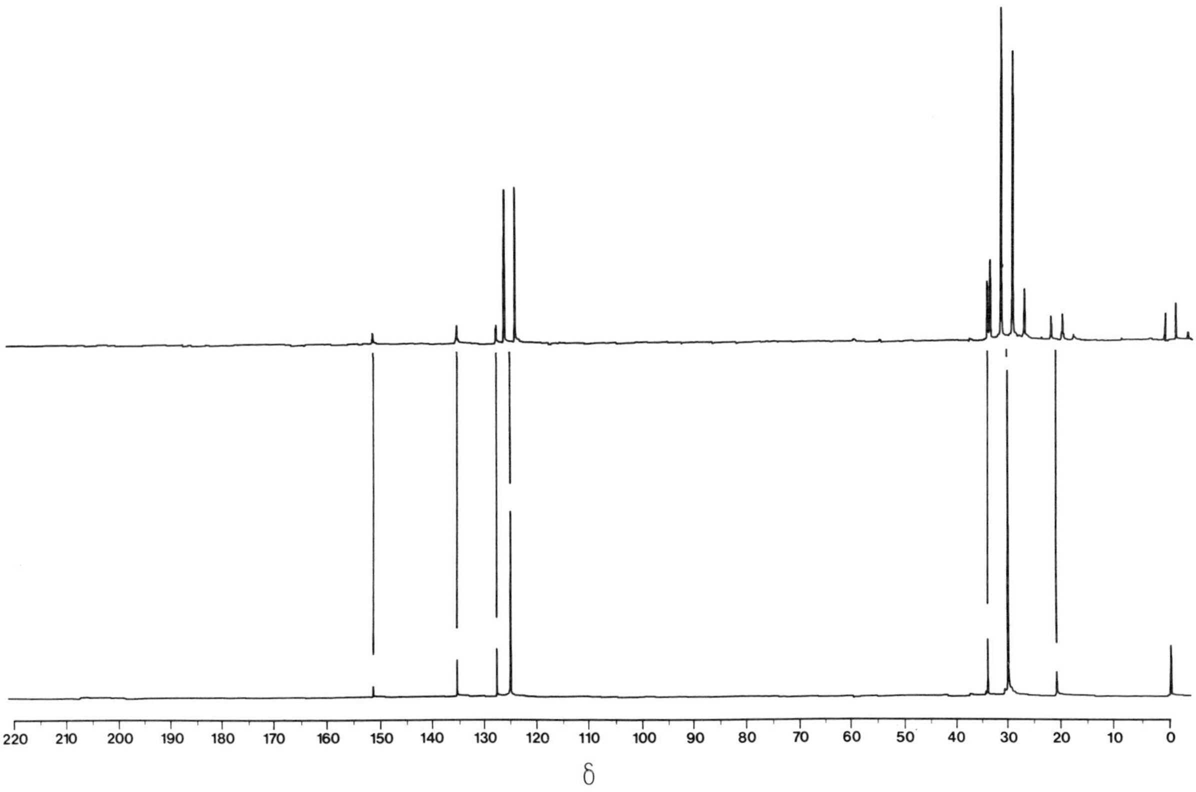
220
210
200
190
180
170
160
150
140
130
120
110
100
90
80
70
60
50
40
30
20
10
0
δ

Problem 1-25

1-25

$C_6H_4N_2O_4$

MS: m168, 123, 78

IR: 2900 (broad), 1730 cm^{-1}

UV: Strong absorption

PMR: d_6-DMSO + TMS: Upper scan offset 360 Hz

CMR: d_6-DMSO + dioxane; 5 sec delay

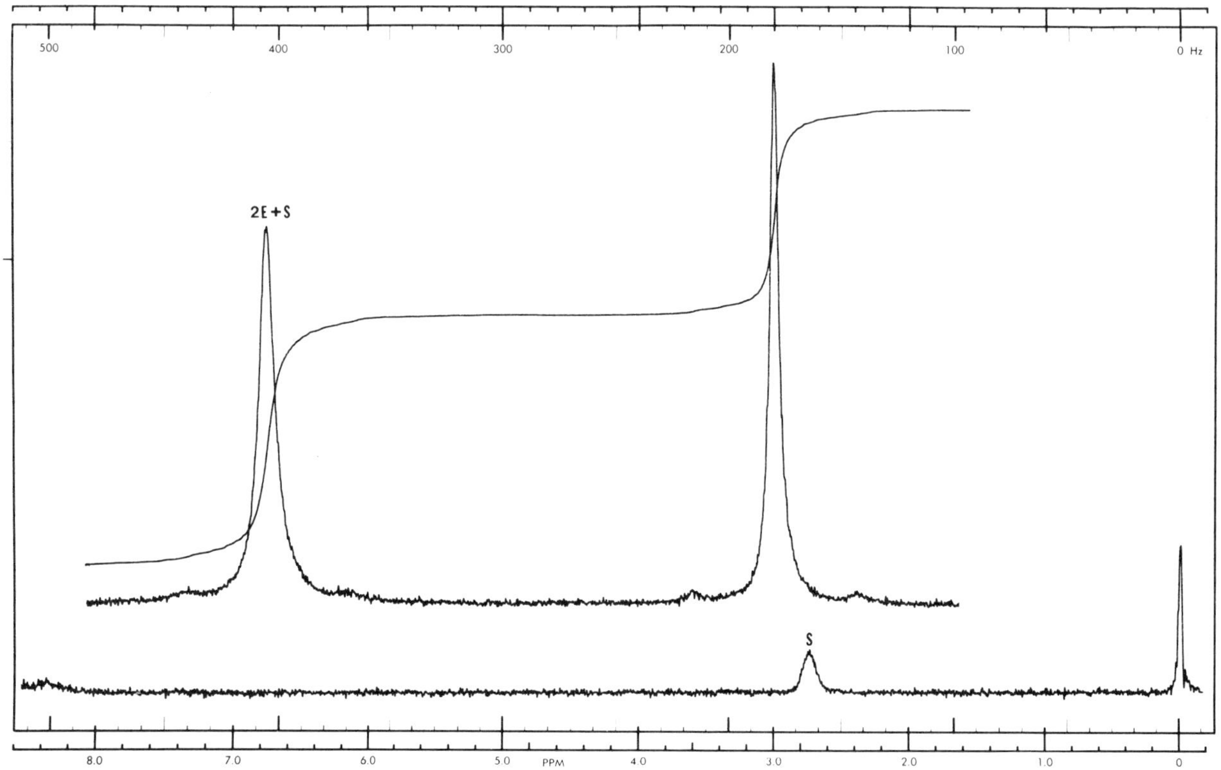

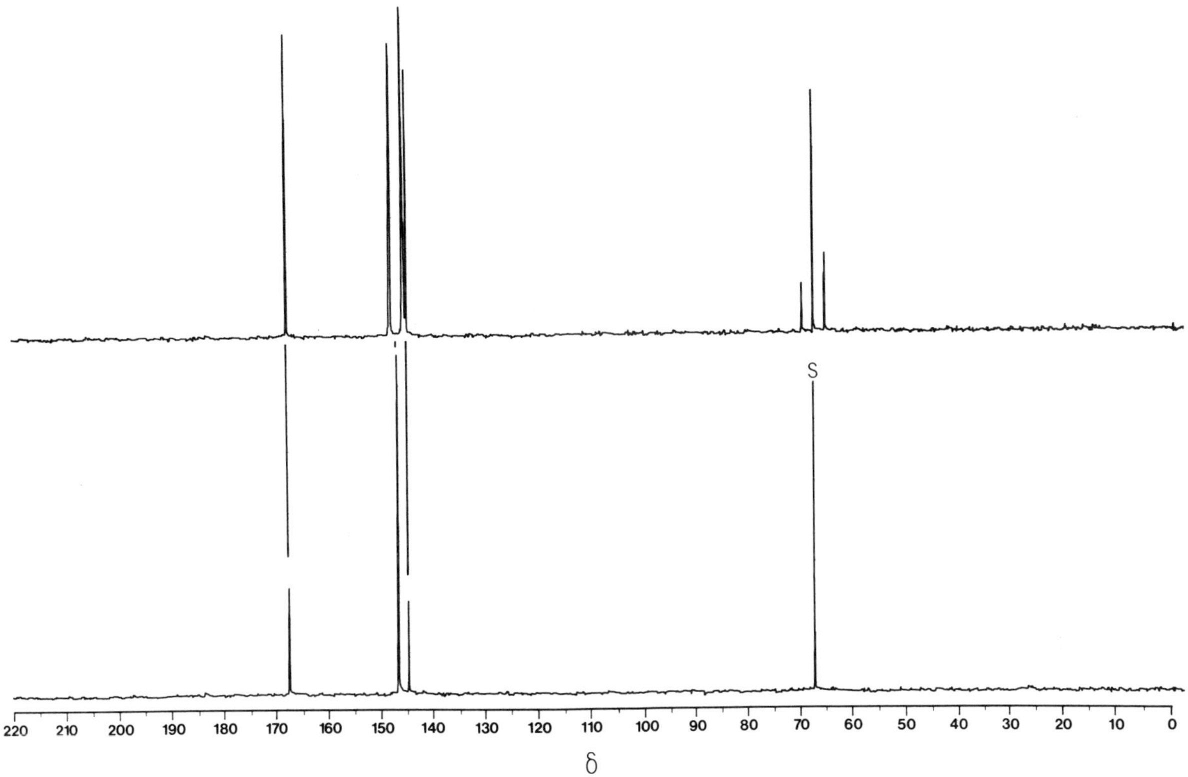
S
220
210
200
190
180
170
160
150
140
130
120
110
100
90
80
70
60
50
40
30
20
10
0
δ

Problem 1-26

1-26

$C_{13}H_{21}N$

MS: m191, 176, 134, 57

IR: -

UV: Strong absorption

PMR: Upper scan amplitude X10

CMR: 15 sec delay

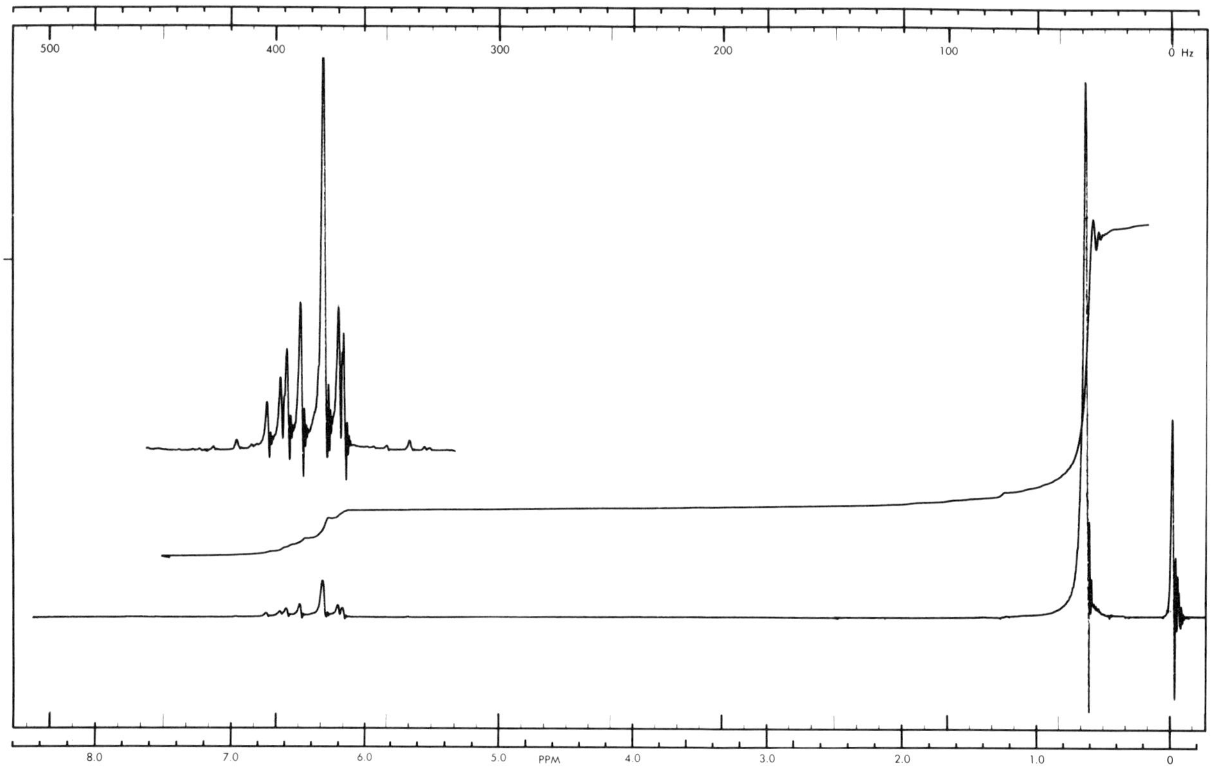

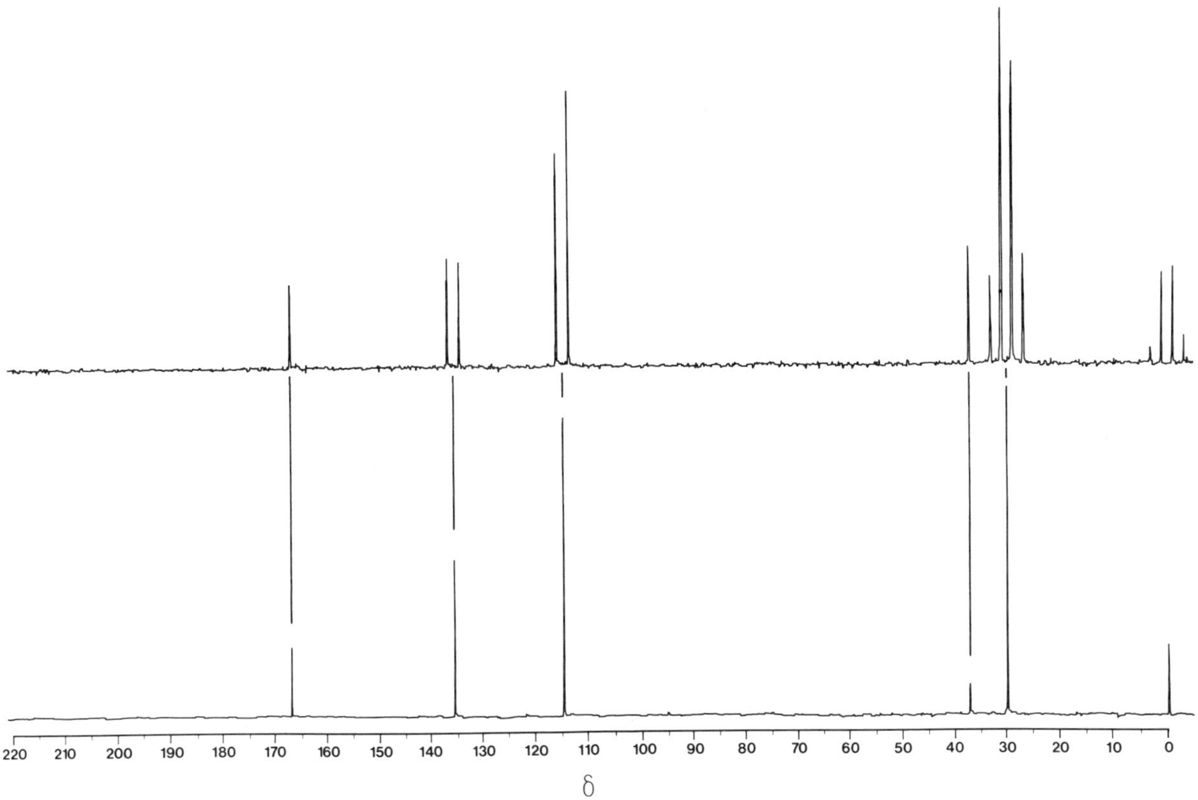
220 210 200 190 180 170 160 150 140 130 120 110 100 90 80 70 60 50 40 30 20 10 0
δ

Problem 1-27

1-27

$C_6H_{24}CoN_6^{3+}3I^-$

MS: Not volatile

IR: 3400 cm^{-1}

UV: Strong absorption

PMR: D_2O + DSS

CMR: D_2O + dioxane

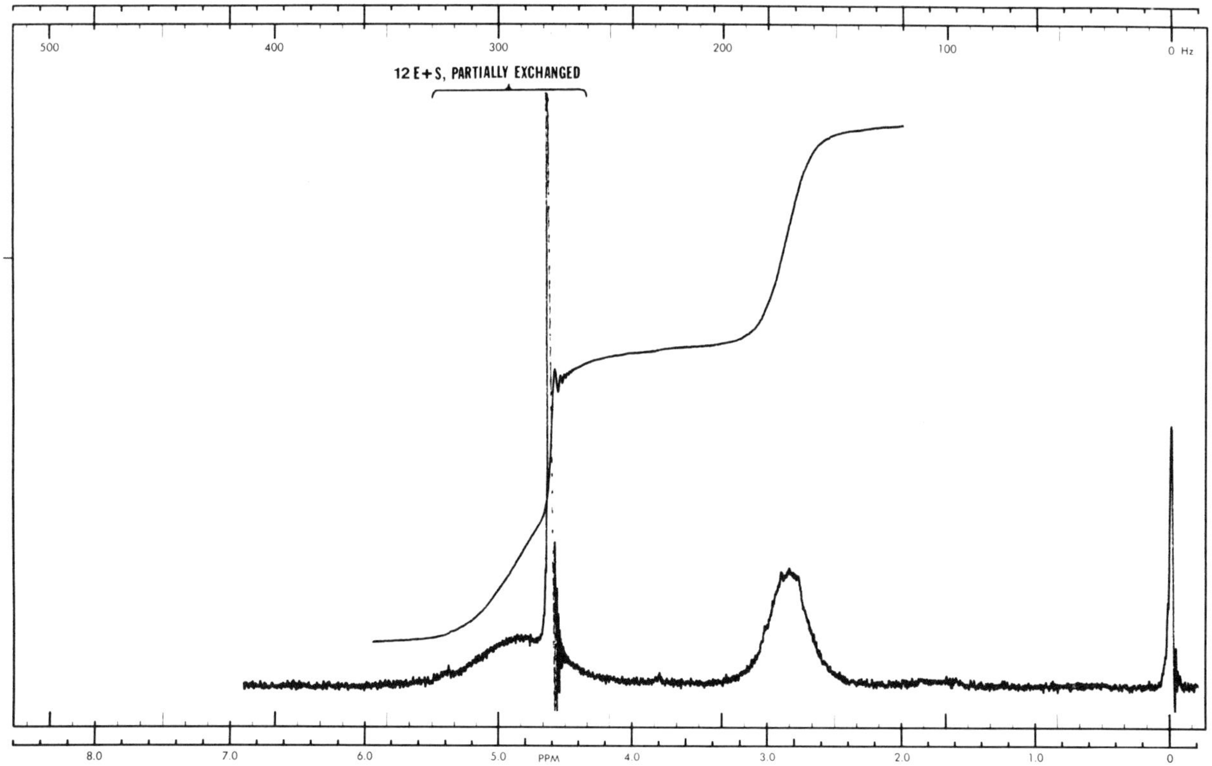

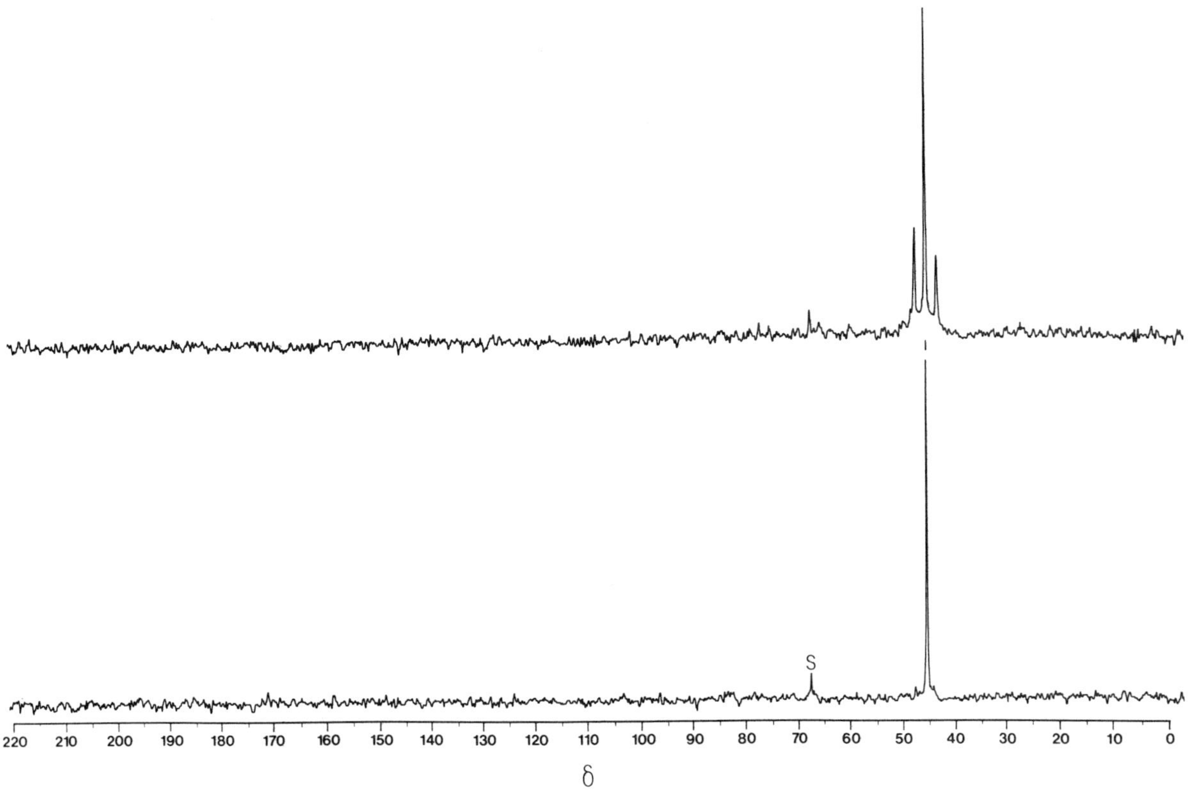
S
220
210
200
190
180
170
160
150
140
130
120
110
100
90
80
70
60
50
40
30
20
10
0
δ

Problem 1-28

1-28

$C_{10}H_{18}O_2$

MS: m170, 153, 152, 134

IR: 3300 cm^{-1} (approaches 3600 cm^{-1} at infinite dilution)

UV: none

CMR: 5 sec delay

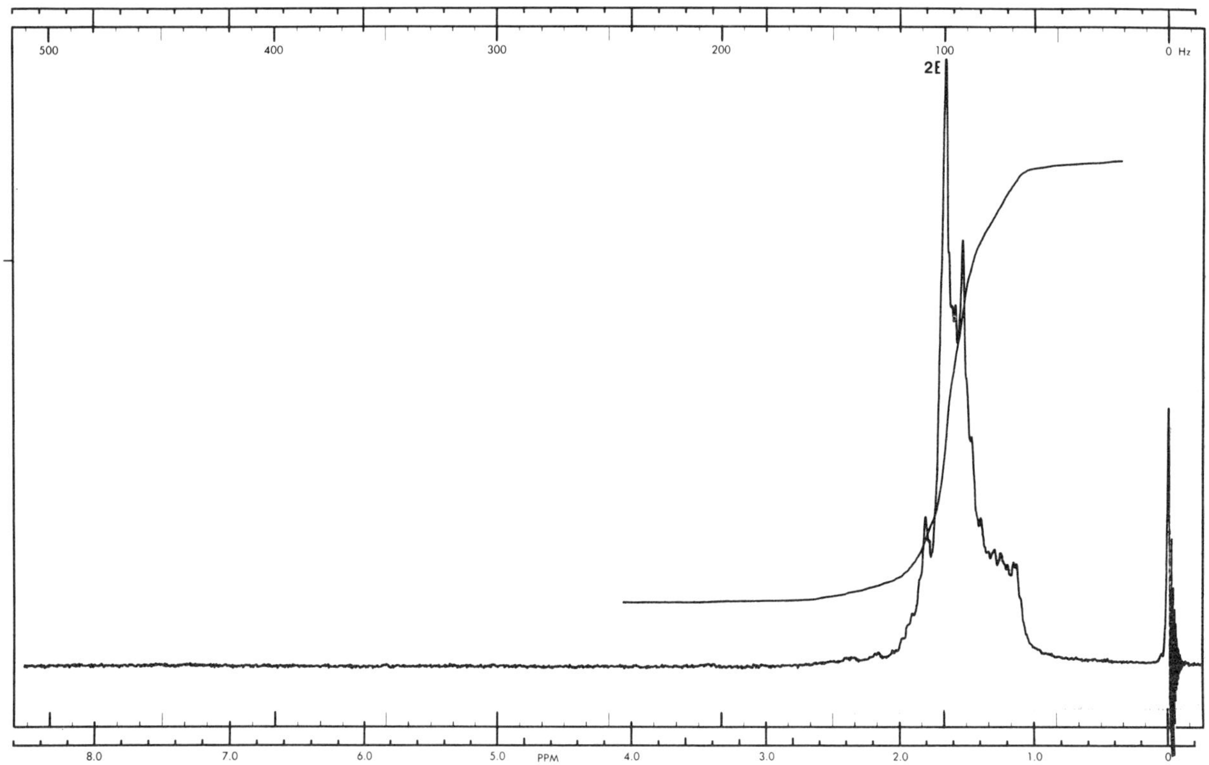

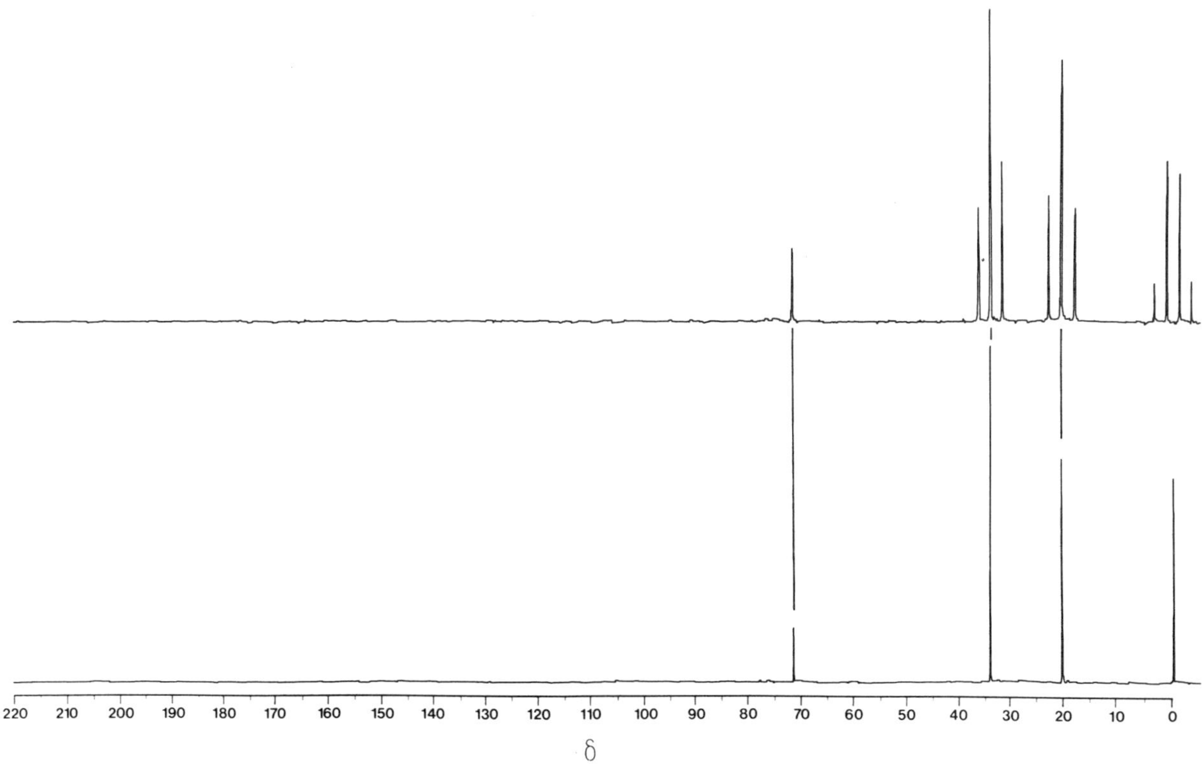
220
210
200
190
180
170
160
150
140
130
120
110
100
90
80
70
60
50
40
30
20
10
0
δ

Problem 1-29

1-29 — $C_7H_{10}O$

MS: m110, 82, 68, 42

IR: 1750 cm^{-1}

UV: 298 (ε 20)

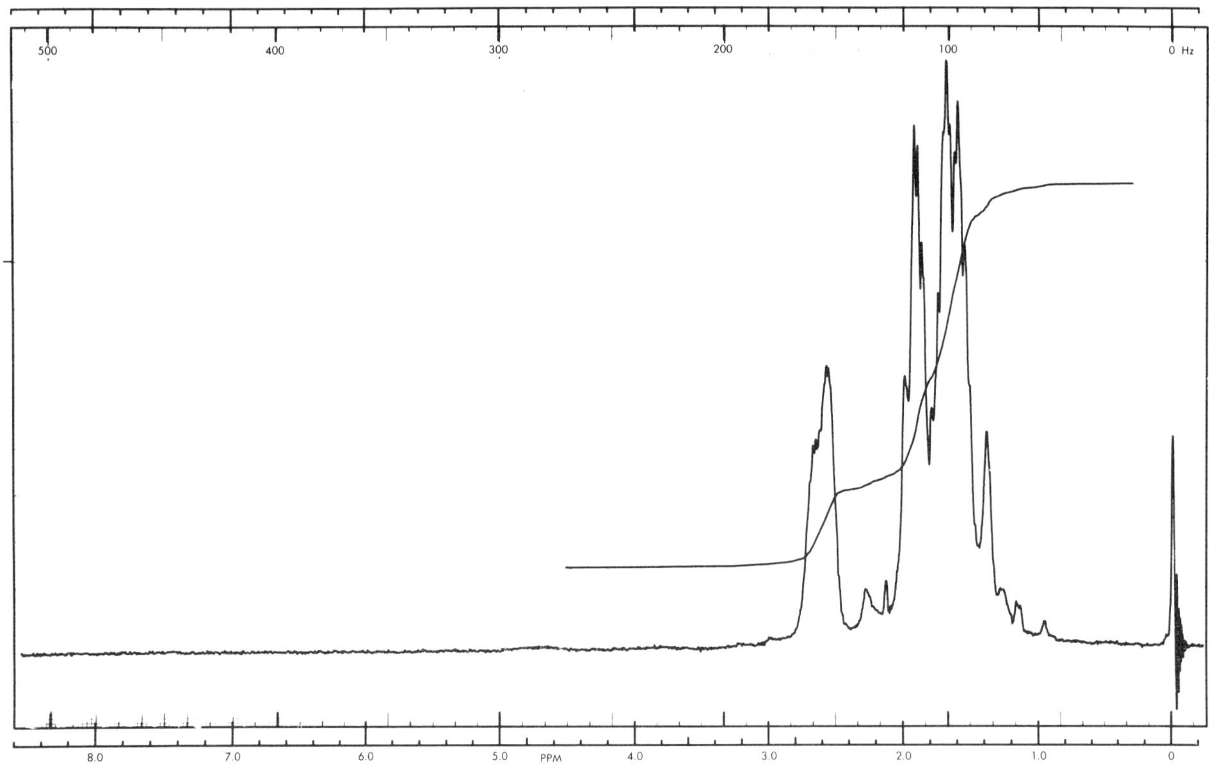

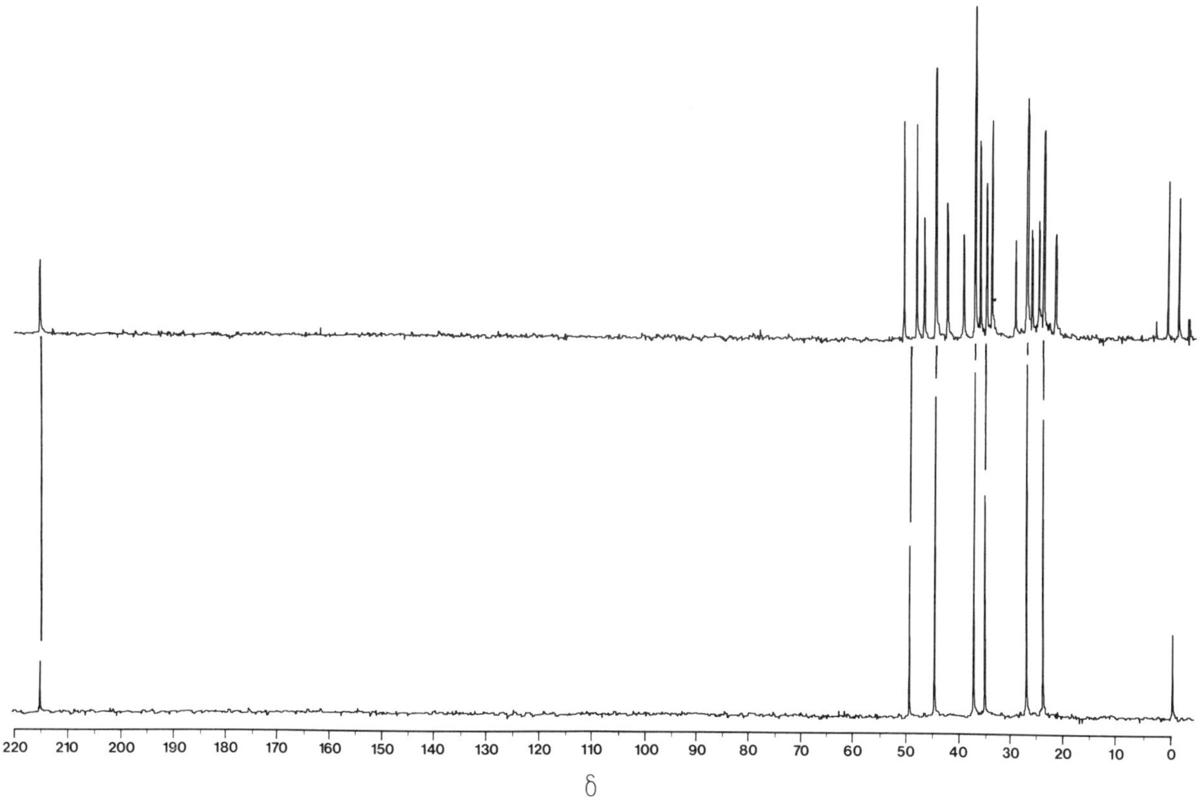
220
210
200
190
180
170
160
150
140
130
120
110
100
90
80
70
60
50
40
30
20
10
0
δ

Problem 1-30

1-30

$C_7H_{10}O$

MS: m110(w), 93, 92, 85, 68

IR: 3400, 3300, 2110 cm^{-1}

UV: none

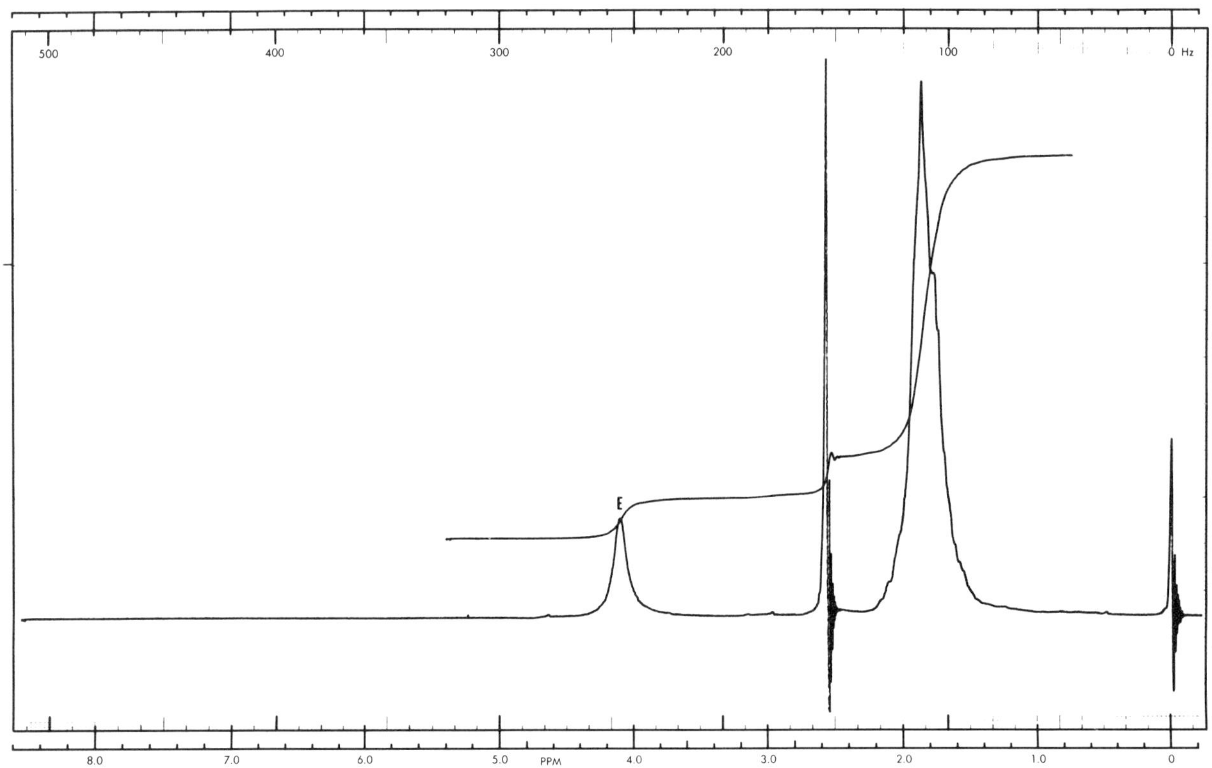

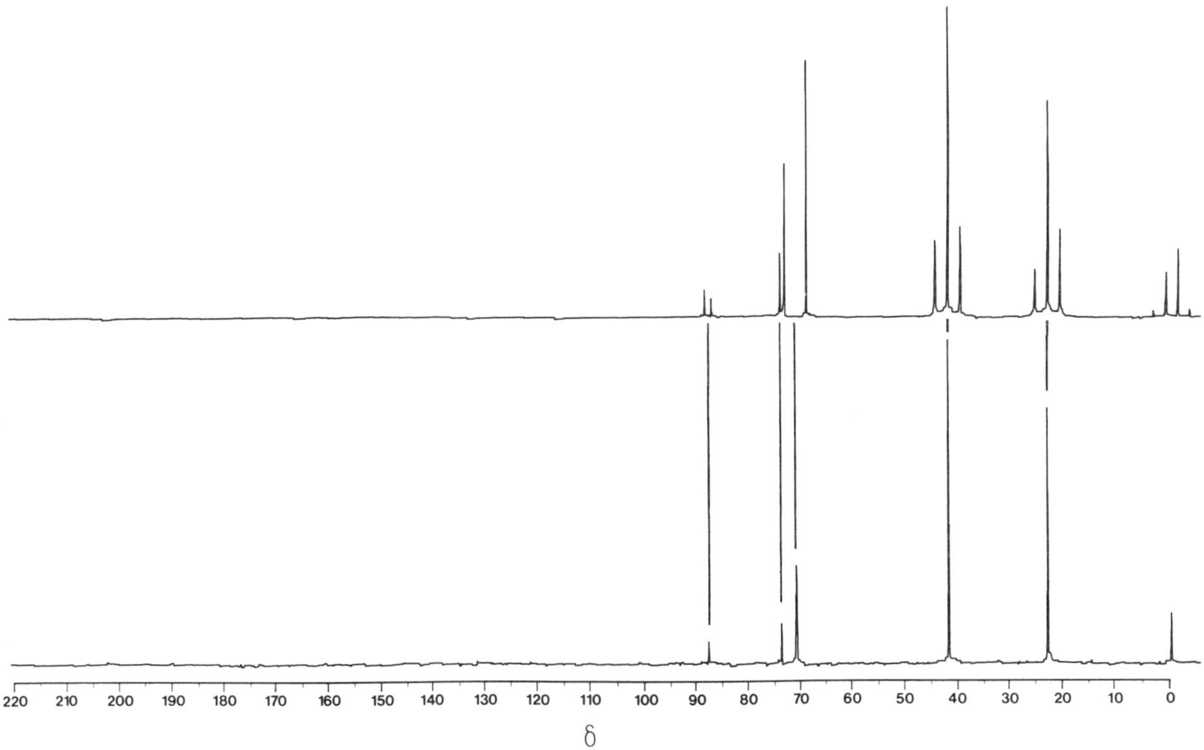
220
210
200
190
180
170
160
150
140
130
120
110
100
90
80
70
60
50
40
30
20
10
0
δ

Problem 1-31

1-31	MS:	m158(s), 143, 117
$C_{10}H_{10}N_2$	IR:	-
	UV:	Strong absorption

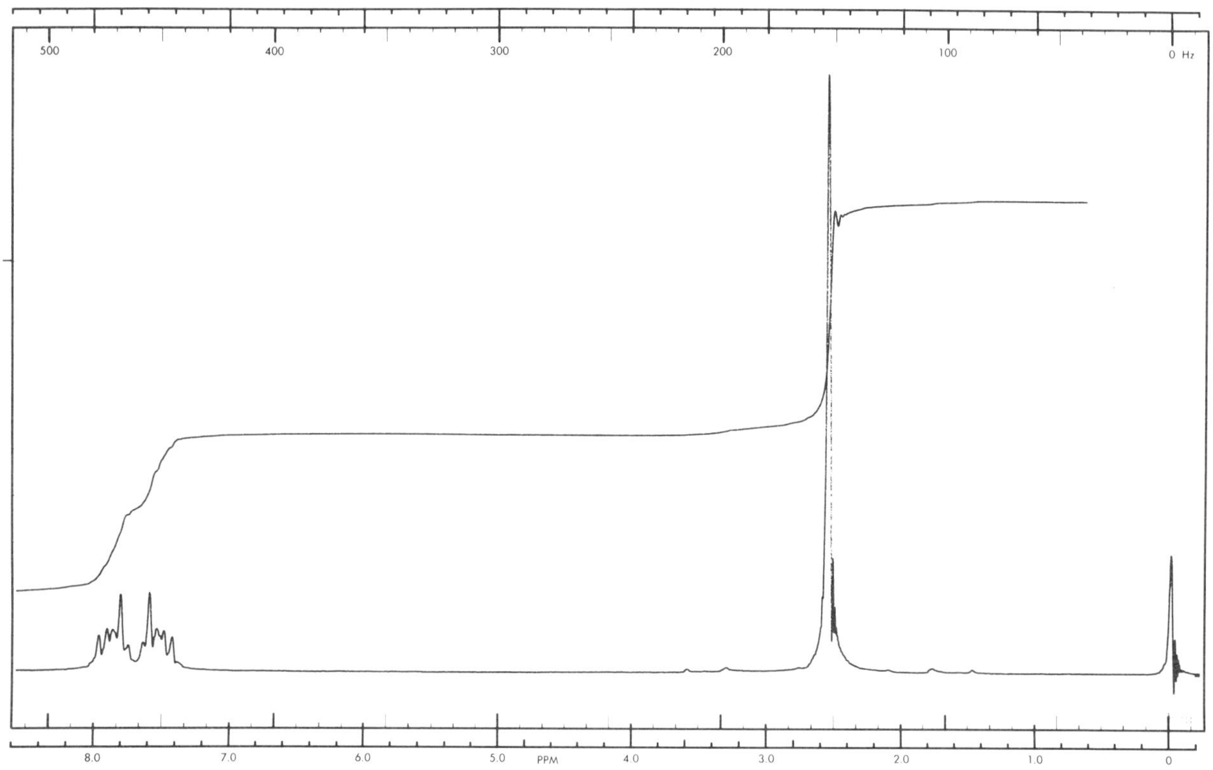

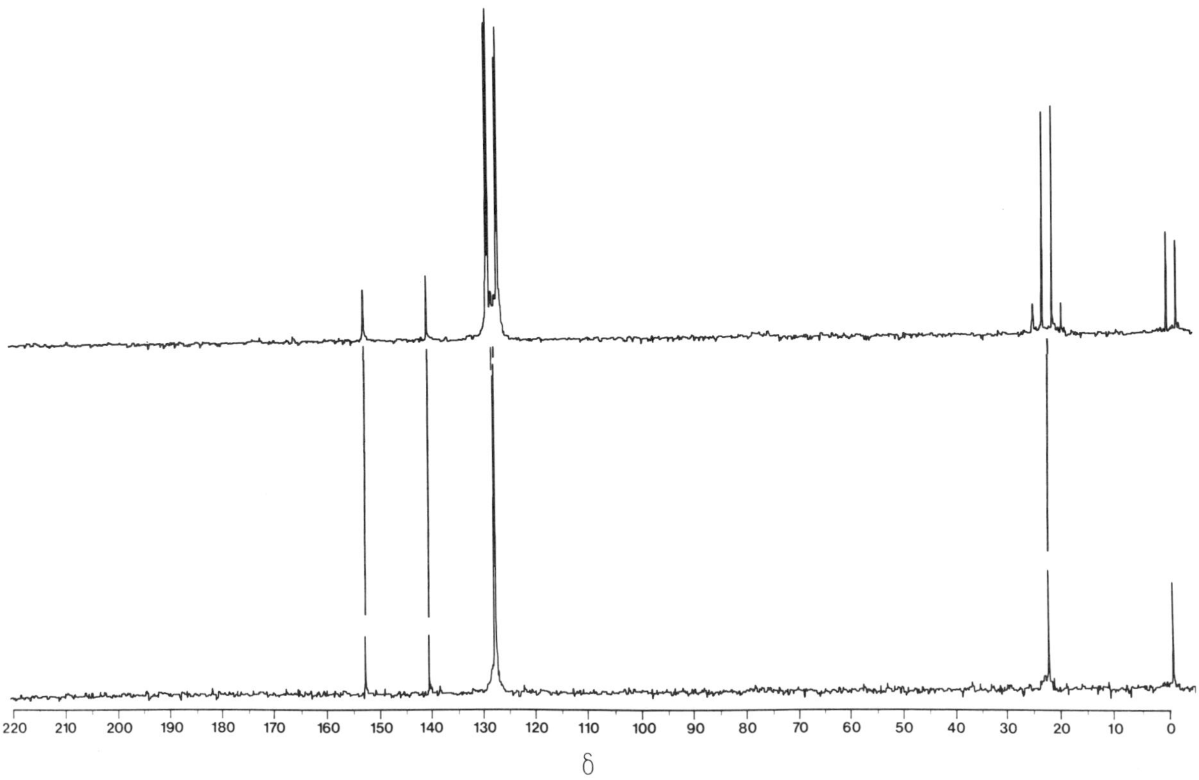
220 210 200 190 180 170 160 150 140 130 120 110 100 90 80 70 60 50 40 30 20 10 0
δ

Problem 1-32

1-32

$C_6H_2Cl_2N_2O_4$

MS: m236(s), 206, 190

IR: -

UV: Strong absorption

PMR: Upper scan offset 200 Hz

CMR: d_6-Me_2CO + TMS

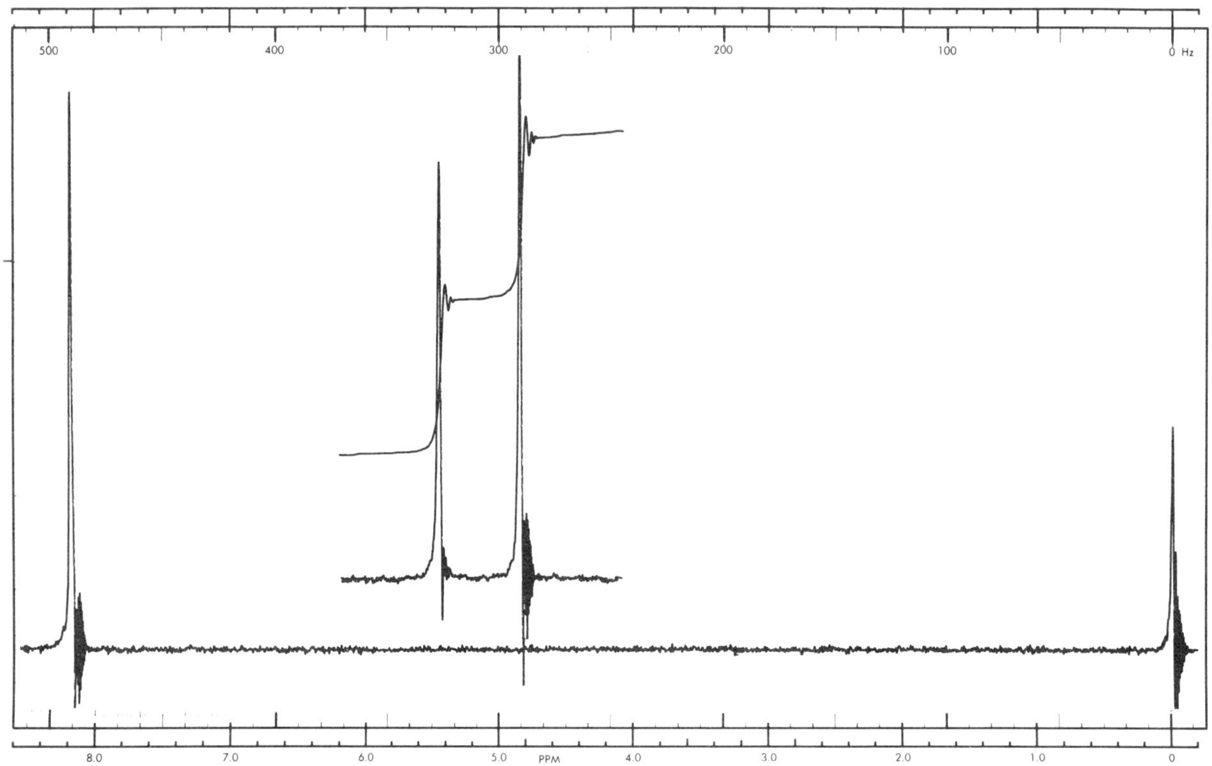

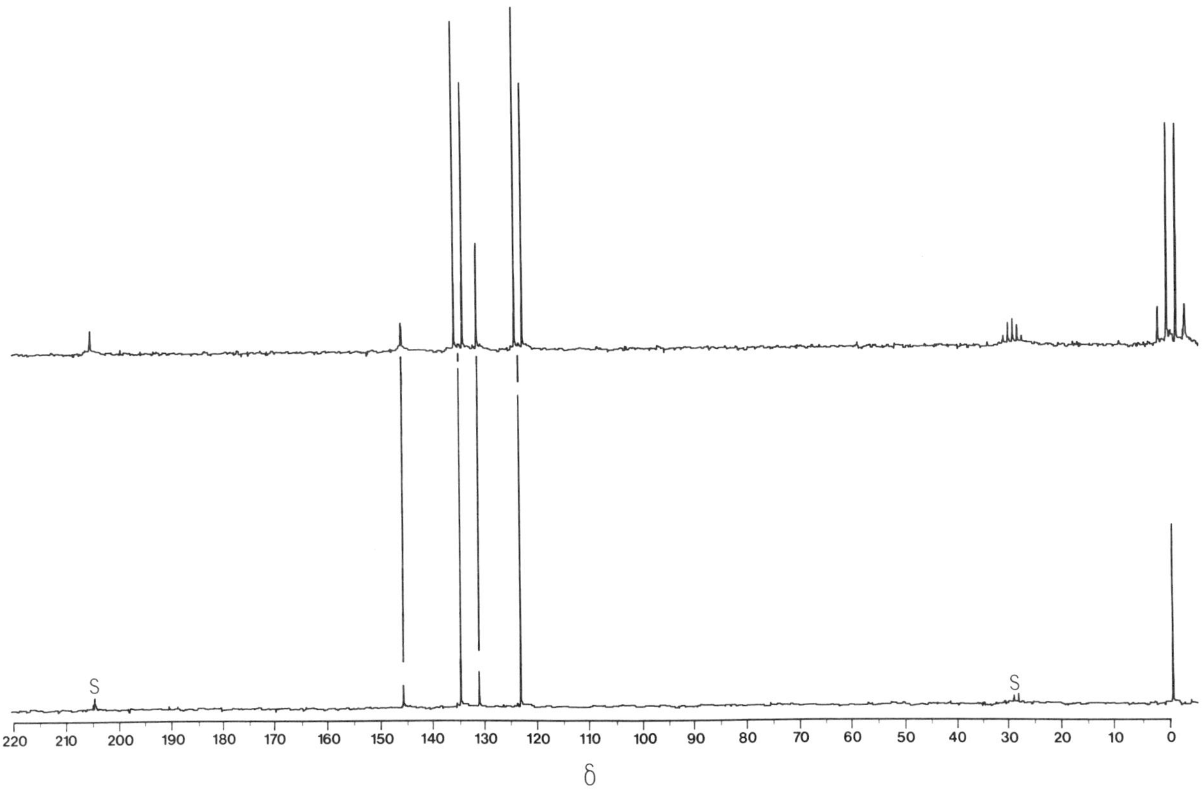
S
S
220
210
200
190
180
170
160
150
140
130
120
110
100
90
80
70
60
50
40
30
20
10
0
δ

Problem 1-33

1-33 MS: m112, 97, 84

$C_7H_{12}O$ IR: 1710 cm^{-1}

UV: 287 (ε 15)

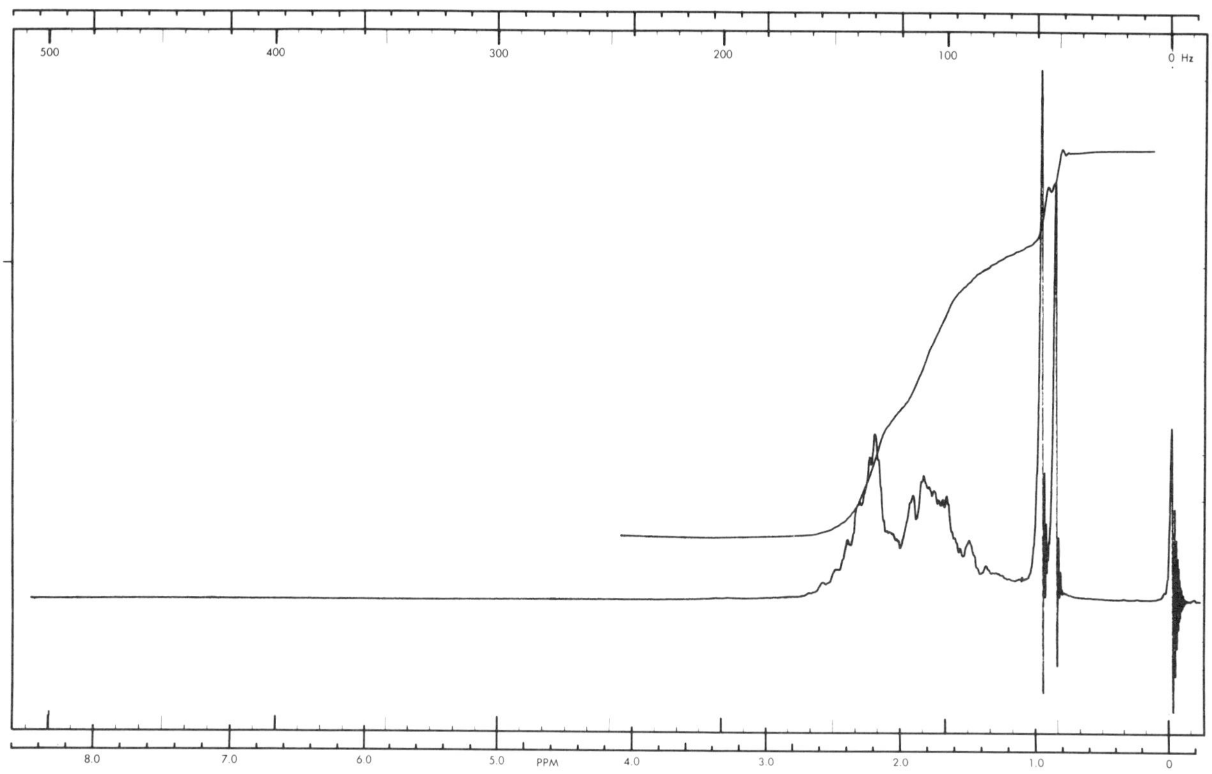

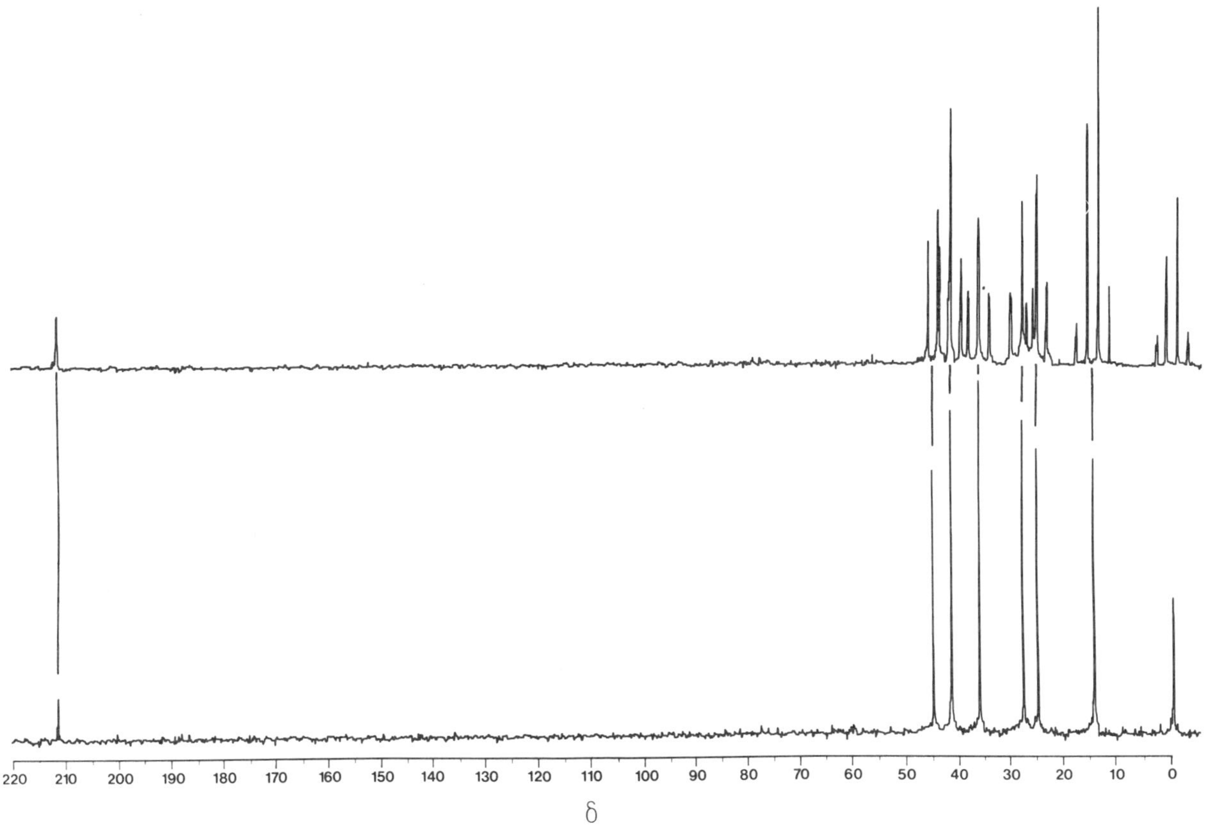
220
210
200
190
180
170
160
150
140
130
120
110
100
90
80
70
60
50
40
30
20
10
0
δ

Problem 1-34

1-34

$C_8H_{16}O_2$

MS: m144(w), 126, 108, 31

IR: 3300 cm^{-1}

UV: none

PMR and CMR: d_6-Me_2CO + TMS

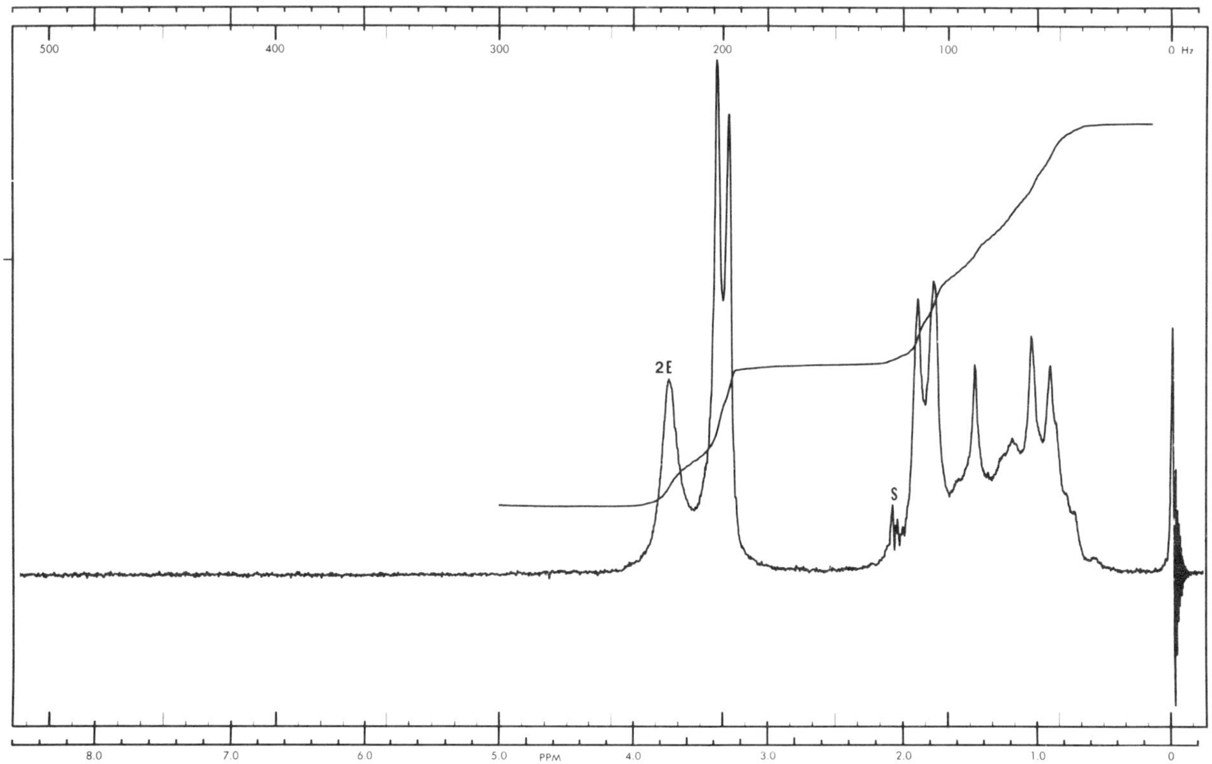

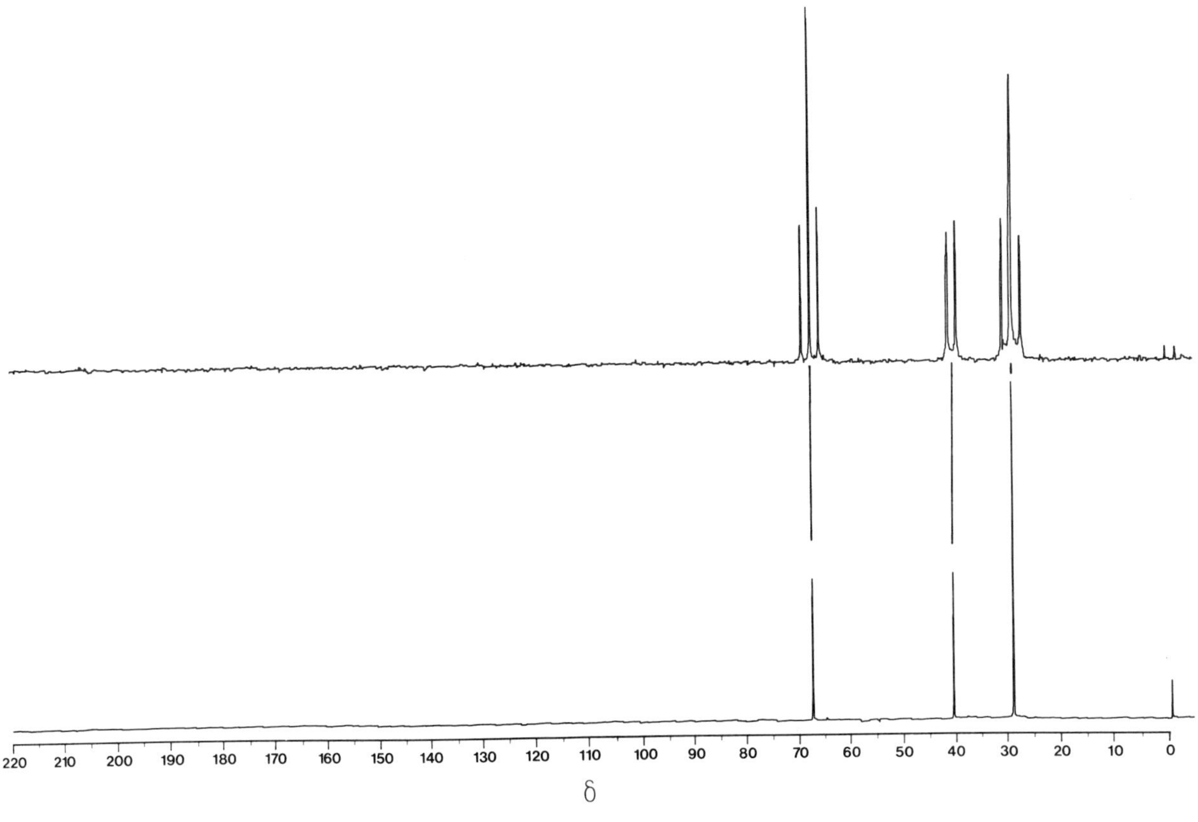
220
210
200
190
180
170
160
150
140
130
120
110
100
90
80
70
60
50
40
30
20
10
0
δ

Problem 1-35

1-35 MS: m198(s), 185, 183

$C_9H_{11}Br$ IR: -

UV: Strong absorption

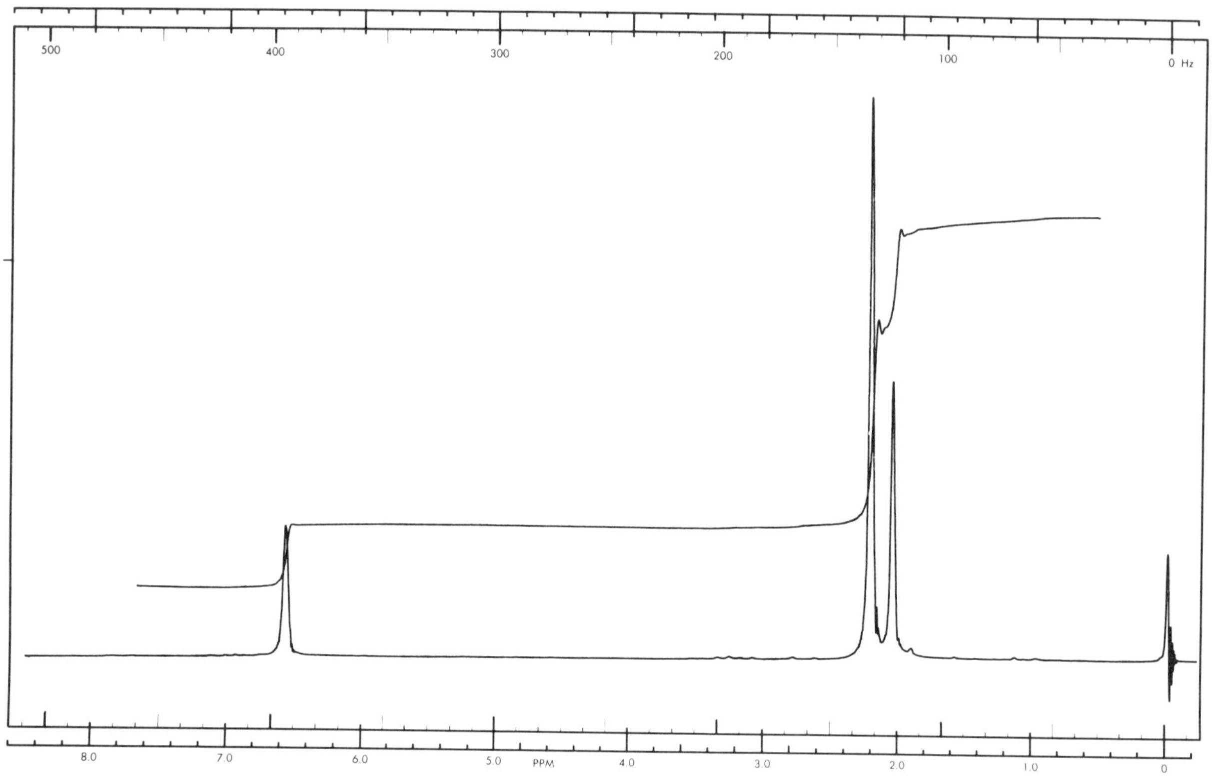

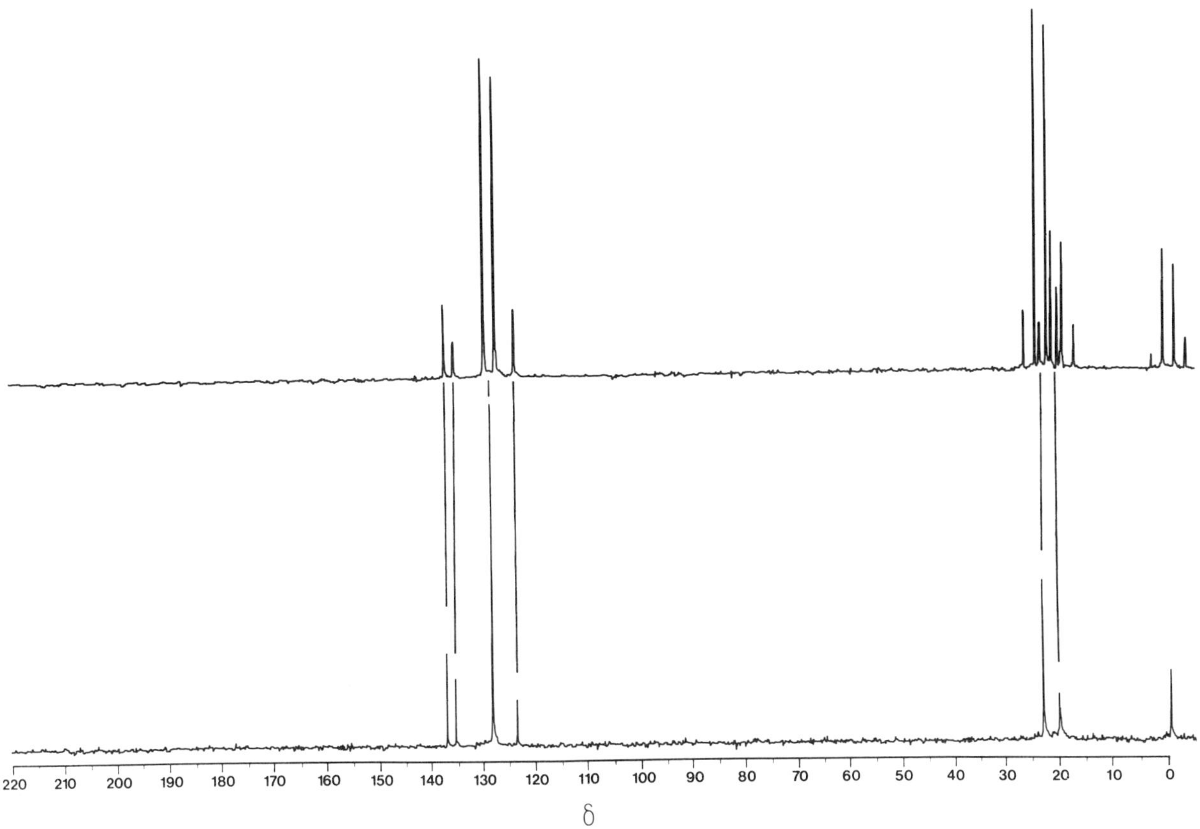
220
210
200
190
180
170
160
150
140
130
120
110
100
90
80
70
60
50
40
30
20
10
0
δ

Problem 1-36

1-36

$C_9H_{23}N_3$

MS: m173(w), 158, 115, 58

IR: -

UV: none

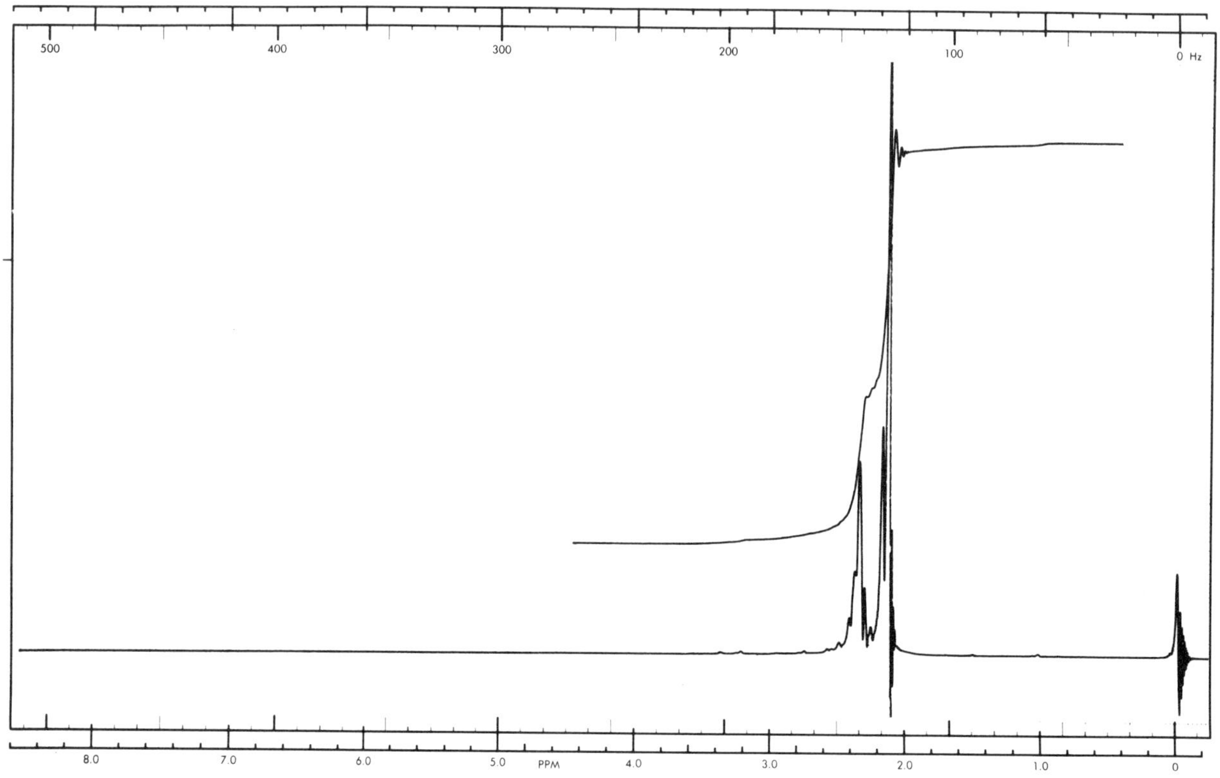

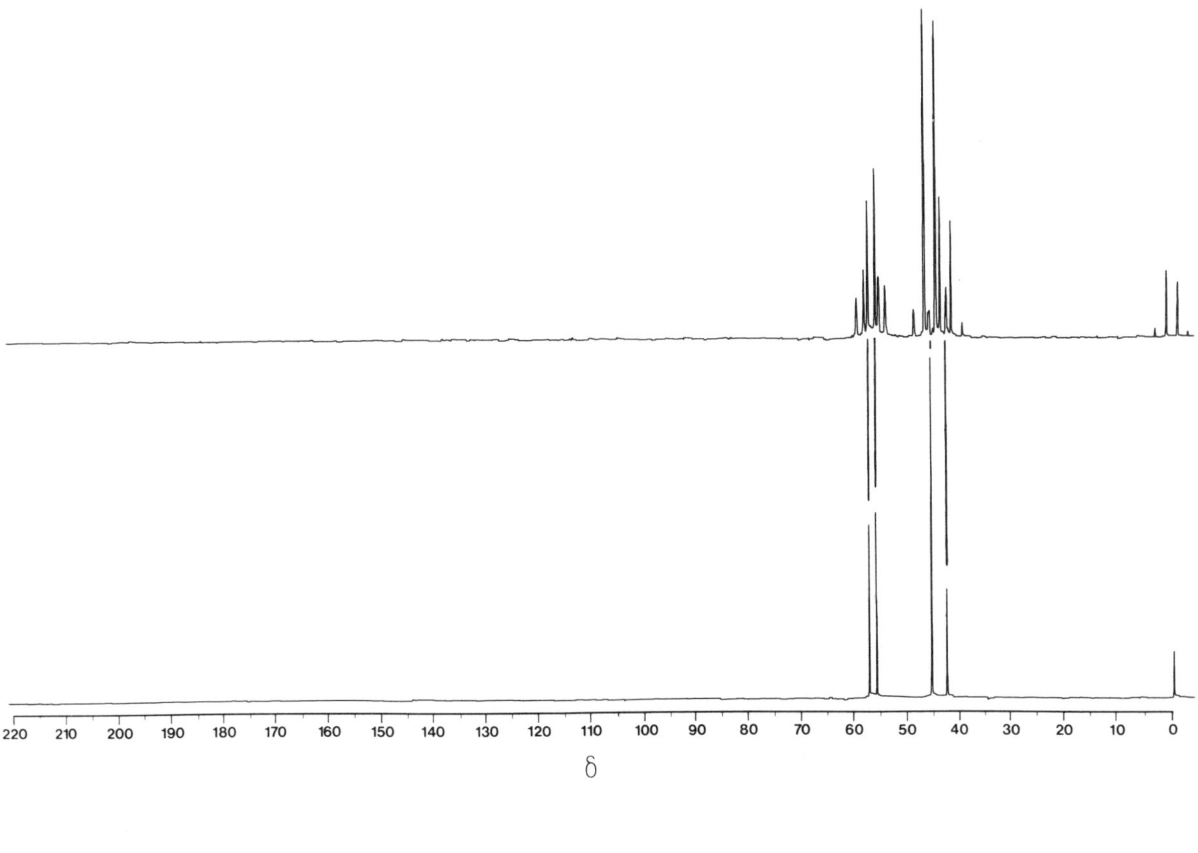
220
210
200
190
180
170
160
150
140
130
120
110
100
90
80
70
60
50
40
30
20
10
0
δ

Problem 1-37

1-37 MS: m122, 107, 95, 93, 66

C_9H_{14} IR: 1640 cm^{-1}

UV: none

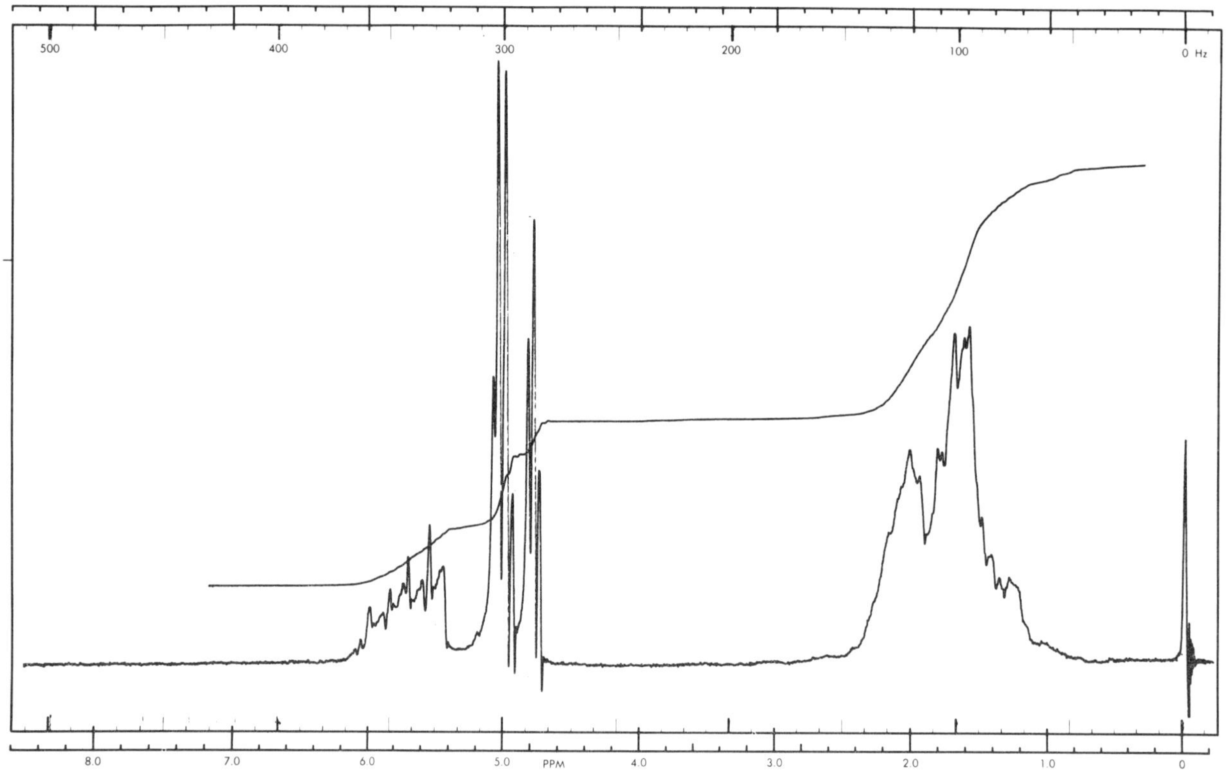

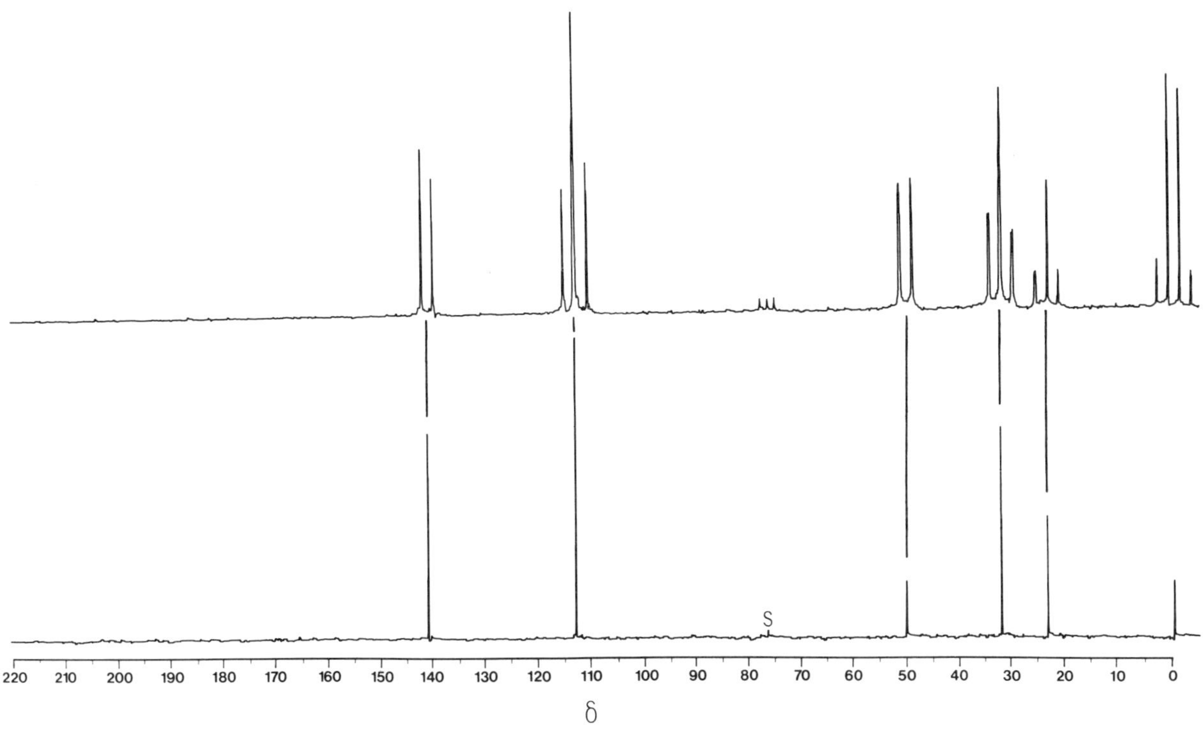
S
220 210 200 190 180 170 160 150 140 130 120 110 100 90 80 70 60 50 40 30 20 10 0
δ

Problem 1-38

1-38

$C_{17}H_{16}O_2$

MS: m252, 105, 77

IR: 1700 cm^{-1}

UV: Strong absorption

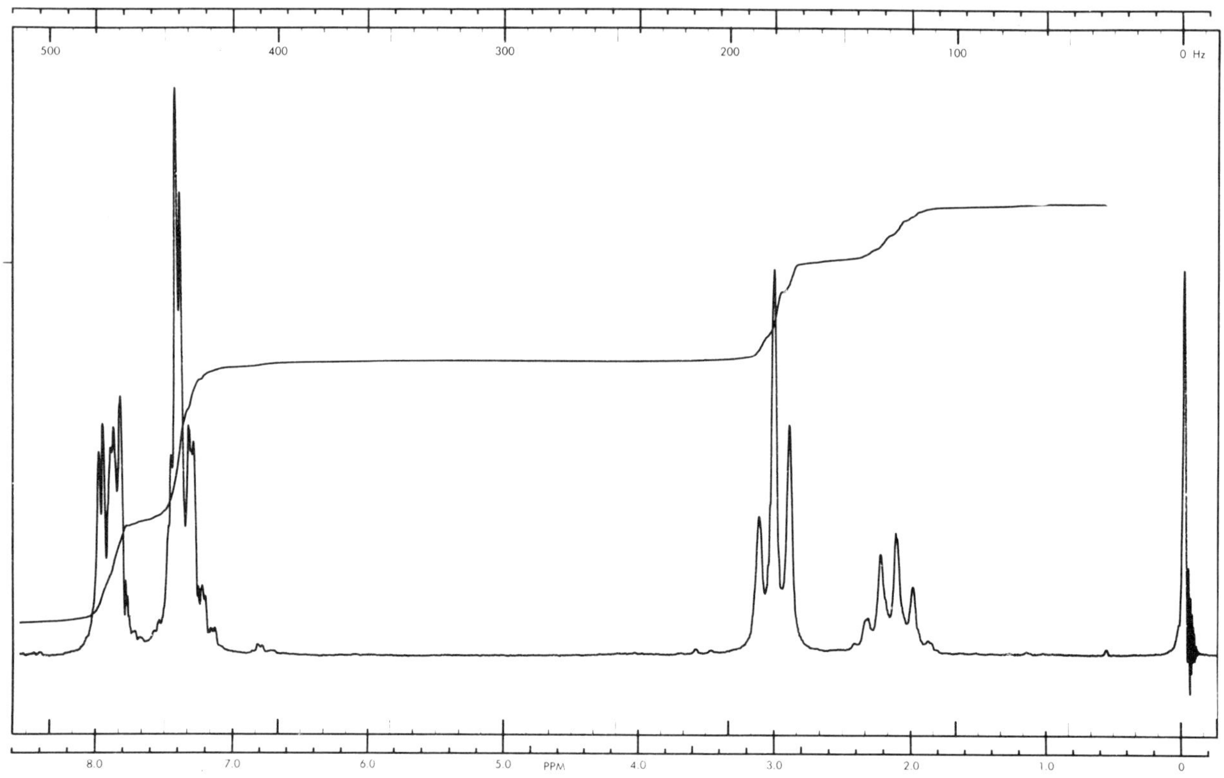

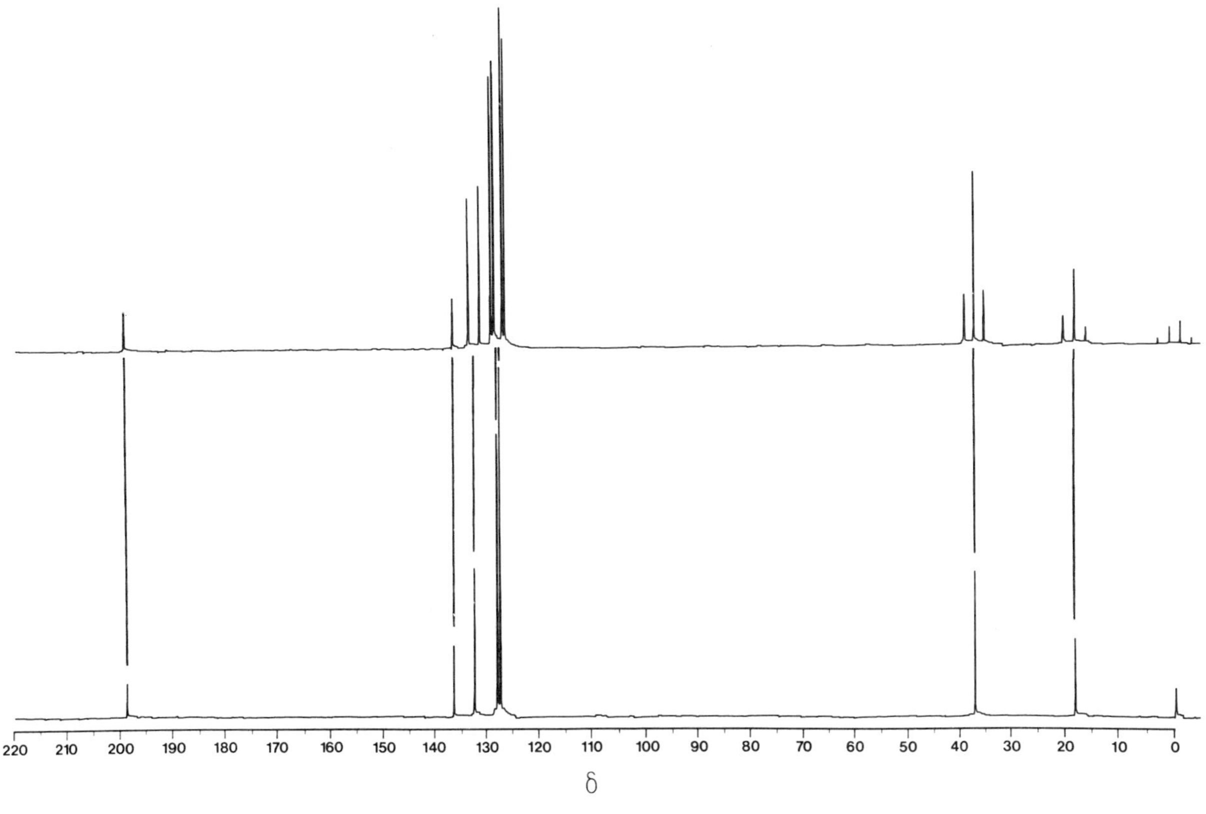
220
210
200
190
180
170
160
150
140
130
120
110
100
90
80
70
60
50
40
30
20
10
0
δ

Problem 1-39

1-39

$C_6H_5N_3$

MS: m119(s), 118, 91, 90

IR: 3300-2700 cm^{-1} (broad)

UV: Strong absorption

PMR: Upper scan offset 600 Hz

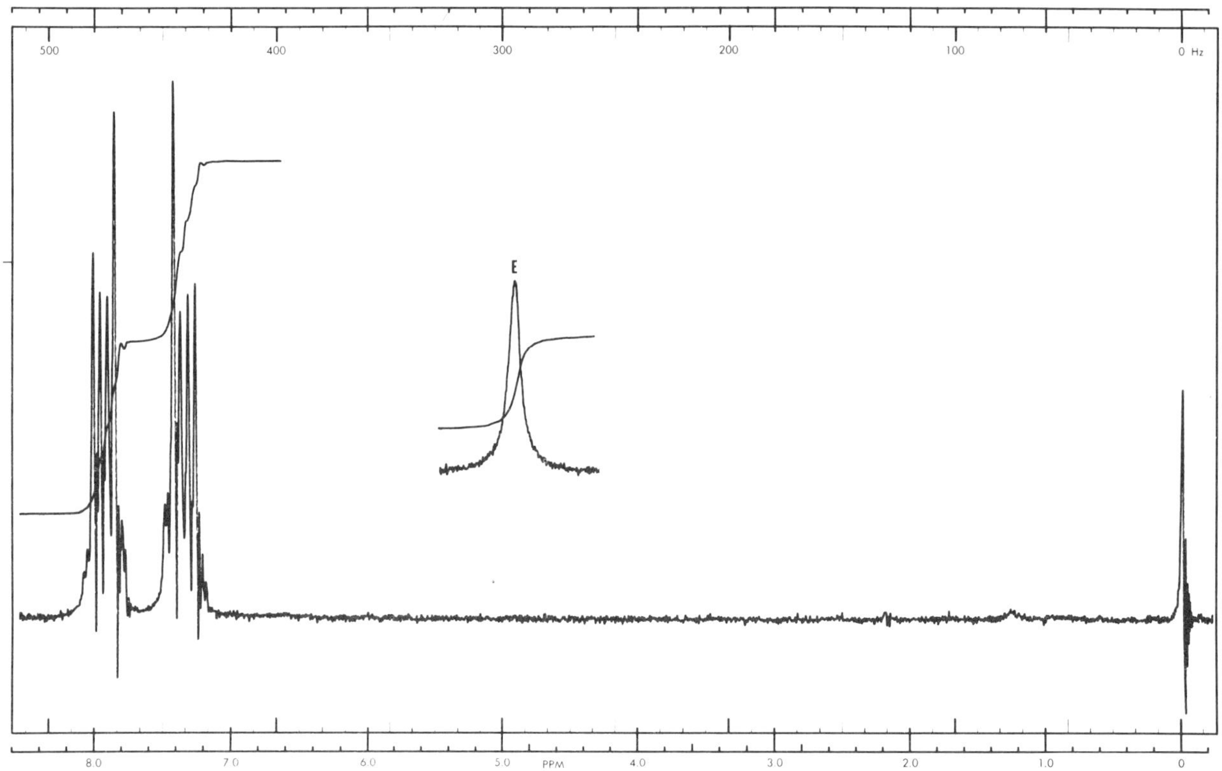

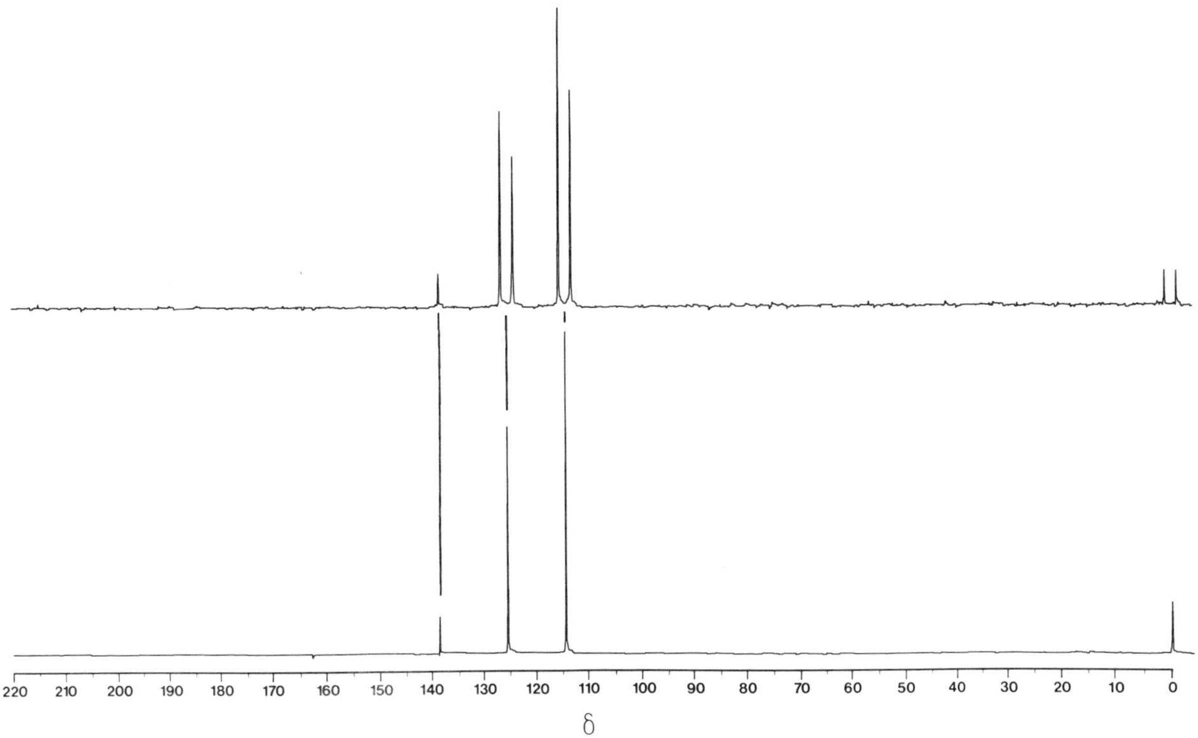
220
210
200
190
180
170
160
150
140
130
120
110
100
90
80
70
60
50
40
30
20
10
0
δ

Problem 1-40

1-40

$C_{14}H_{22}O_8$

MS: m318(w), 273, 245, 159

IR: 1724 cm^{-1}

UV: none

PMR: d_6-Me_2CO + TMS

CMR: d_6-Me_2CO + TMS; 5 sec delay

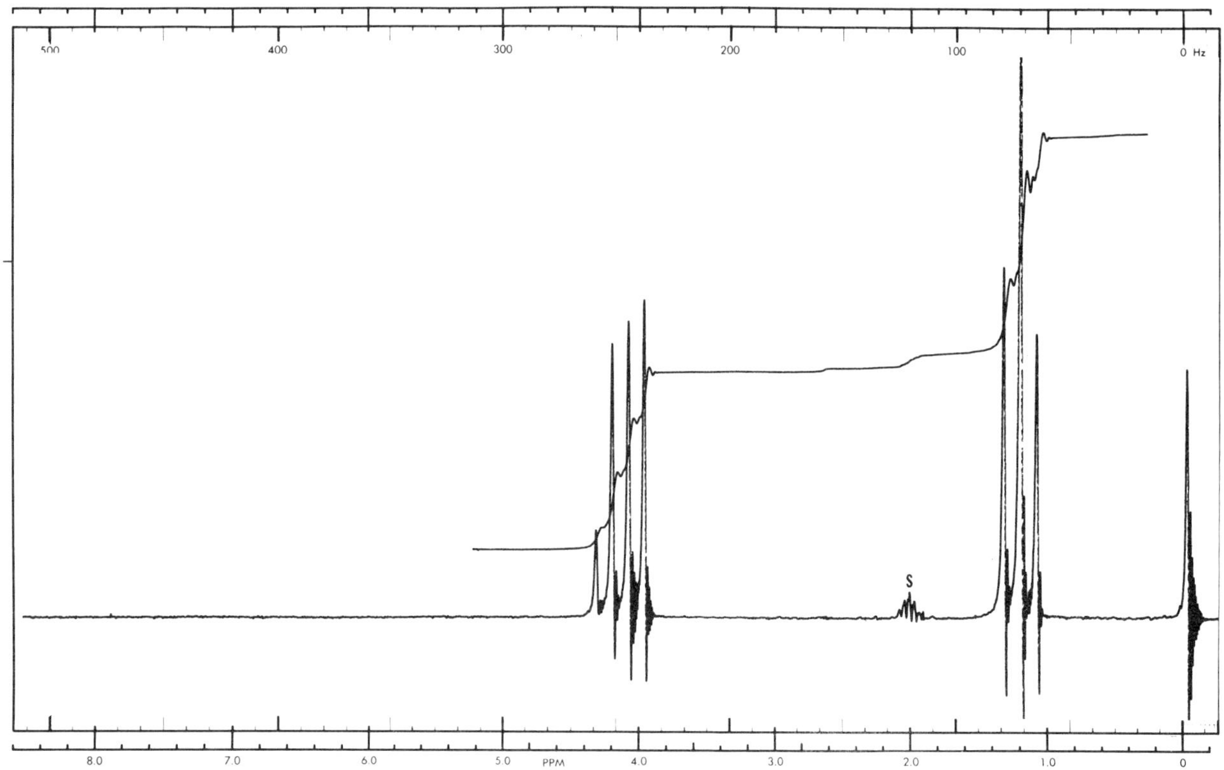

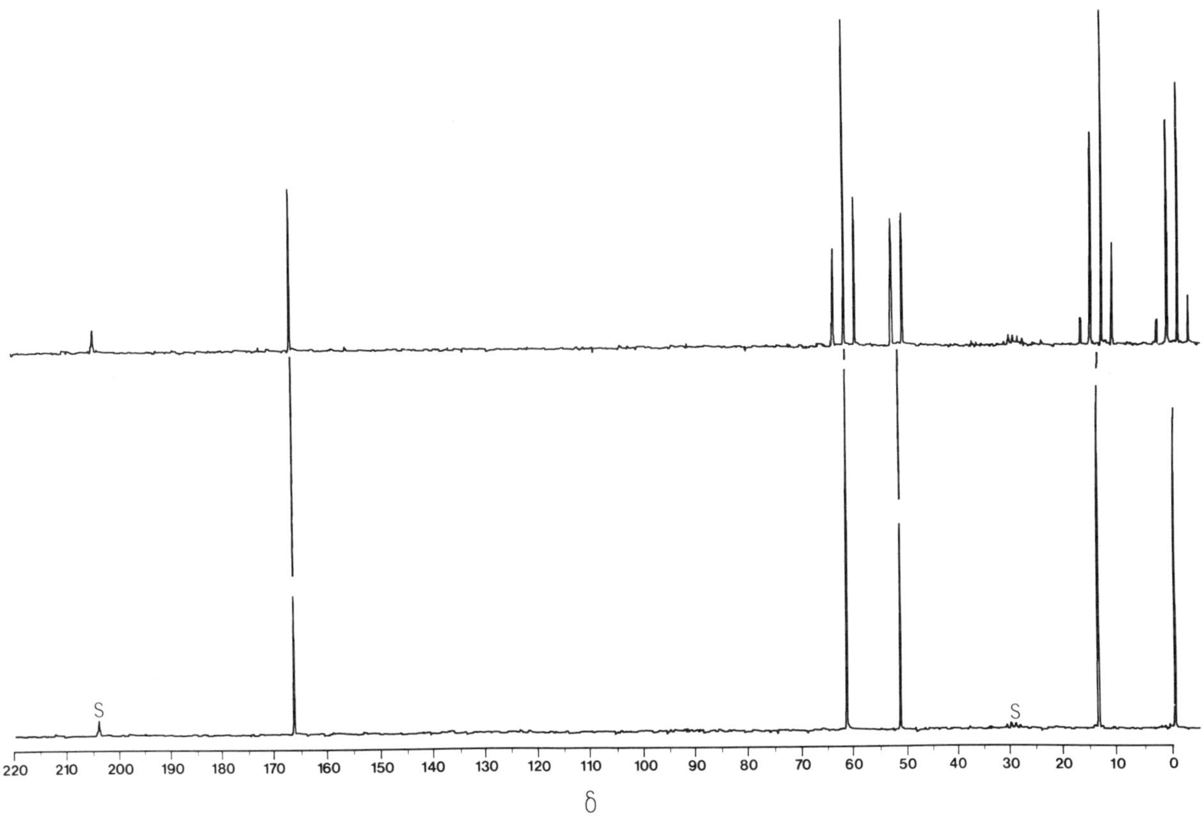
S
S
220
210
200
190
180
170
160
150
140
130
120
110
100
90
80
70
60
50
40
30
20
10
0
δ

Problem 1-41

1-41

$C_{10}H_{20}O_2$

MS: m172(w), 157, 154, 59

IR: 3300, 1650 cm^{-1}

UV: none

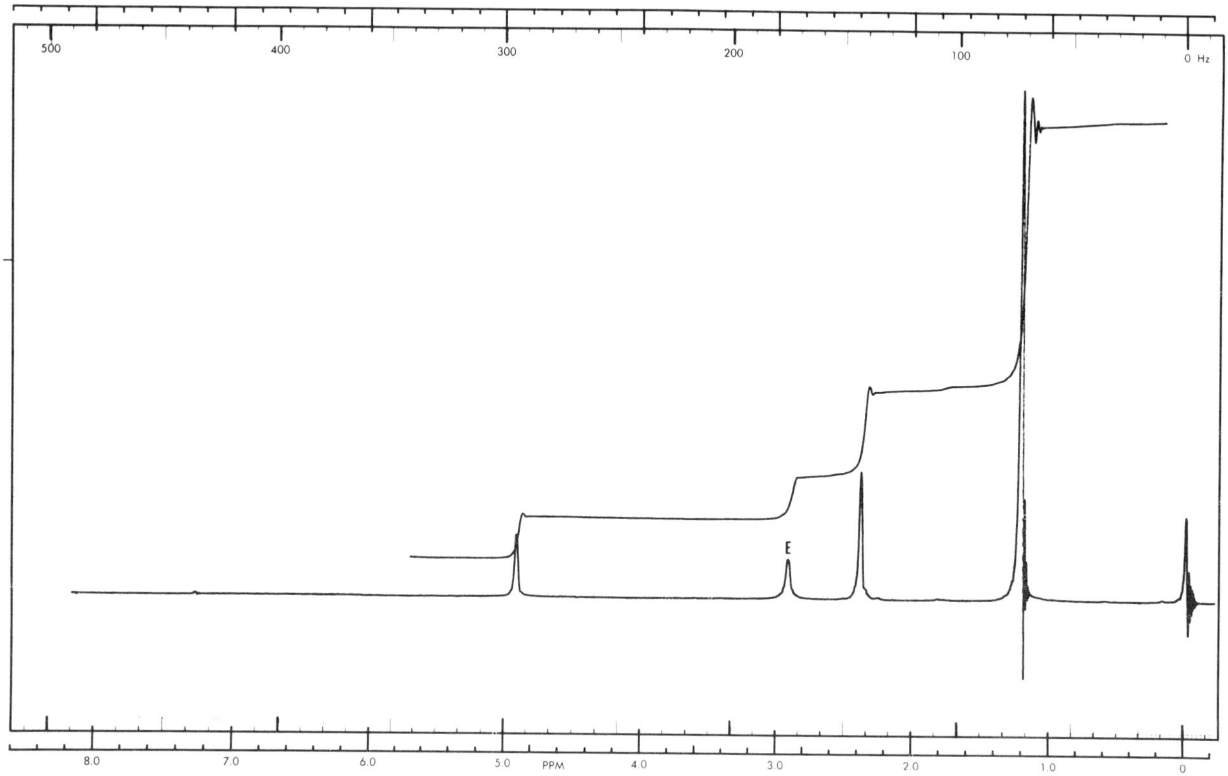

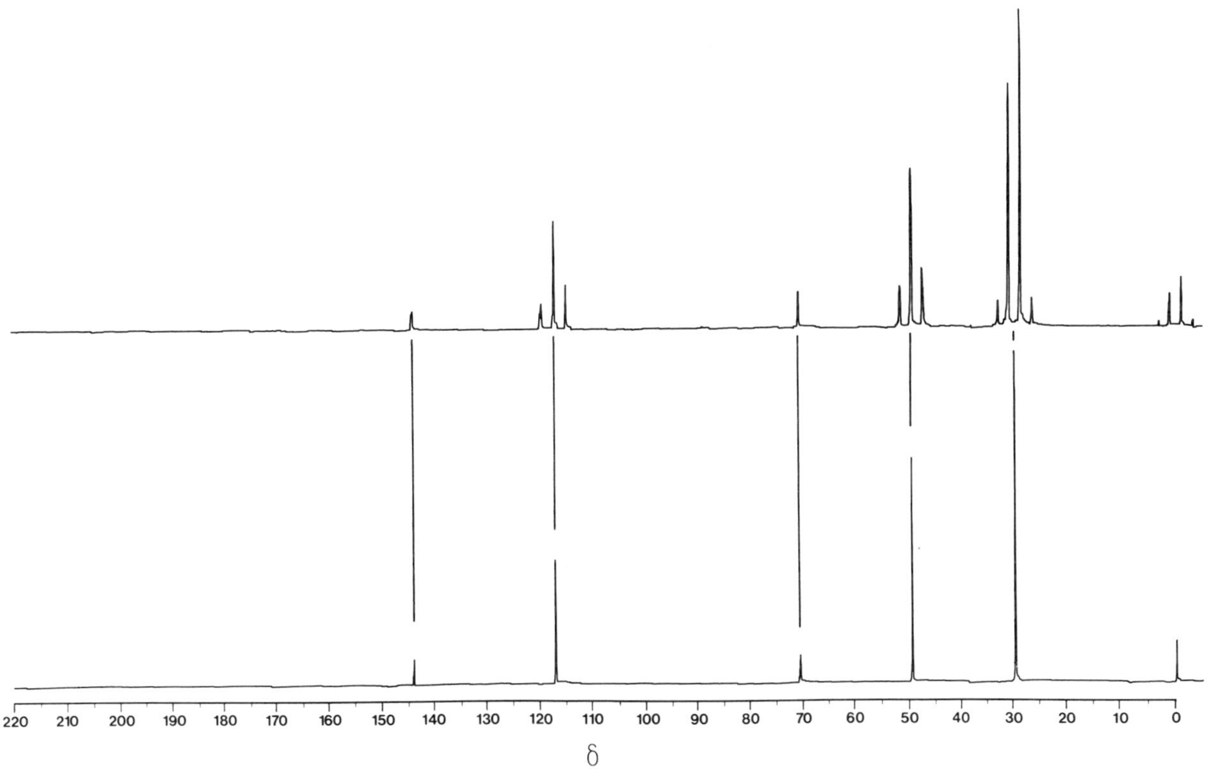
220
210
200
190
180
170
160
150
140
130
120
110
100
90
80
70
60
50
40
30
20
10
0
δ

Problem 1-42

1-42

$C_9H_{10}O_4$

MS: m182, 181, 153, 152

IR: 3200 (broad), 1620 cm^{-1}

UV: Strong absorption

PMR: Upper scan offset 200 Hz

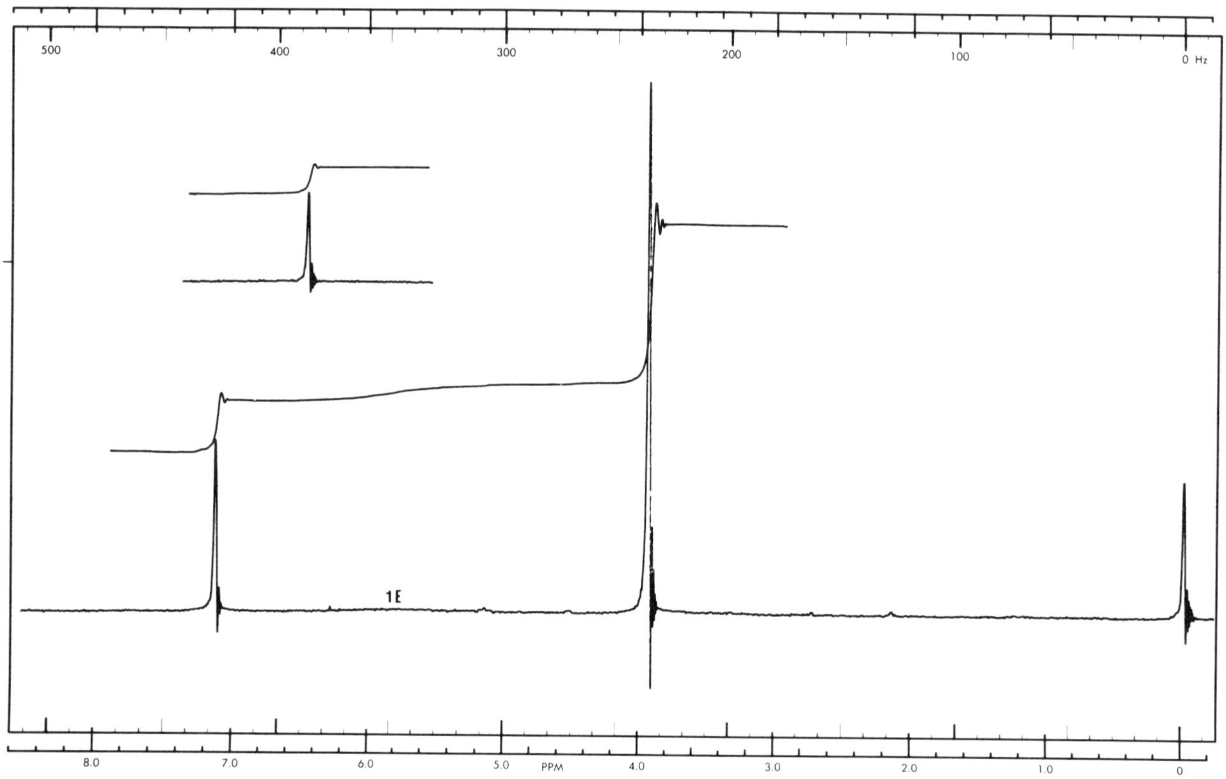

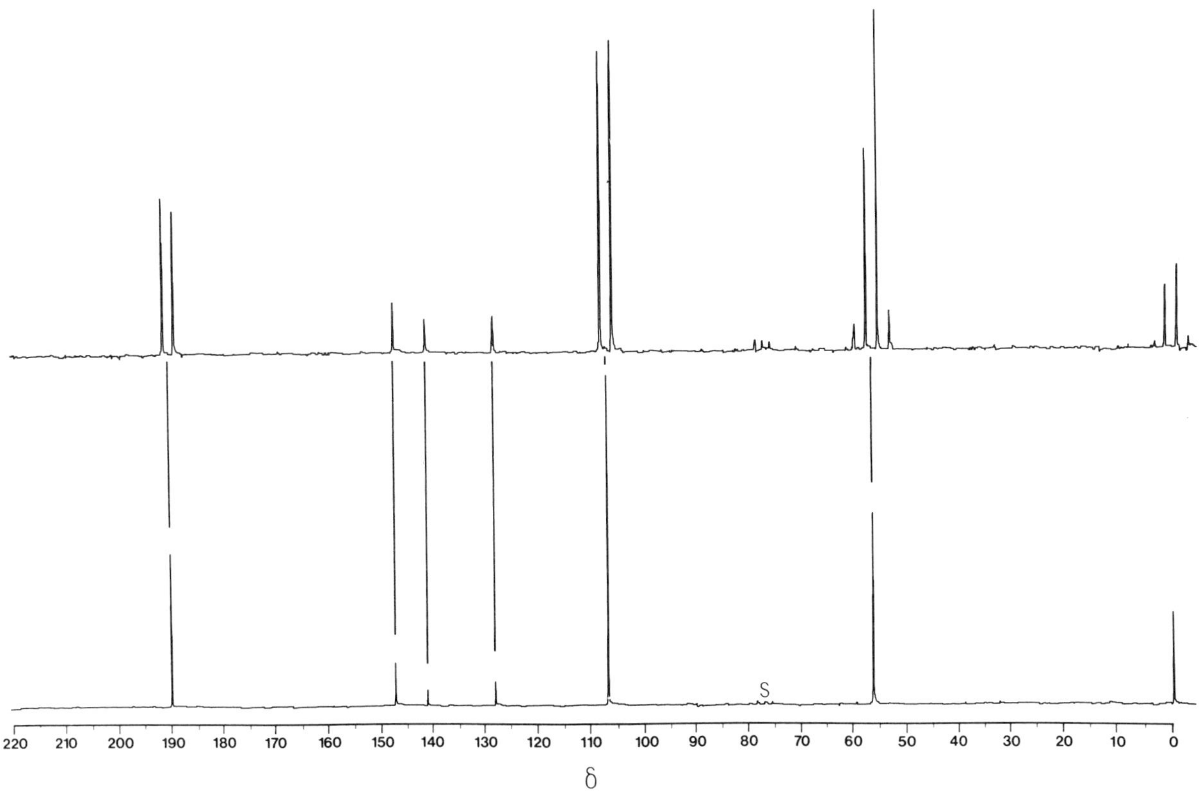
S
220 210 200 190 180 170 160 150 140 130 120 110 100 90 80 70 60 50 40 30 20 10 0
δ

Problem 1-43

1-43

$C_{10}H_{14}O$

MS: m150, 135, 122, 121

IR: 3450 cm^{-1}

UV: Strong absorption

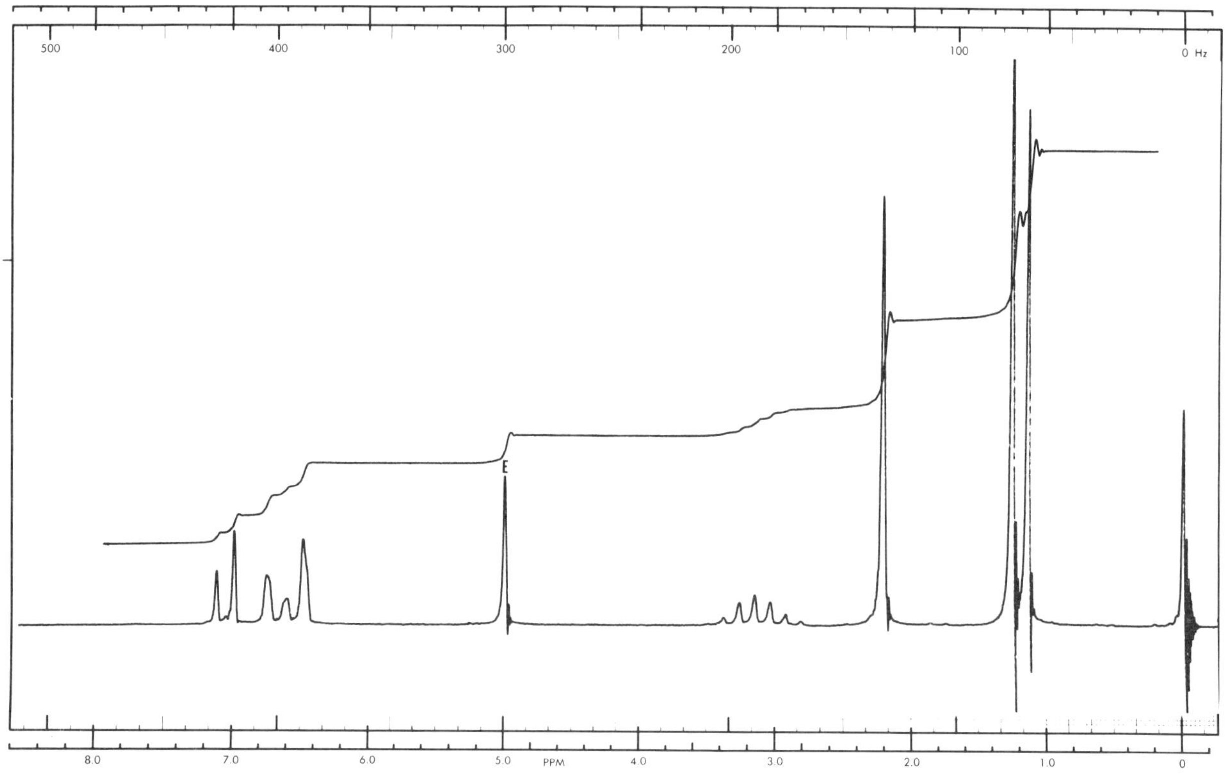

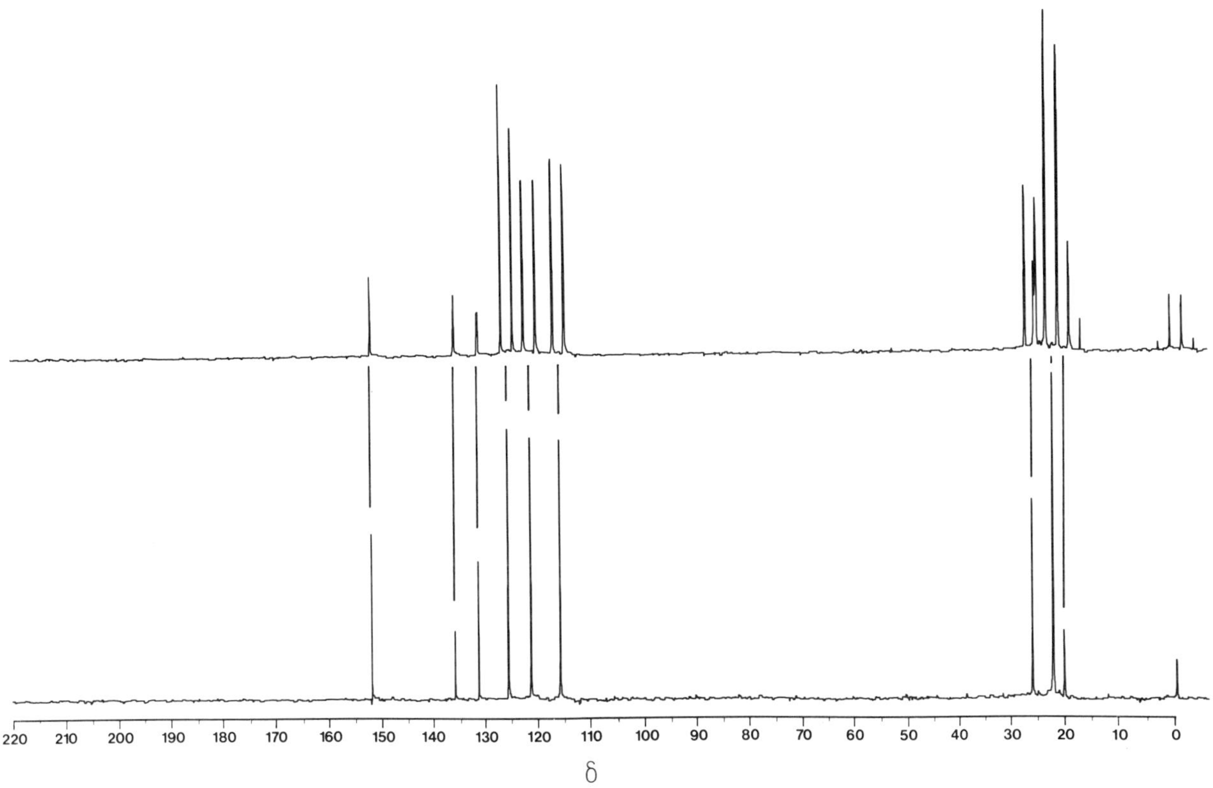
220 210 200 190 180 170 160 150 140 130 120 110 100 90 80 70 60 50 40 30 20 10 0
δ

Problem 1-44

1-44	MS:	m207(s), 180, 130, 105
$C_{14}H_9NO$	IR:	2230, 1660 cm^{-1}
	UV:	Strong absorption

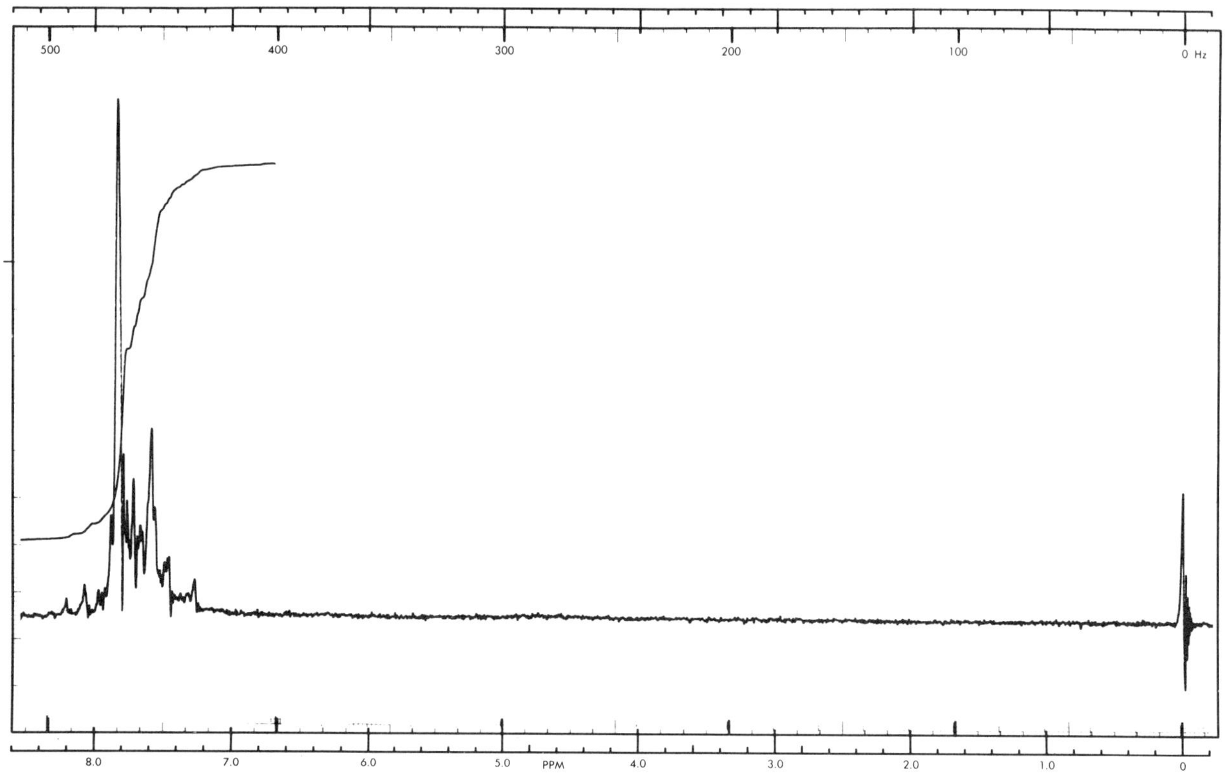

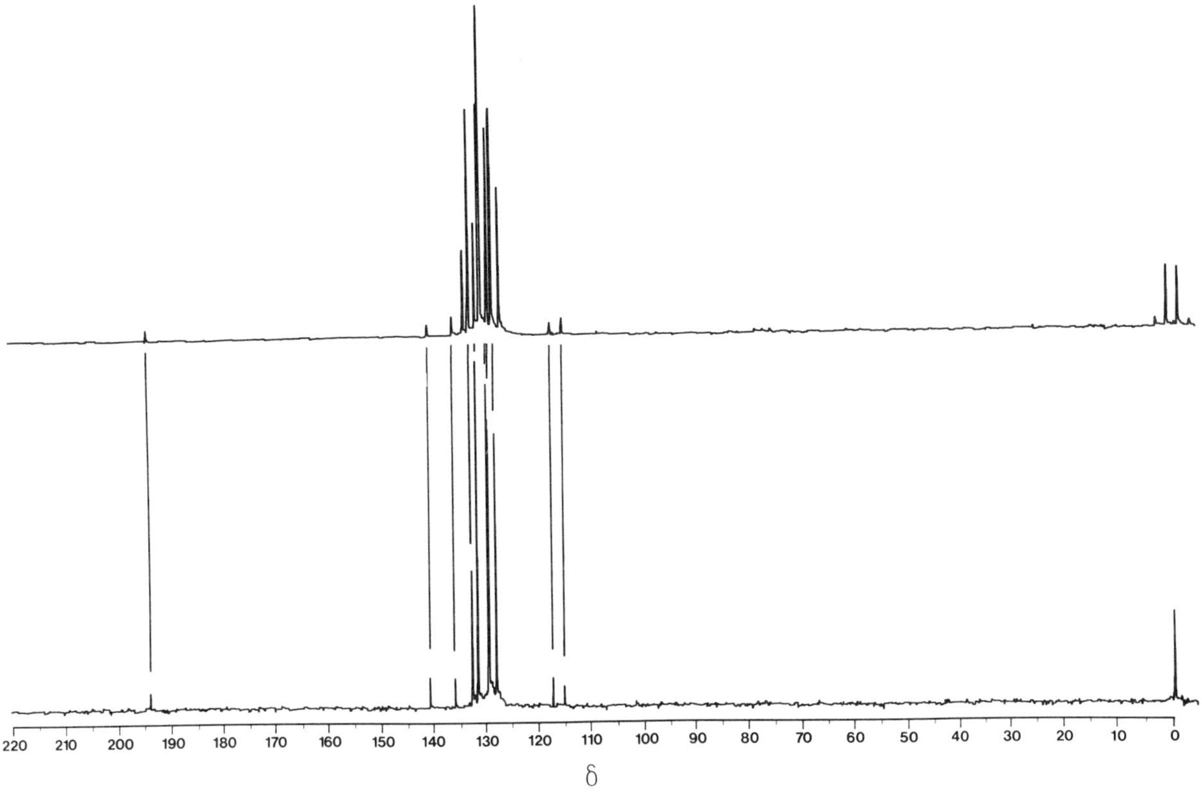
220
210
200
190
180
170
160
150
140
130
120
110
100
90
80
70
60
50
40
30
20
10
0
δ

Problem 1-45

1-45

$C_8H_{14}O$

MS: m126, 111, 69, 43

IR: 1710 cm^{-1}

UV: 279 (ε 13)

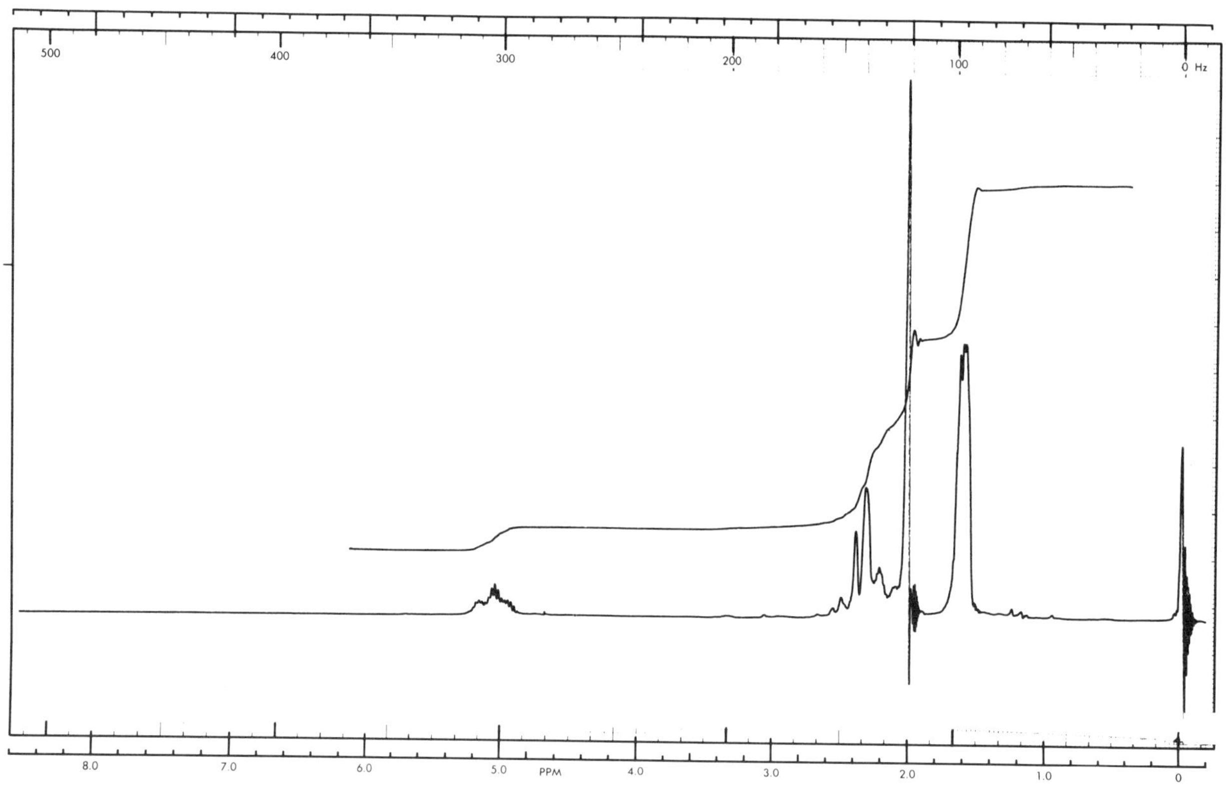

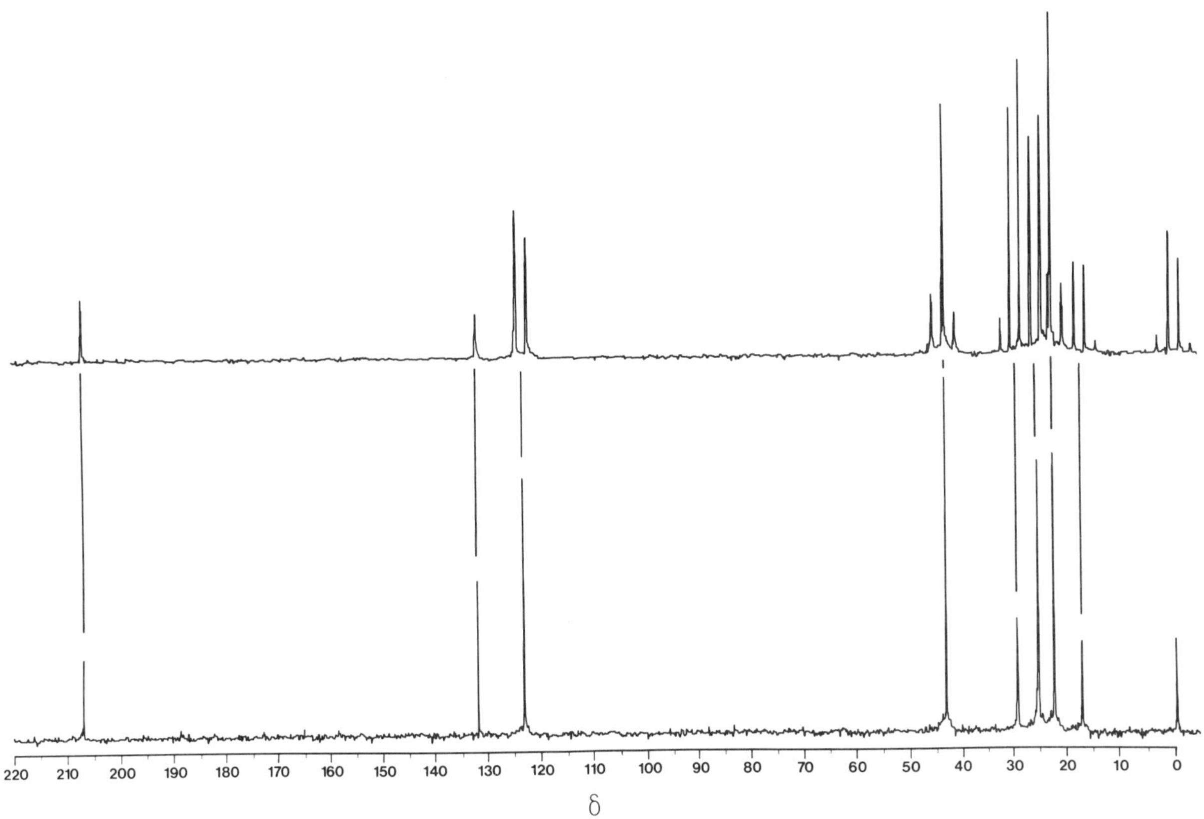
220
210
200
190
180
170
160
150
140
130
120
110
100
90
80
70
60
50
40
30
20
10
0
δ

Problem 1-46

1-46
$C_{10}H_{14}O_2$

MS: m166, 164, 151, 138, 137, 57
IR: 3450 cm^{-1}
UV: Strong absorption

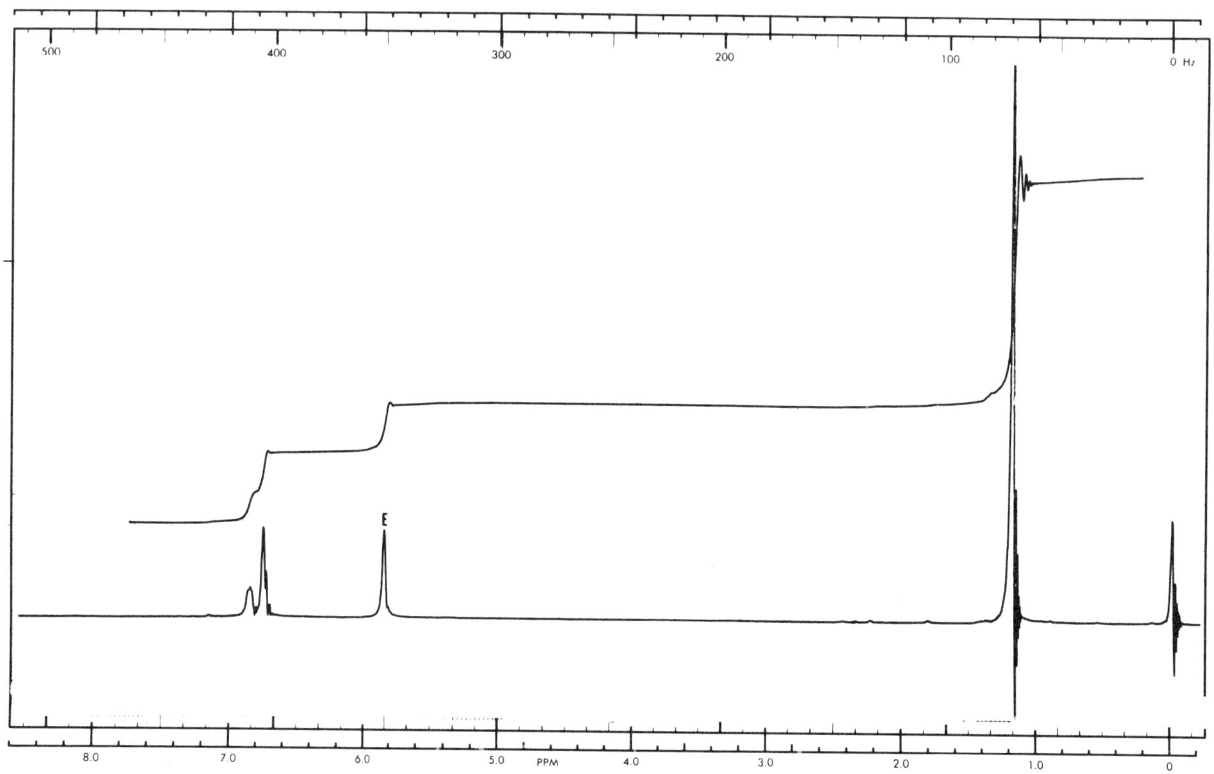

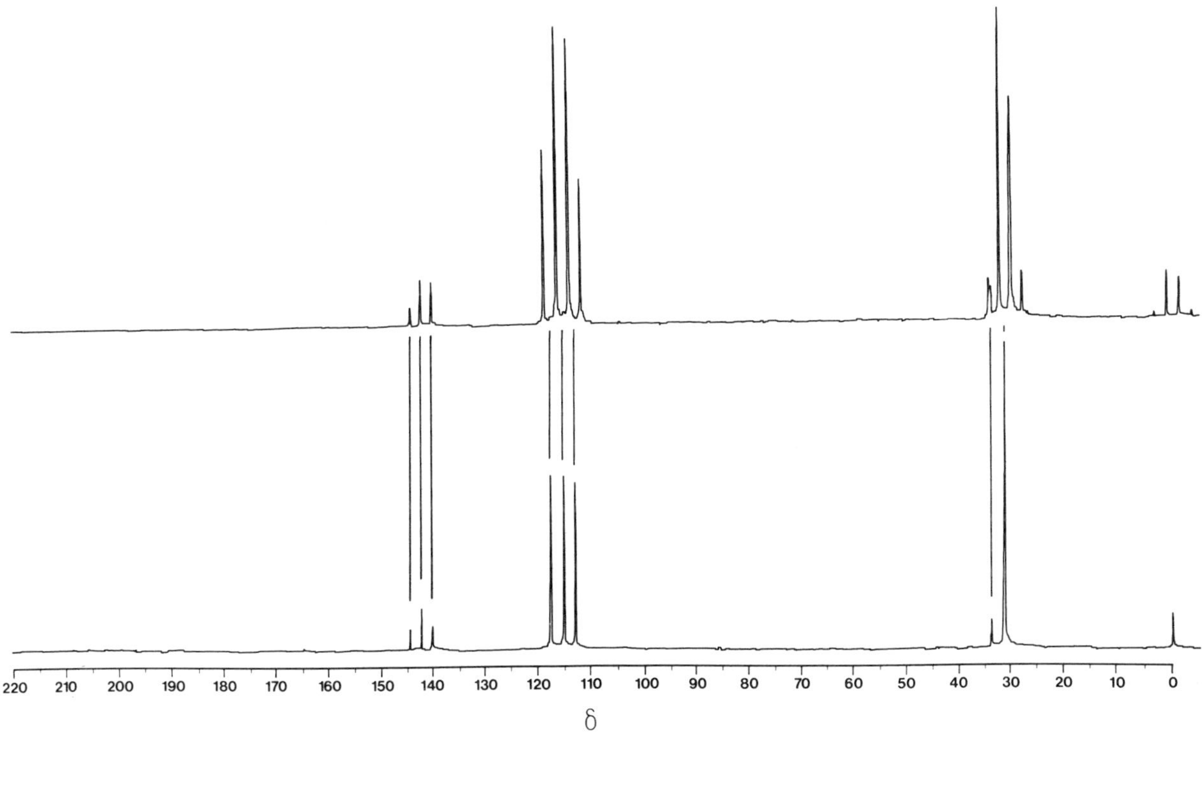
220 210 200 190 180 170 160 150 140 130 120 110 100 90 80 70 60 50 40 30 20 10 0
δ

Problem 1-47

1-47

$C_{10}H_{10}O_2$

MS: m162(s), 161, 132

IR: 1640 cm^{-1}

UV: Strong absorption

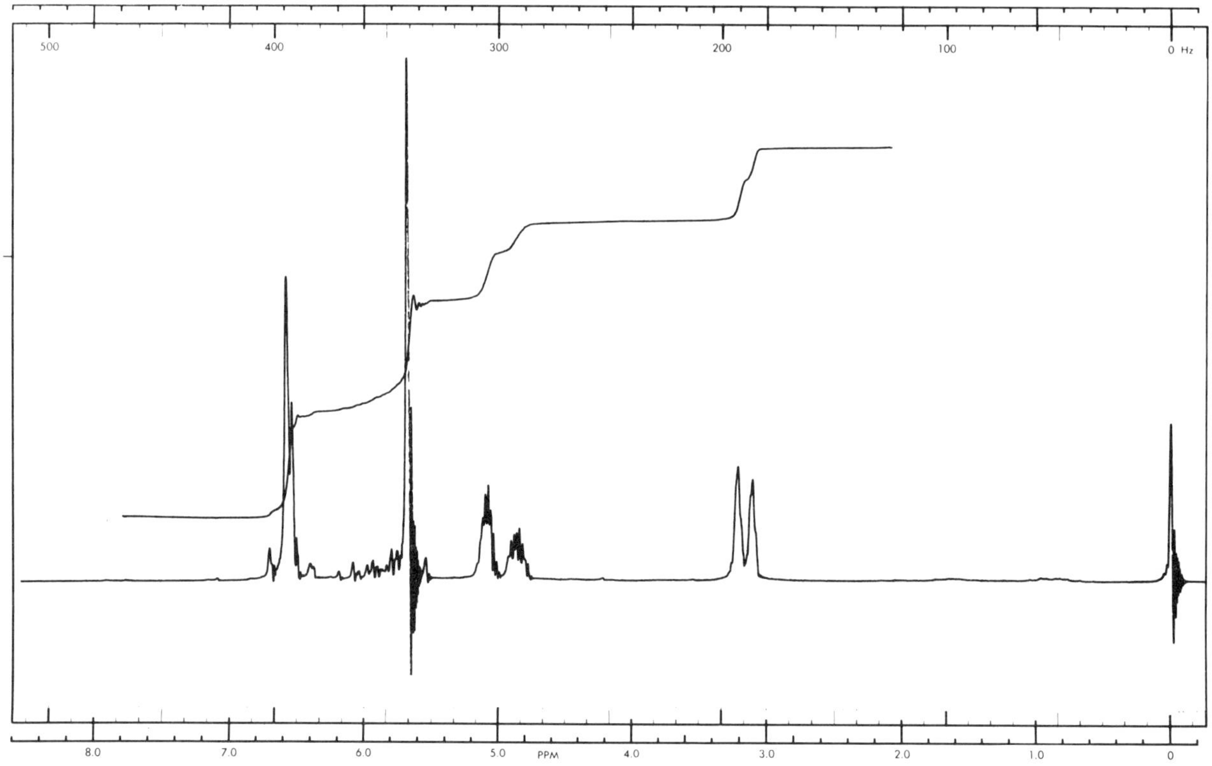

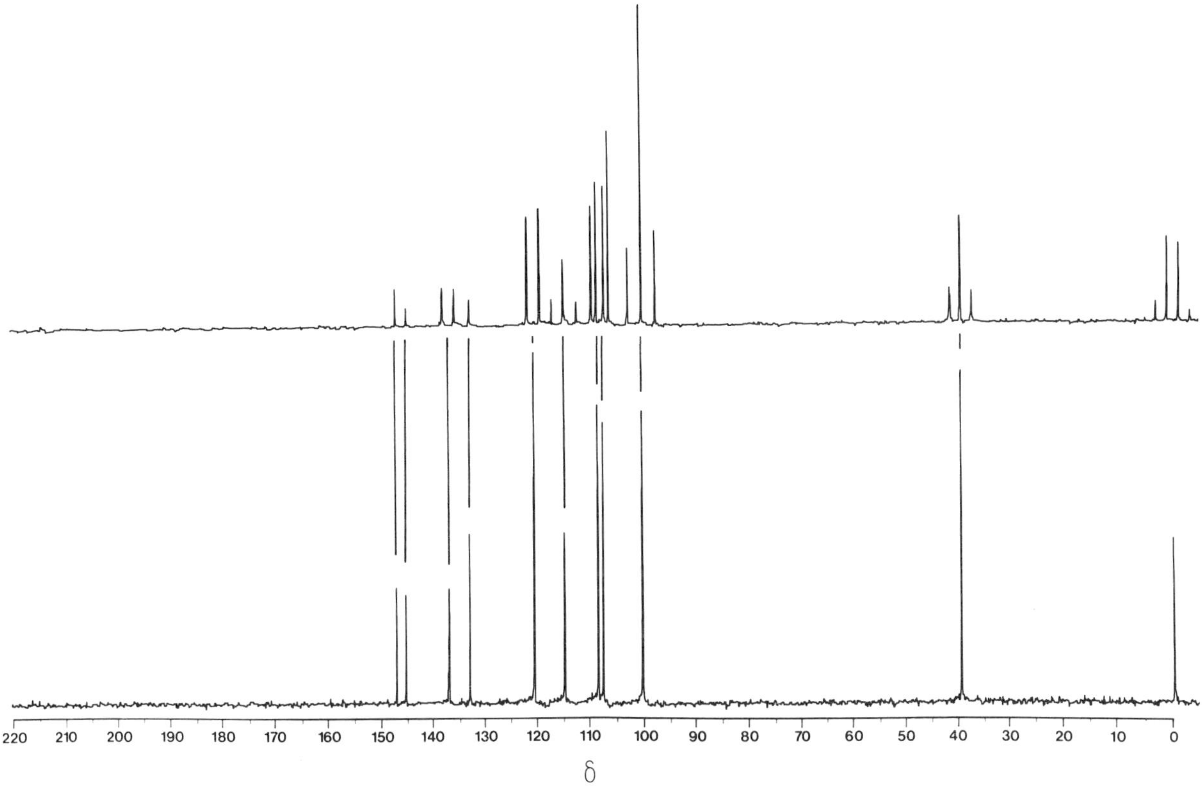
220
210
200
190
180
170
160
150
140
130
120
110
100
90
80
70
60
50
40
30
20
10
0
δ

Problem 1-48

1-48	MS:	m194, 166, 121, 120
$C_{11}H_{14}O_3$	IR:	3190 (persists in dilute solution), 1675 cm^{-1}
	UV:	Strong absorption
	PMR:	Upper scan offset 300 Hz

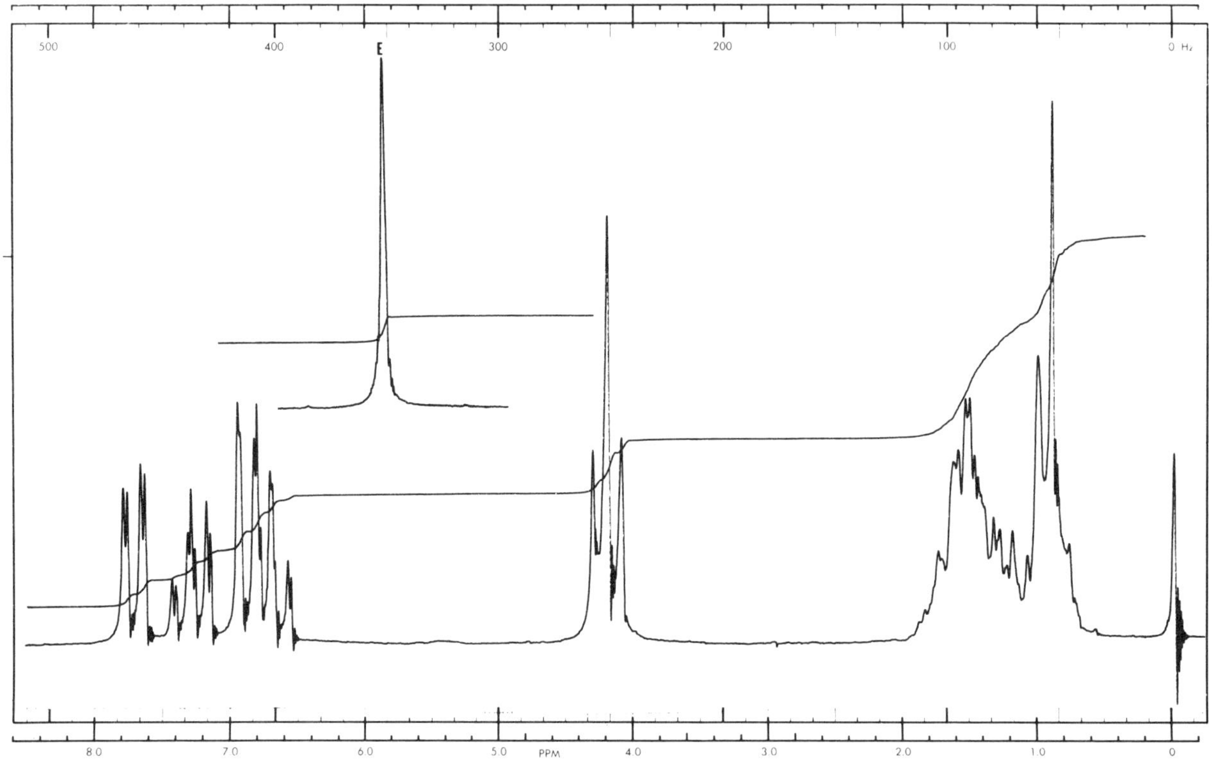

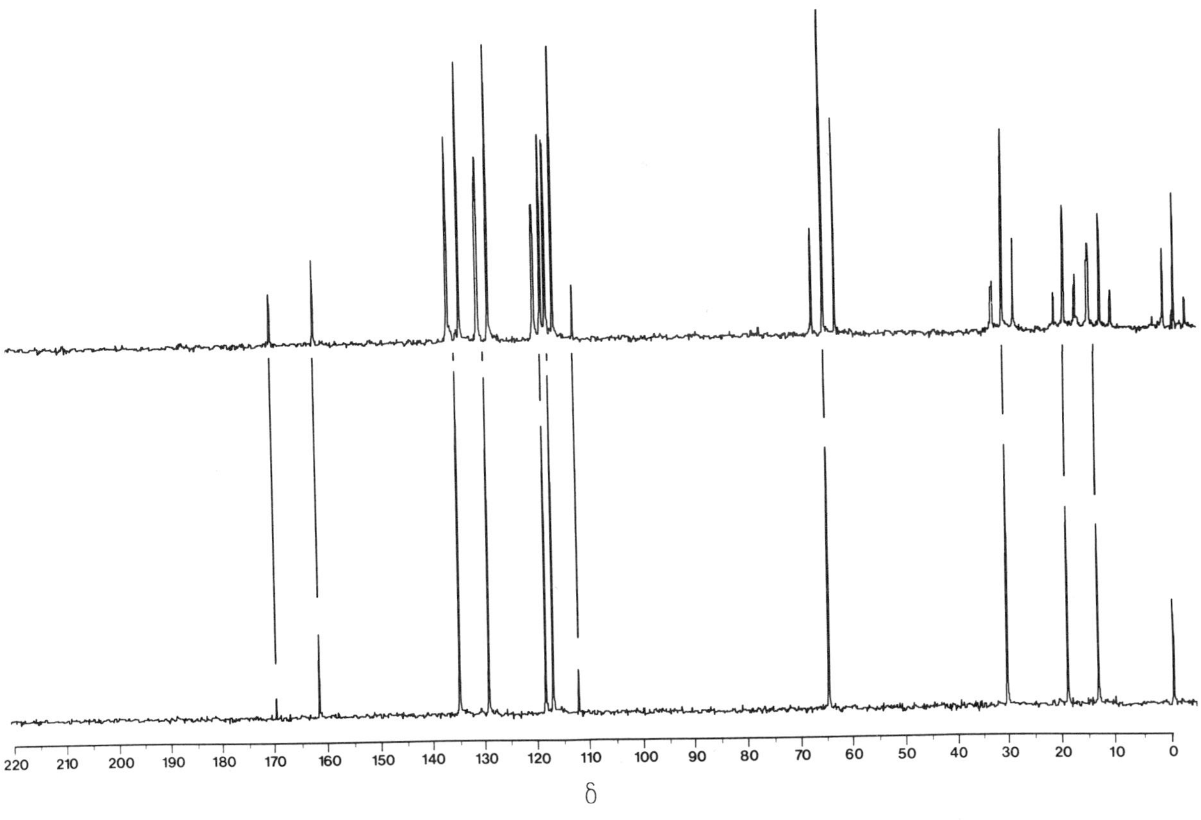
220
210
200
190
180
170
160
150
140
130
120
110
100
90
80
70
60
50
40
30
20
10
0
δ

Problem 1-49

1-49

$C_6H_{12}O_6$

MS: m180(w), 162, 144

IR: 3400, 3200 cm^{-1}

UV: none

PMR: D_2O + DSS

CMR: D_2O + dioxane

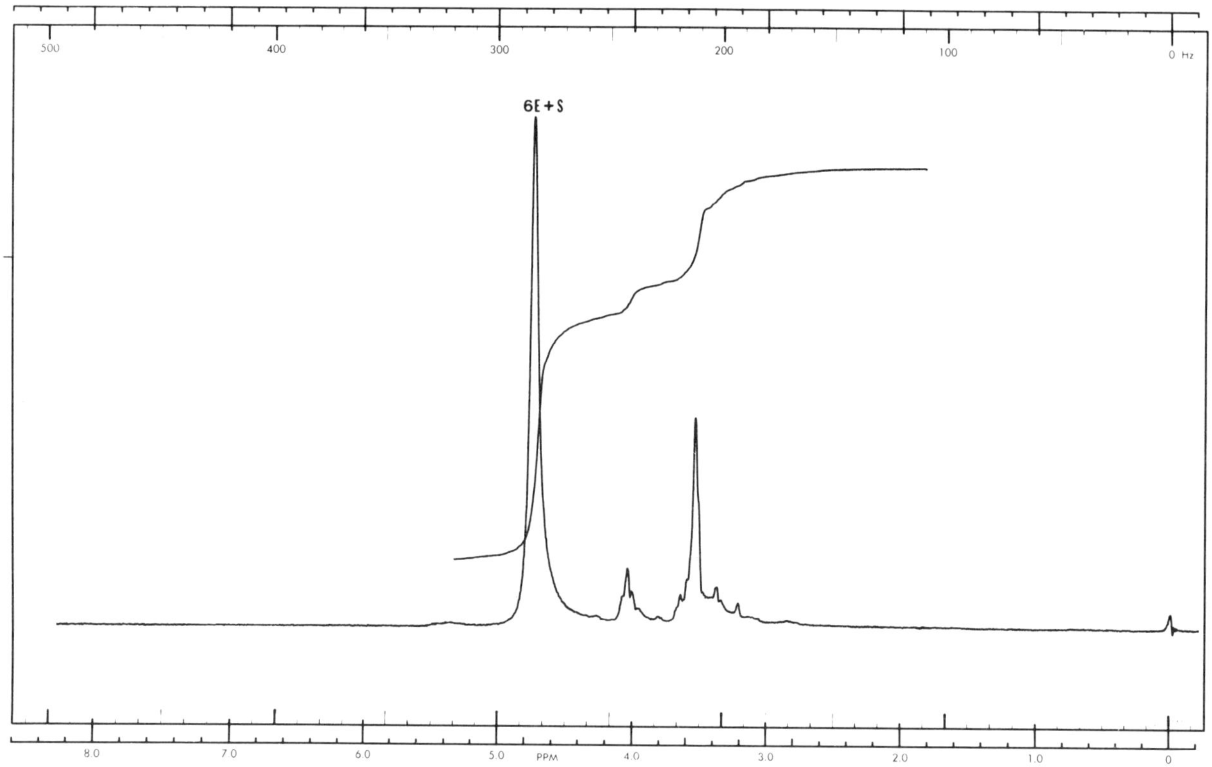

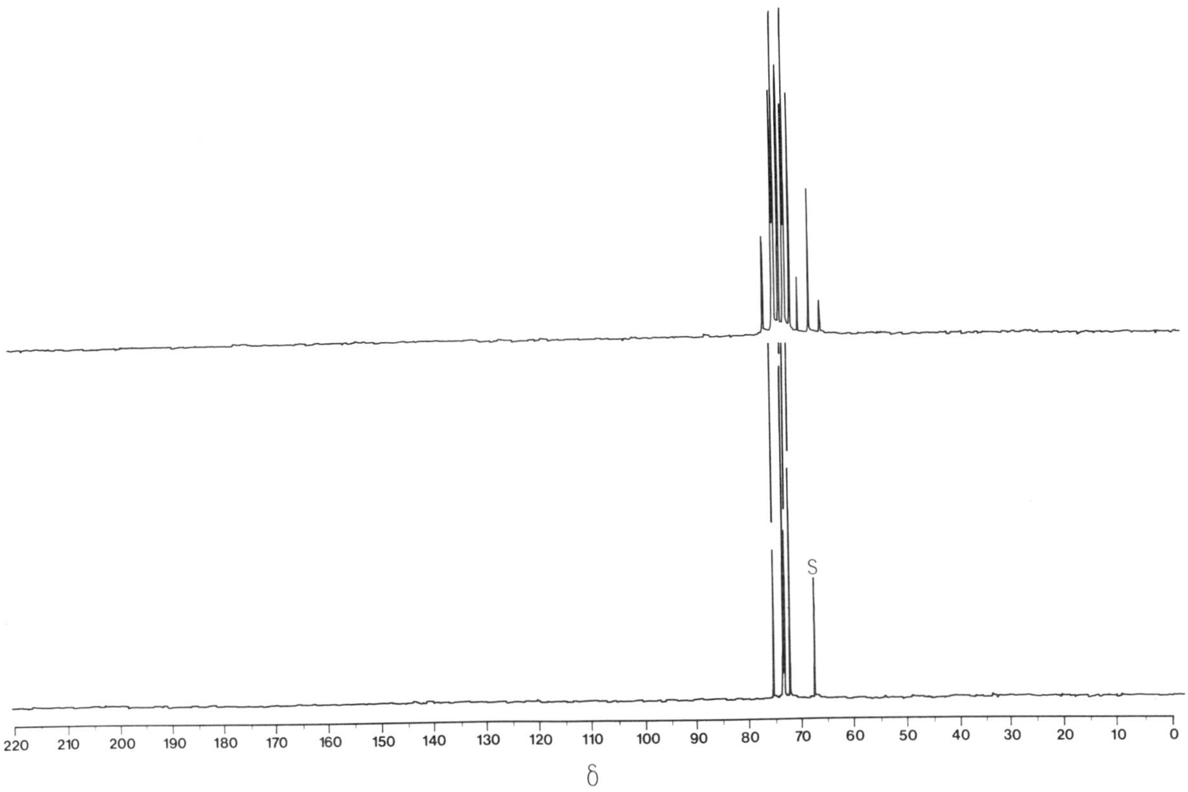
S
220 210 200 190 180 170 160 150 140 130 120 110 100 90 80 70 60 50 40 30 20 10 0
δ

Problem 1-50

1-50

$C_{20}H_{14}O_4$

MS: m318, 225, 93

IR: 1740 cm^{-1}

UV: Strong absorption

PMR: Upper scan offset 250 Hz

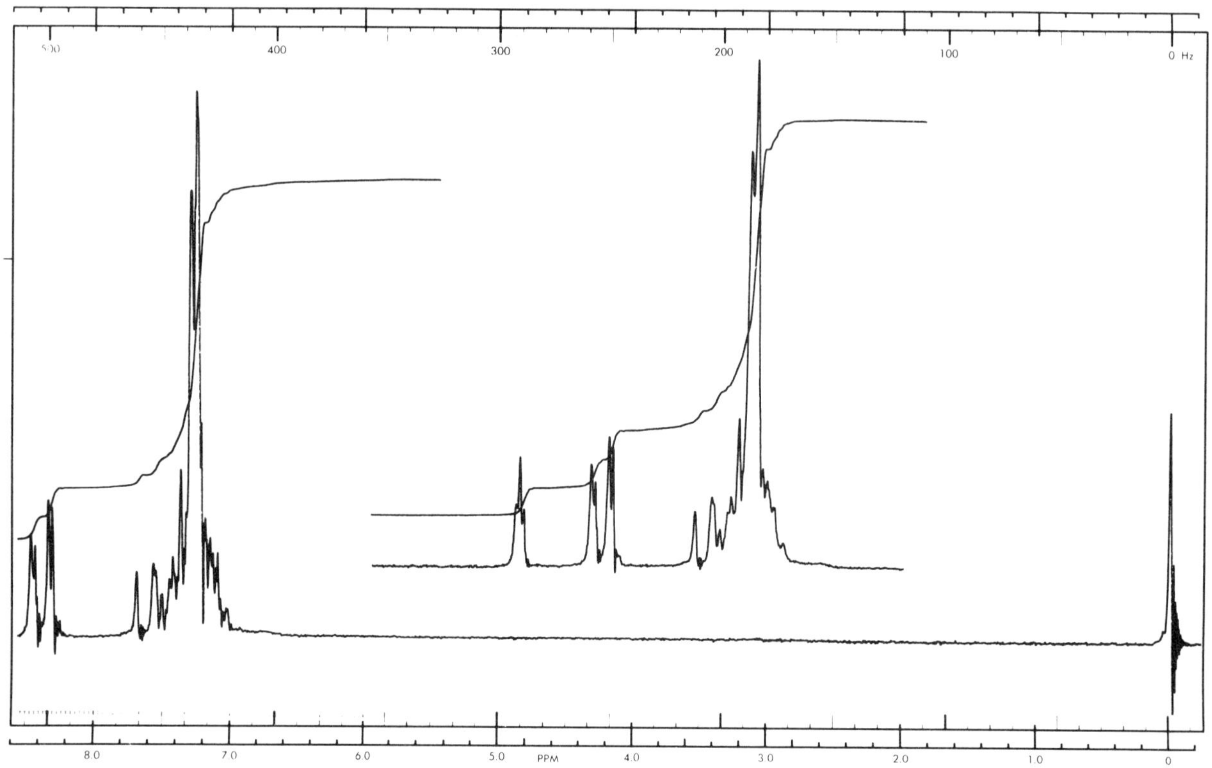

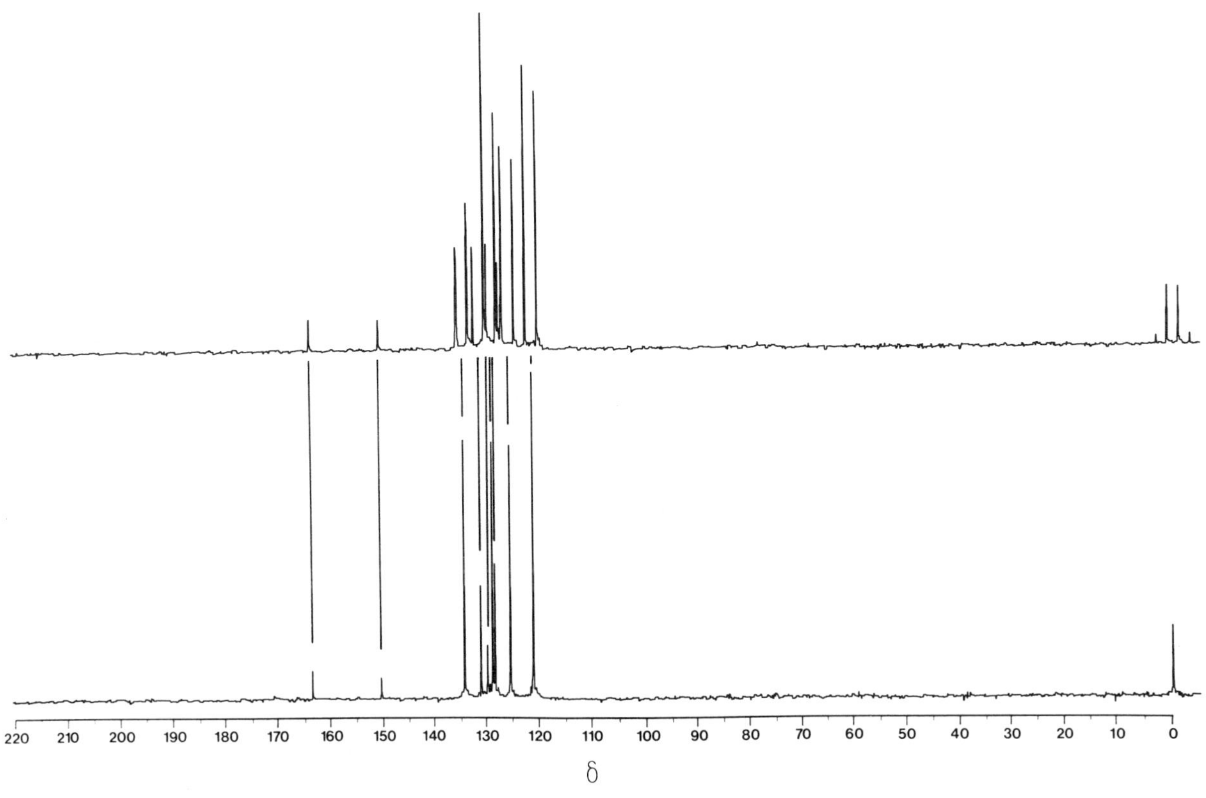
220
210
200
190
180
170
160
150
140
130
120
110
100
90
80
70
60
50
40
30
20
10
0
δ

Problem 1-51

1-51

$C_9H_{14}O$

MS: m138, 123, 110

IR: 1660 cm^{-1}

UV: 238 (ε 11,000)

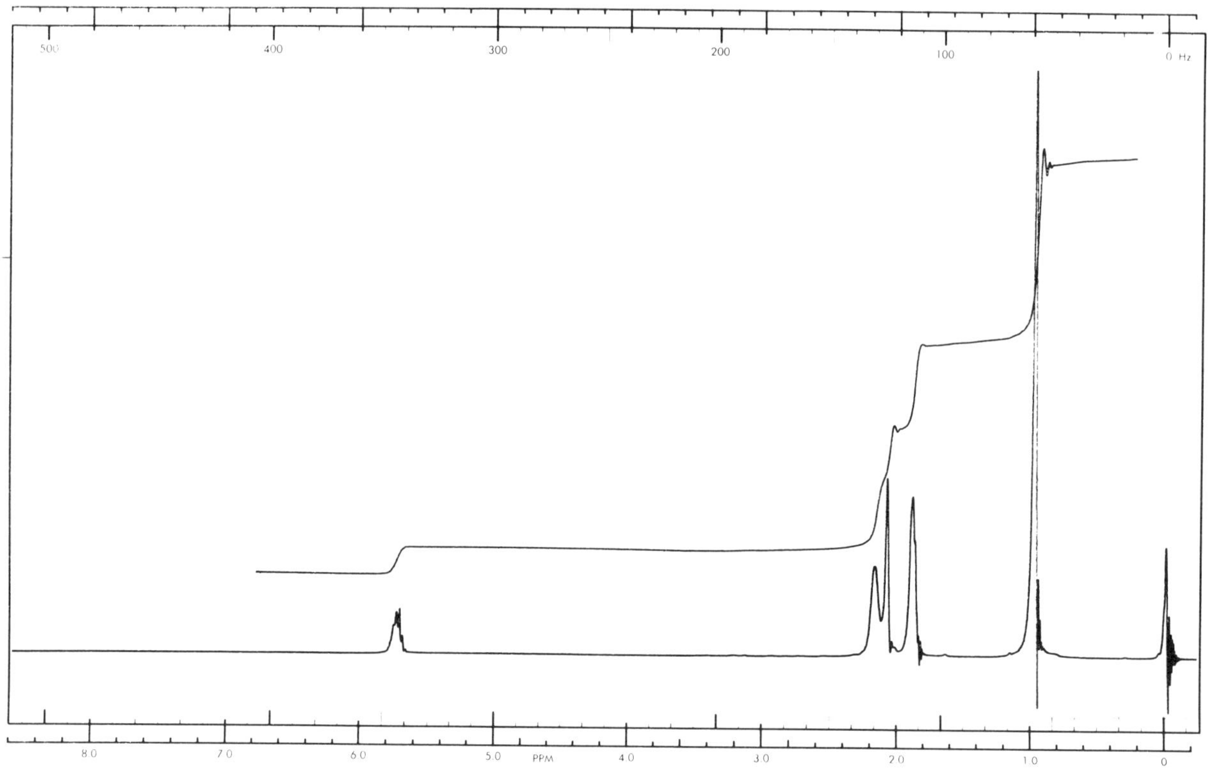

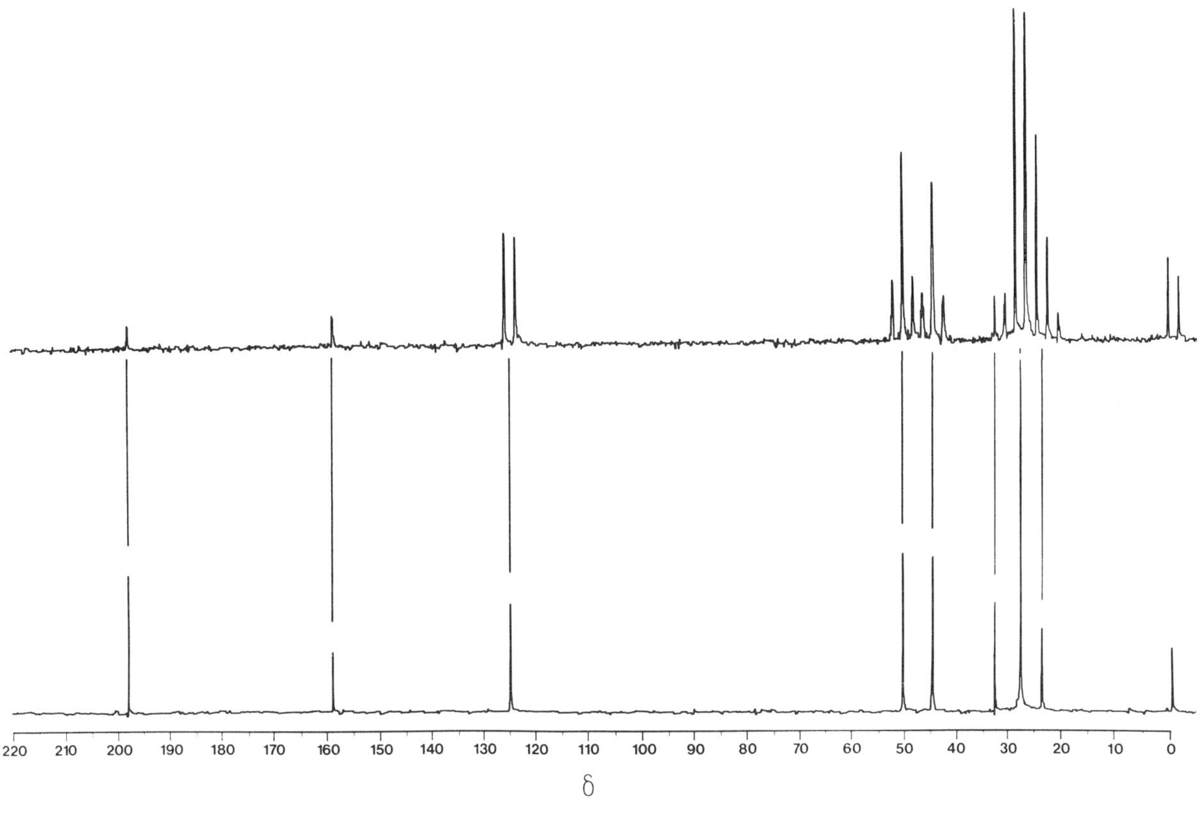
220
210
200
190
180
170
160
150
140
130
120
110
100
90
80
70
60
50
40
30
20
10
0
δ

Problem 1-52

1-52	MS:	m120(s), 118, 94, 91
C_9H_{12}	IR:	-
	UV:	none

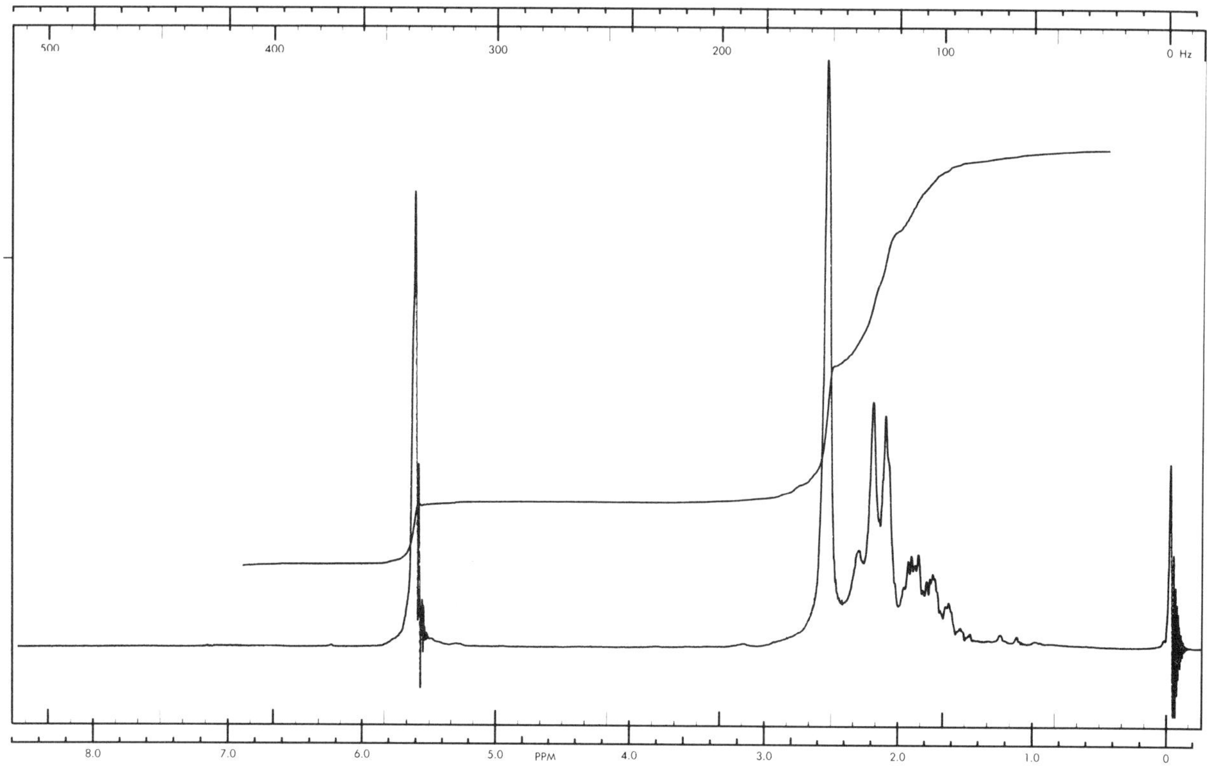

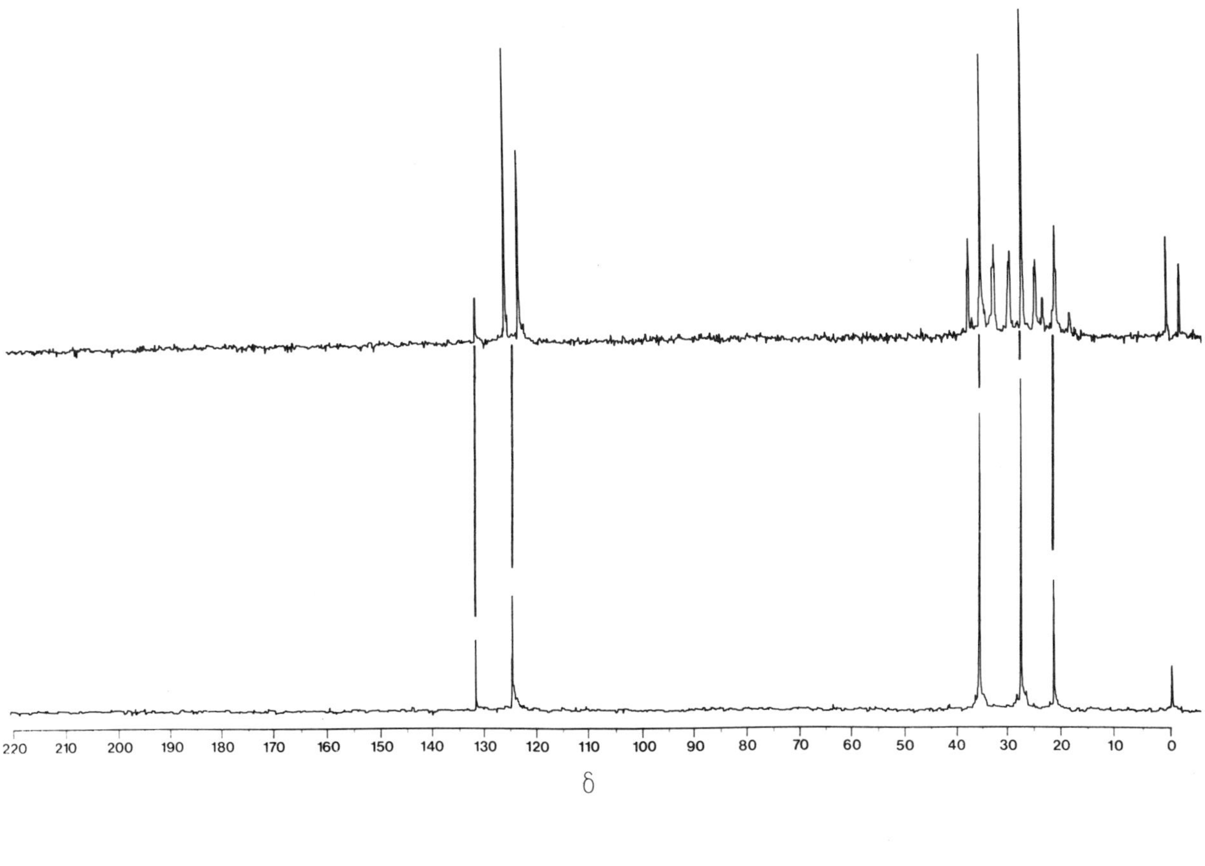
220
210
200
190
180
170
160
150
140
130
120
110
100
90
80
70
60
50
40
30
20
10
0
δ

Problem 1-53

1-53

$C_9H_8O_4$

MS: m180(s), 163, 135, 43

IR: 3000 (broad), 1760, 1700, 1600 cm^{-1}

UV: 234 (ε 6950), 303 (ε 3760)

PMR: Upper scan offset 250 Hz

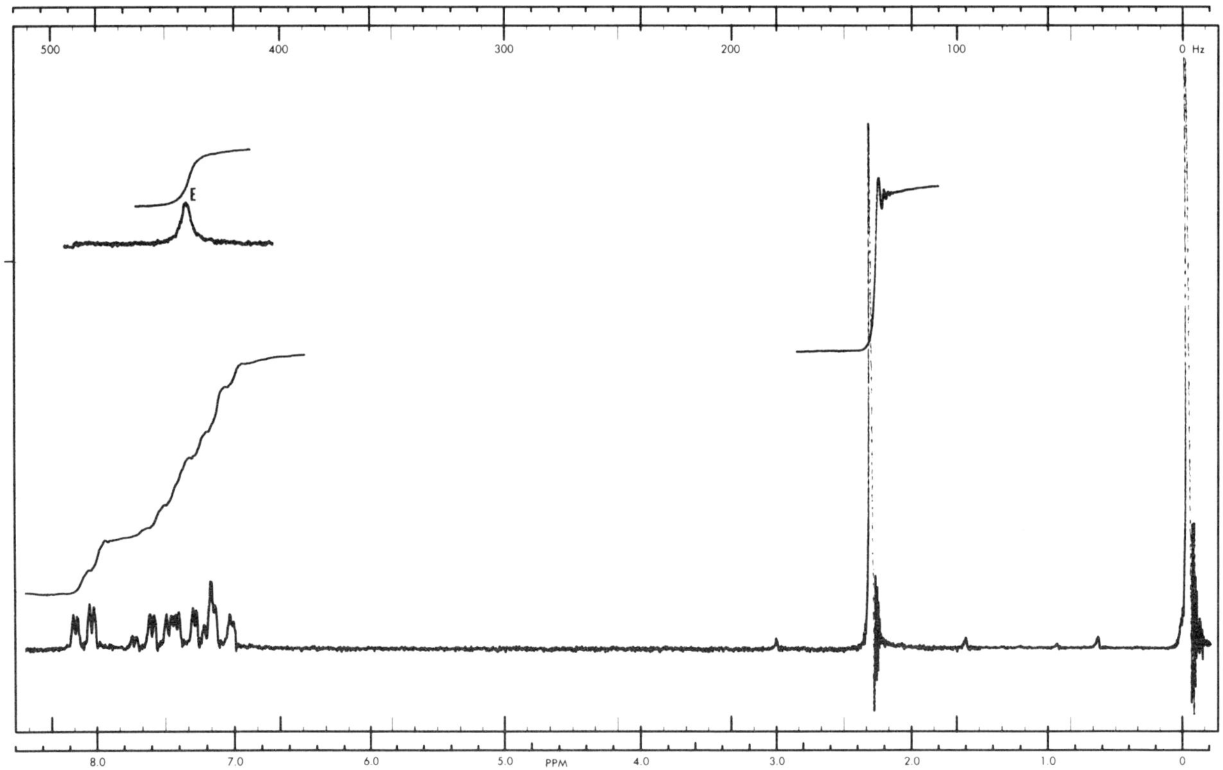

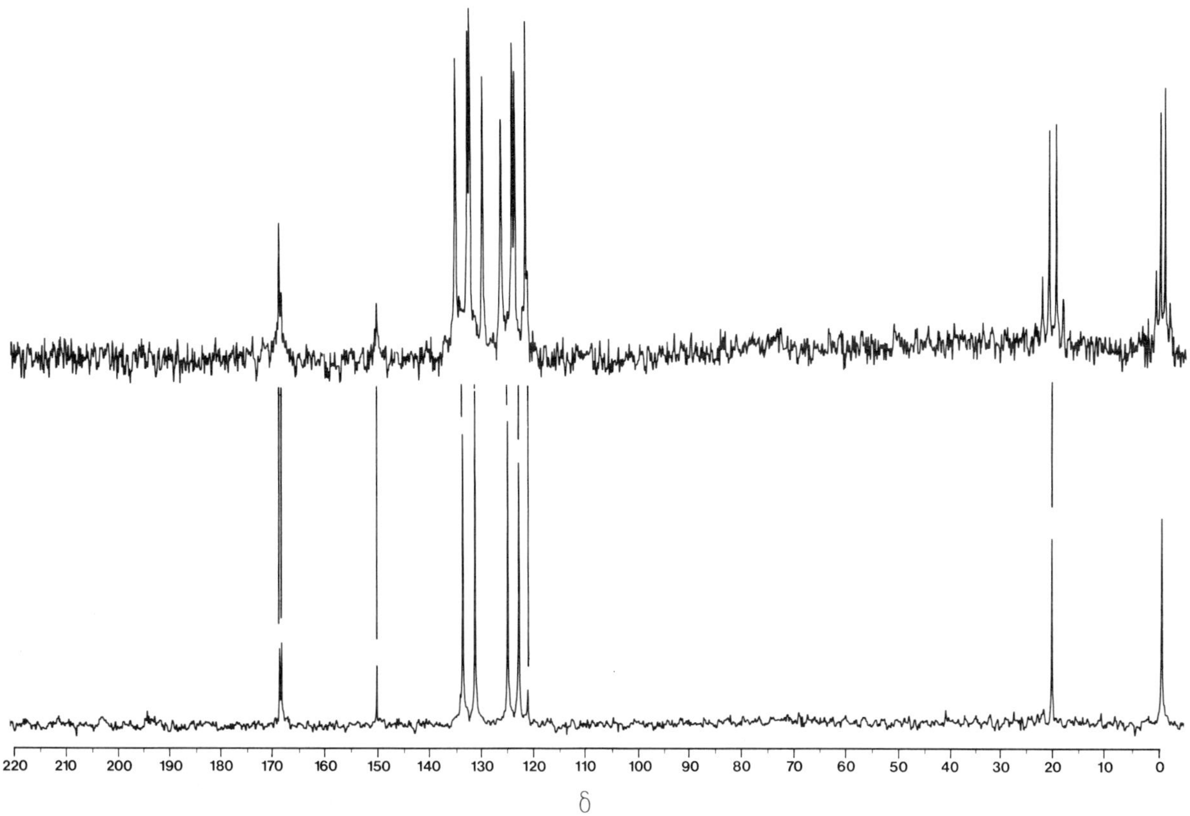
220
210
200
190
180
170
160
150
140
130
120
110
100
90
80
70
60
50
40
30
20
10
0
δ

Problem 1-54

1-54

$C_3H_7NO_2S$

MS: m121(w), 120, 76, 74

IR: 3300-2300 cm^{-1} (broad), 1680 cm^{-1}

UV: End absorption

PMR: D_2O + DSS

CMR: D_2O + dioxane

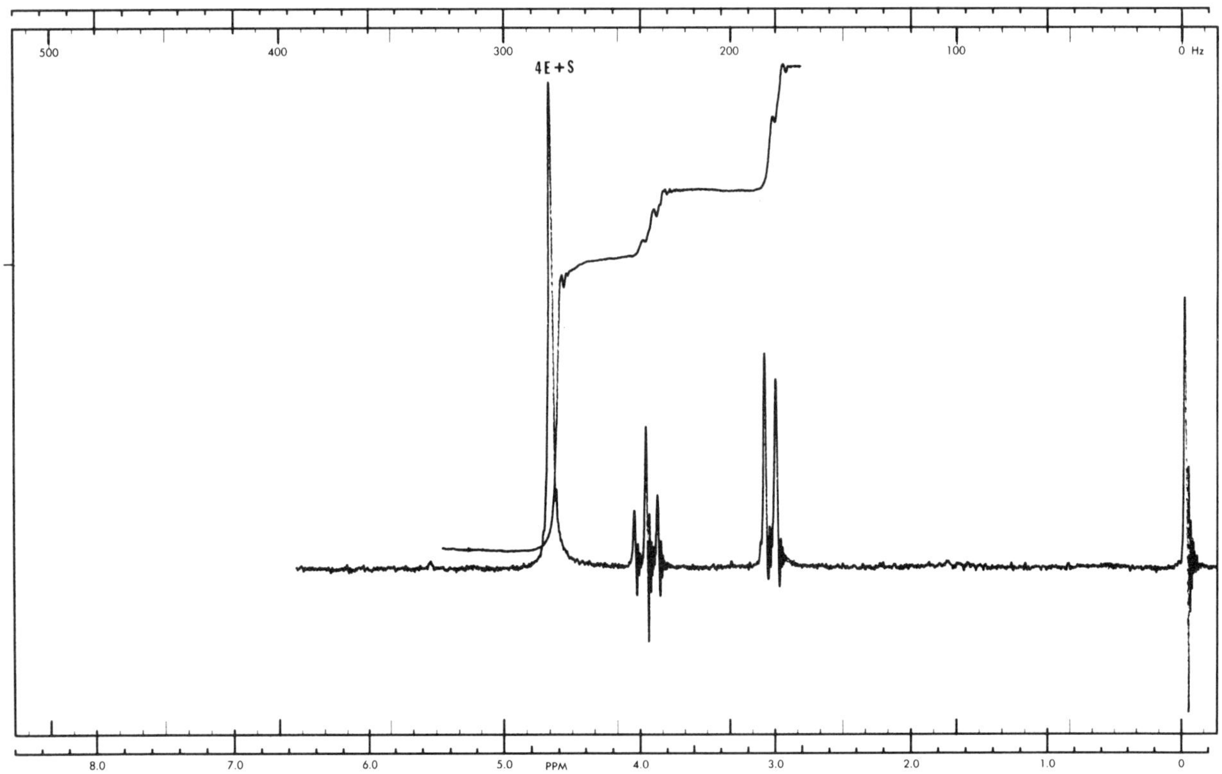

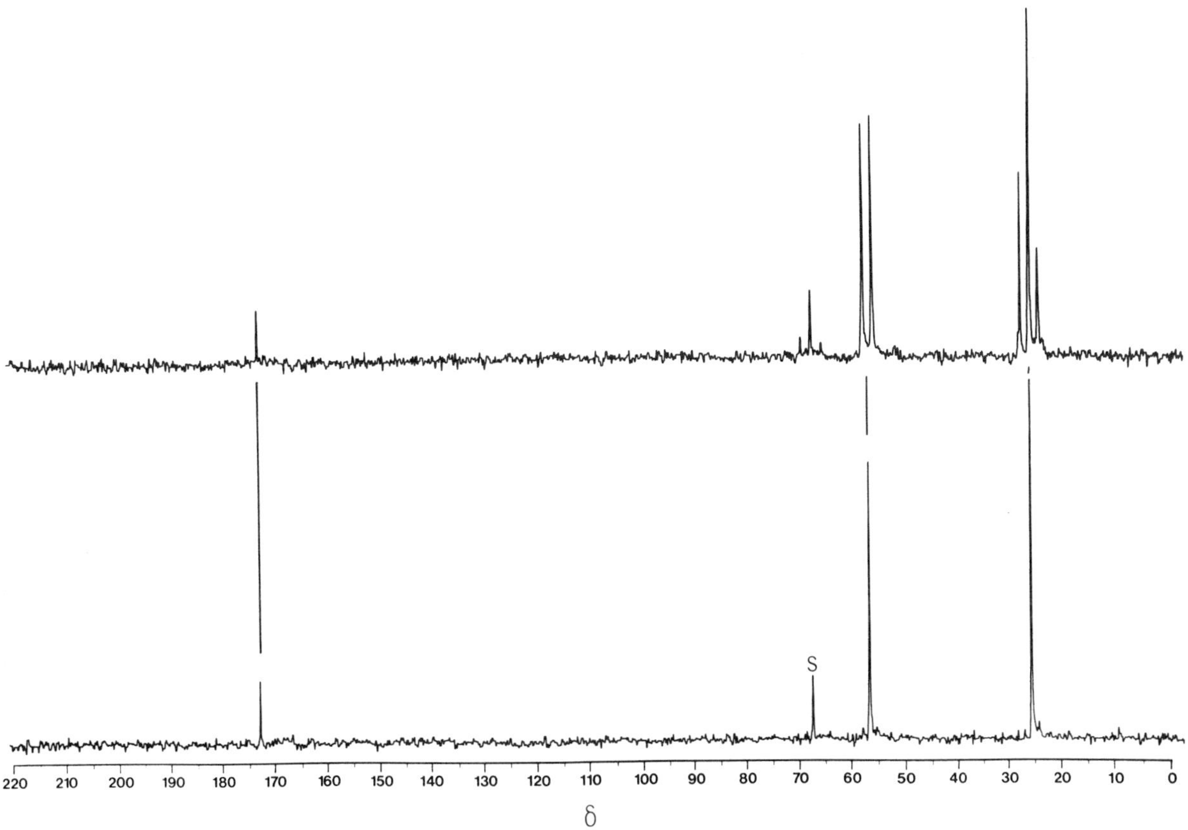
S
220
210
200
190
180
170
160
150
140
130
120
110
100
90
80
70
60
50
40
30
20
10
0
δ

Problem 1-55

1-55	MS:	Not volatile
$C_7H_{16}NO_2^+Cl^-$	IR:	1740 cm^{-1}
	UV:	none
	PMR:	D_2O + DSS
	CMR:	D_2O + dioxane

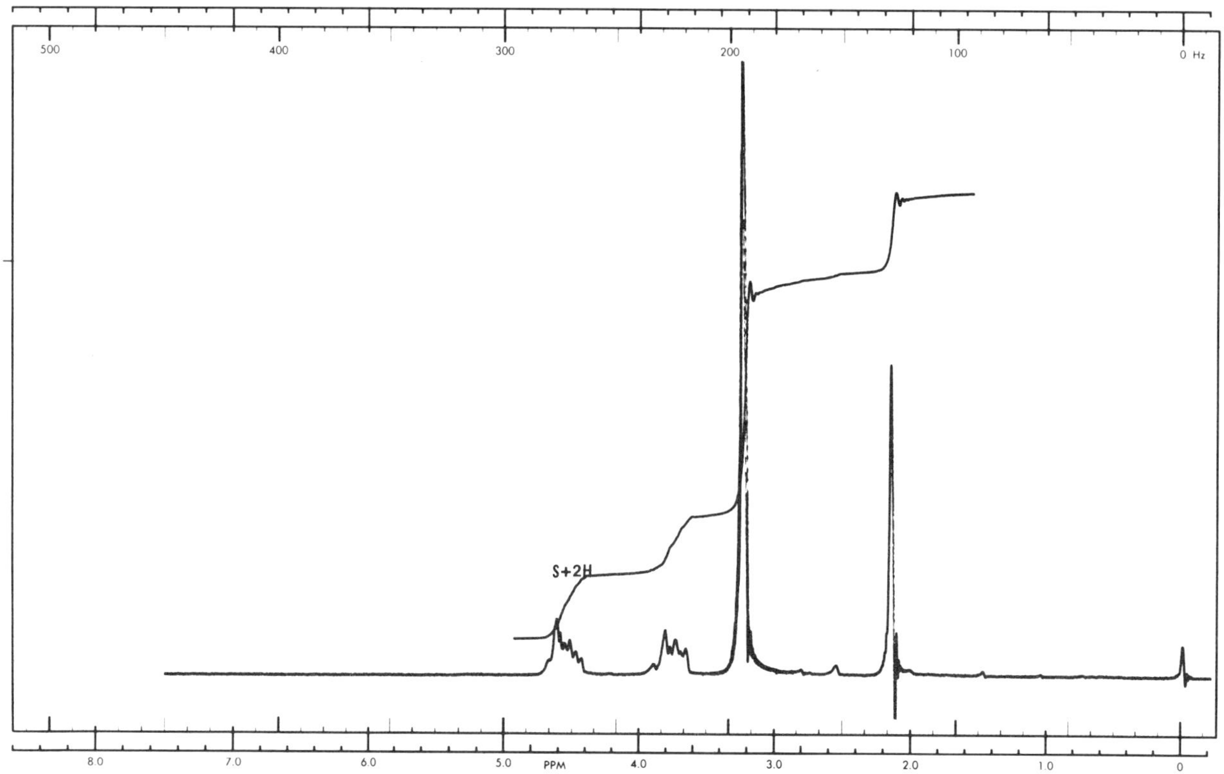

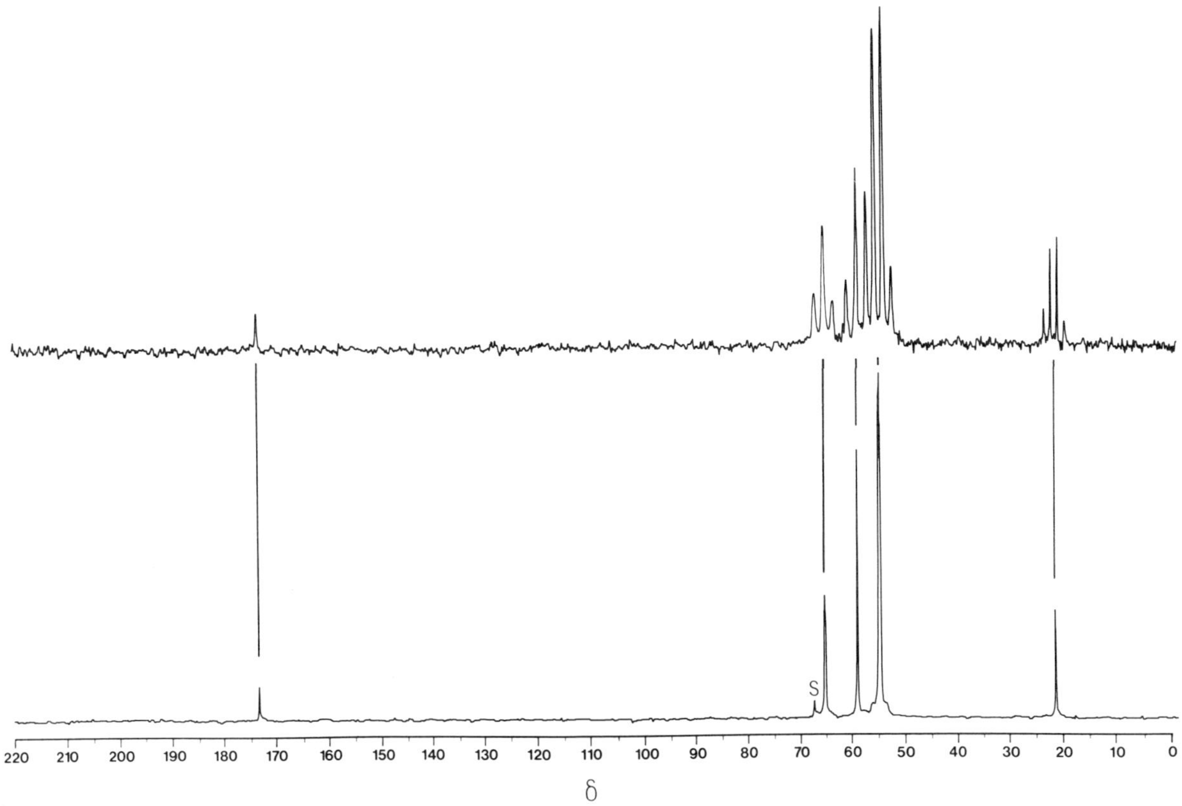
S
220 210 200 190 180 170 160 150 140 130 120 110 100 90 80 70 60 50 40 30 20 10 0
δ

Problem 1-56

1-56 MS: m304, 276, 248, 177, 149

$C_7H_5FeIO_2$ IR: 2043, 2005 cm^{-1}

UV: Strong absorption

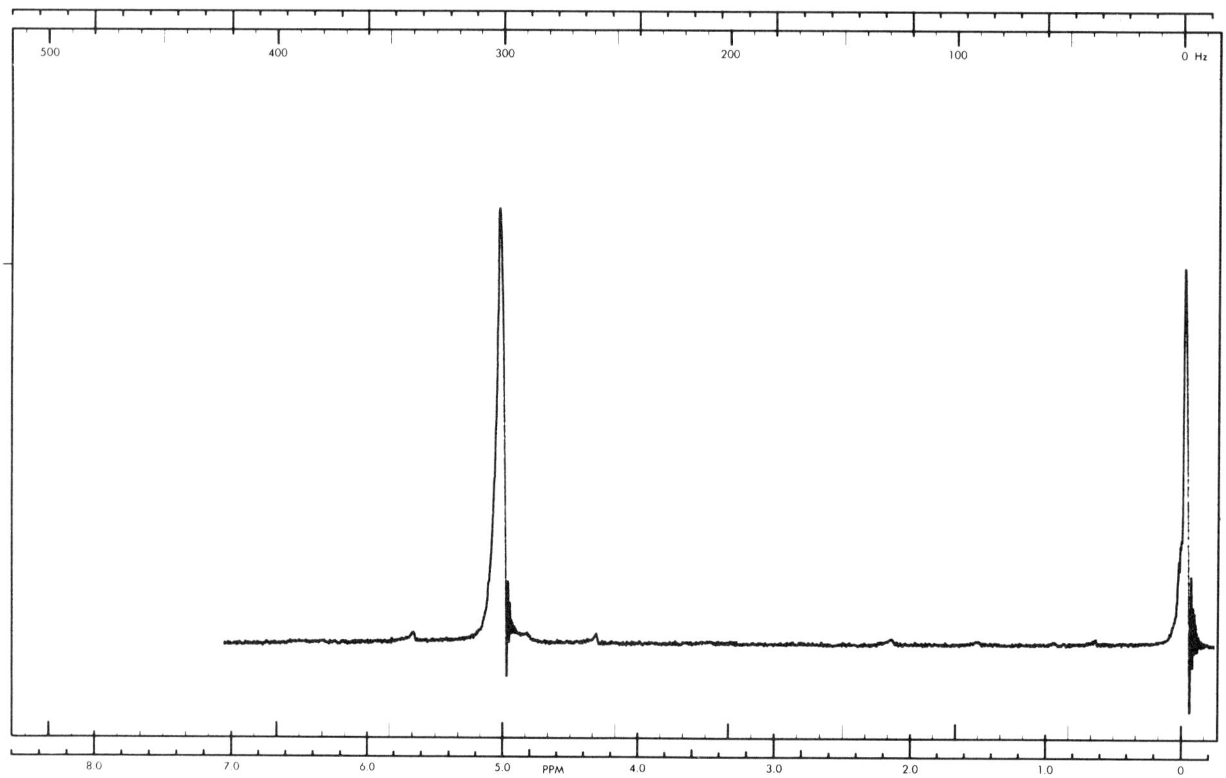

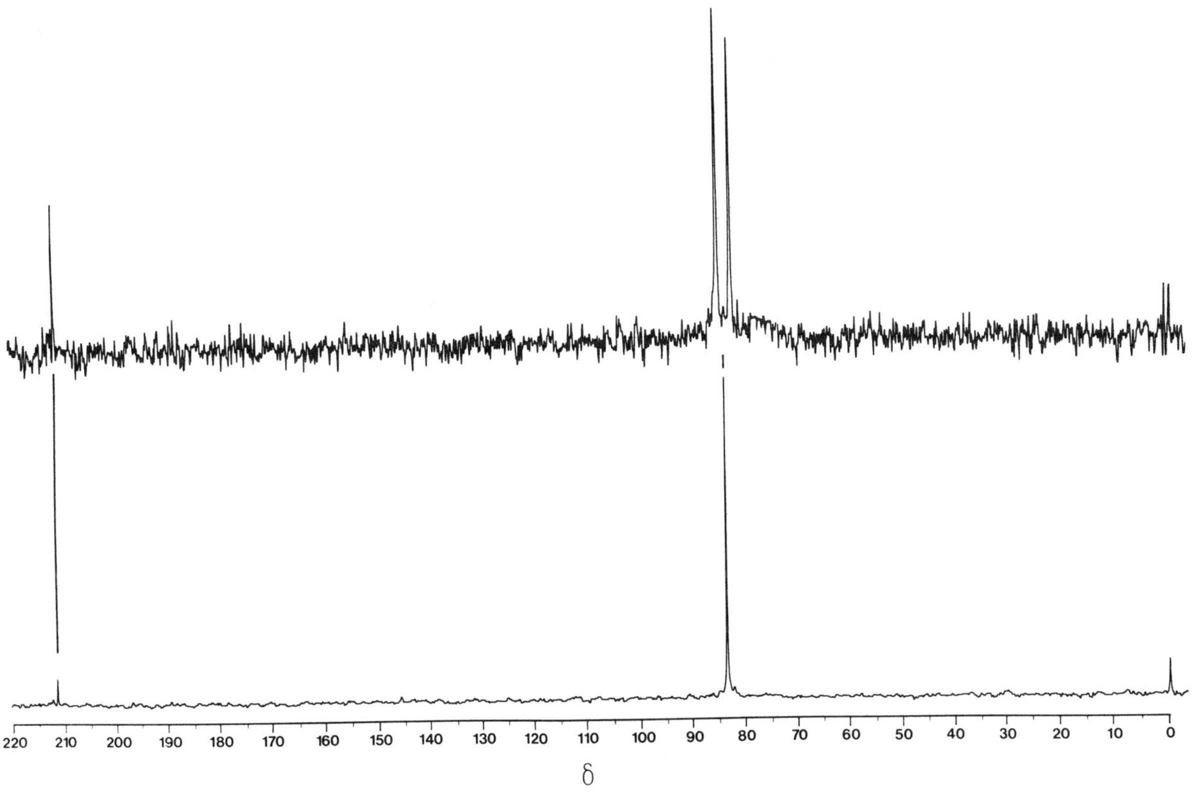
220
210
200
190
180
170
160
150
140
130
120
110
100
90
80
70
60
50
40
30
20
10
0
δ

Problem 1-57

1-57

$C_9H_{15}NO_2$

MS: m169, 141, 112, 71

IR: 1778, 1710 (broad) cm^{-1}

UV: none

CMR: d_6-Me_2CO + TMS

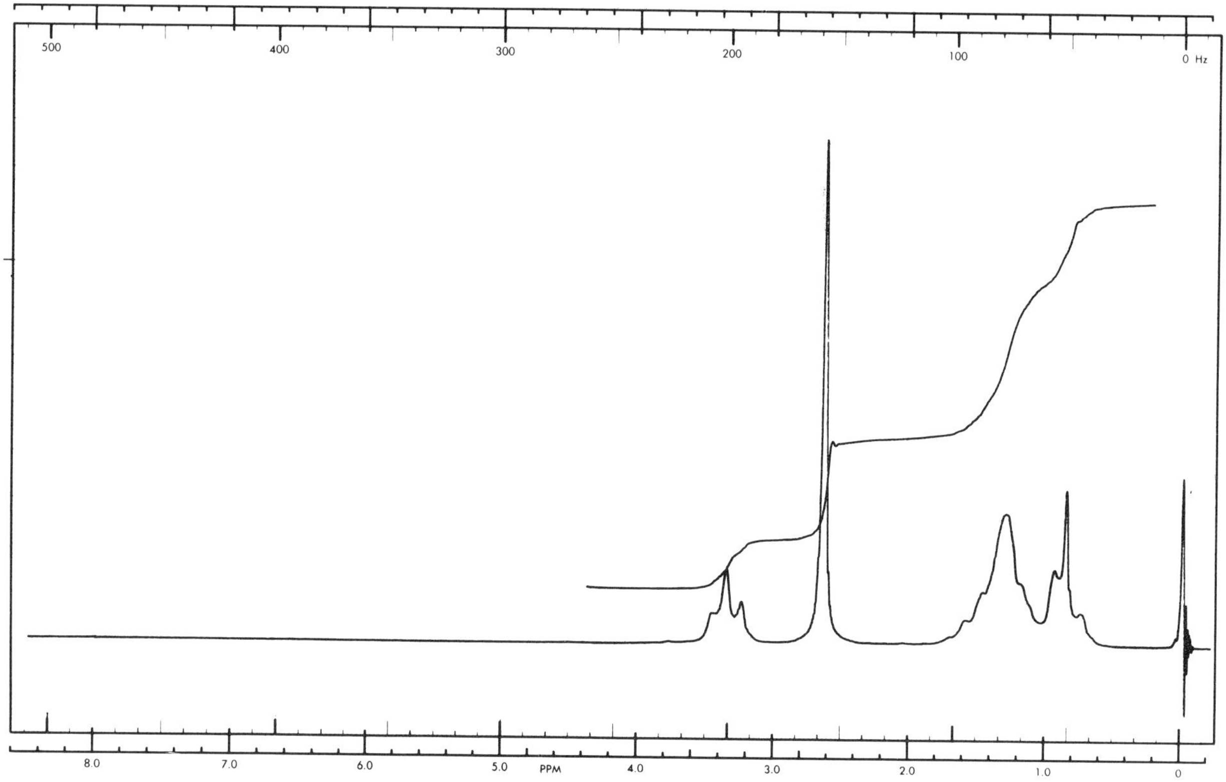

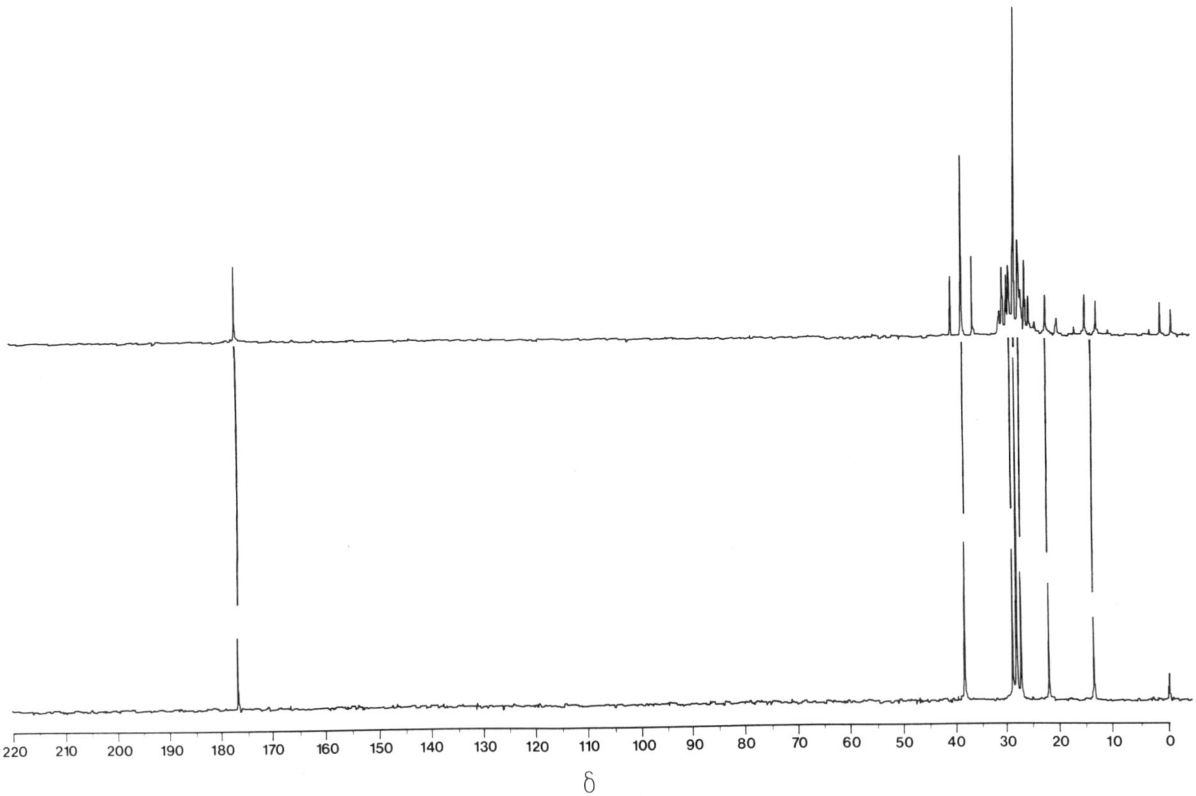
220 210 200 190 180 170 160 150 140 130 120 110 100 90 80 70 60 50 40 30 20 10 0
δ

Problem 1-58

1-58

$C_{12}H_{18}$

MS: m162, 108, 54

IR: 1678, 1650, 970, 700 cm^{-1}

UV: none

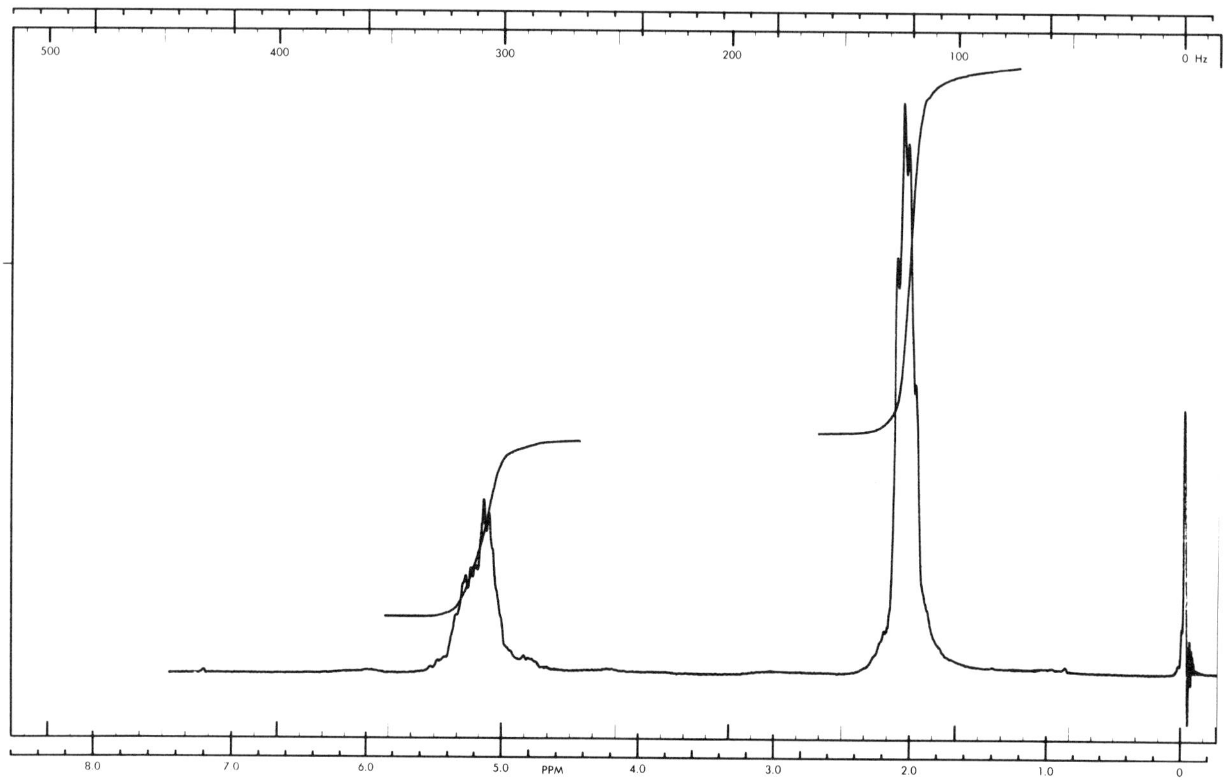

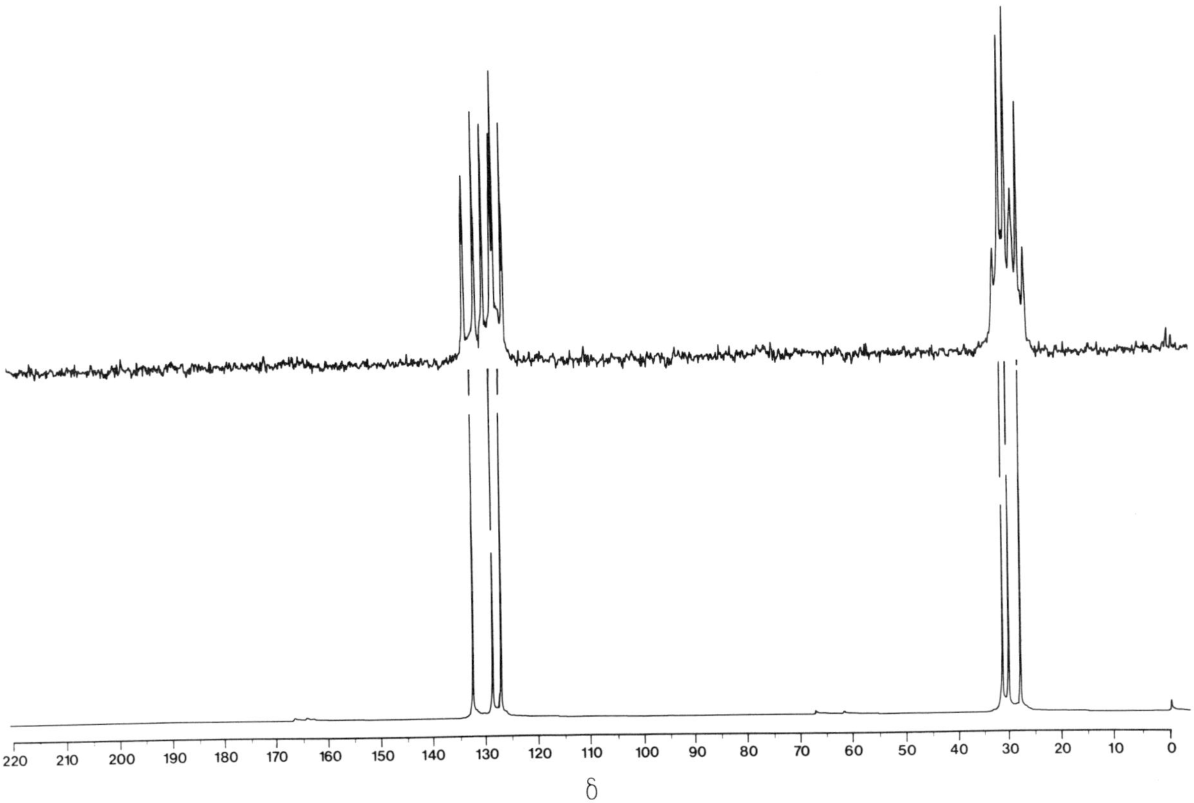
220
210
200
190
180
170
160
150
140
130
120
110
100
90
80
70
60
50
40
30
20
10
0
δ

Problem 1-59

1-59

C_6H_6FN

MS: m111(s), 110, 91

IR: 3475, 3400 cm^{-1}

UV: Strong absorption

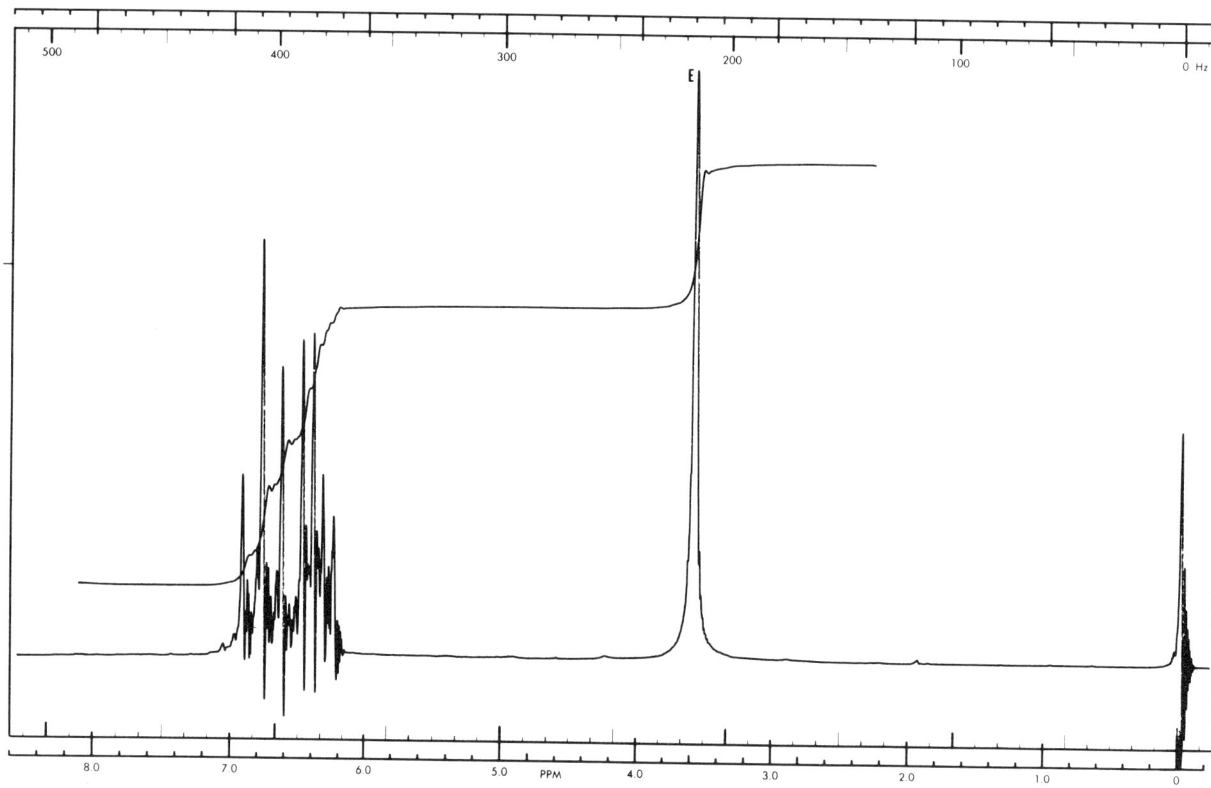

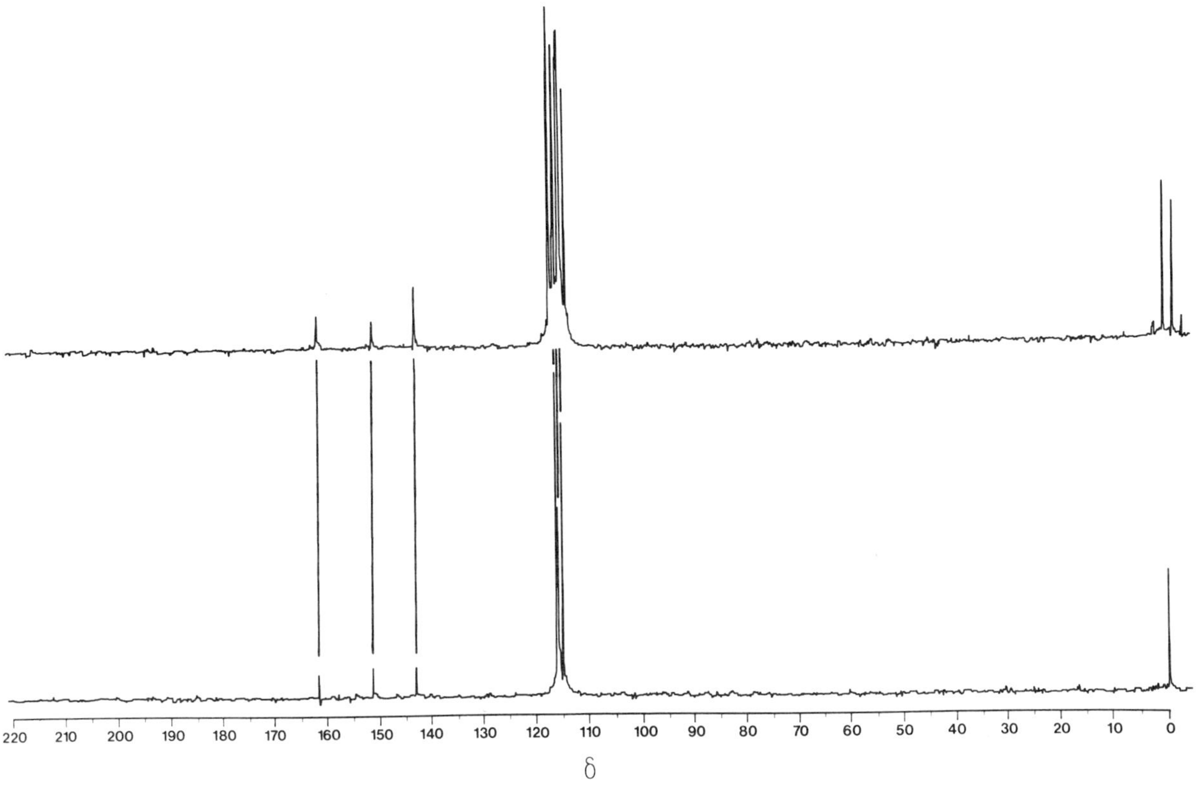
220 210 200 190 180 170 160 150 140 130 120 110 100 90 80 70 60 50 40 30 20 10 0
δ

Problem 1-60

1-60 MS: m218, 190

$C_9H_7MnO_3$ IR: 2024, 1941 cm^{-1}

UV: Strong absorption

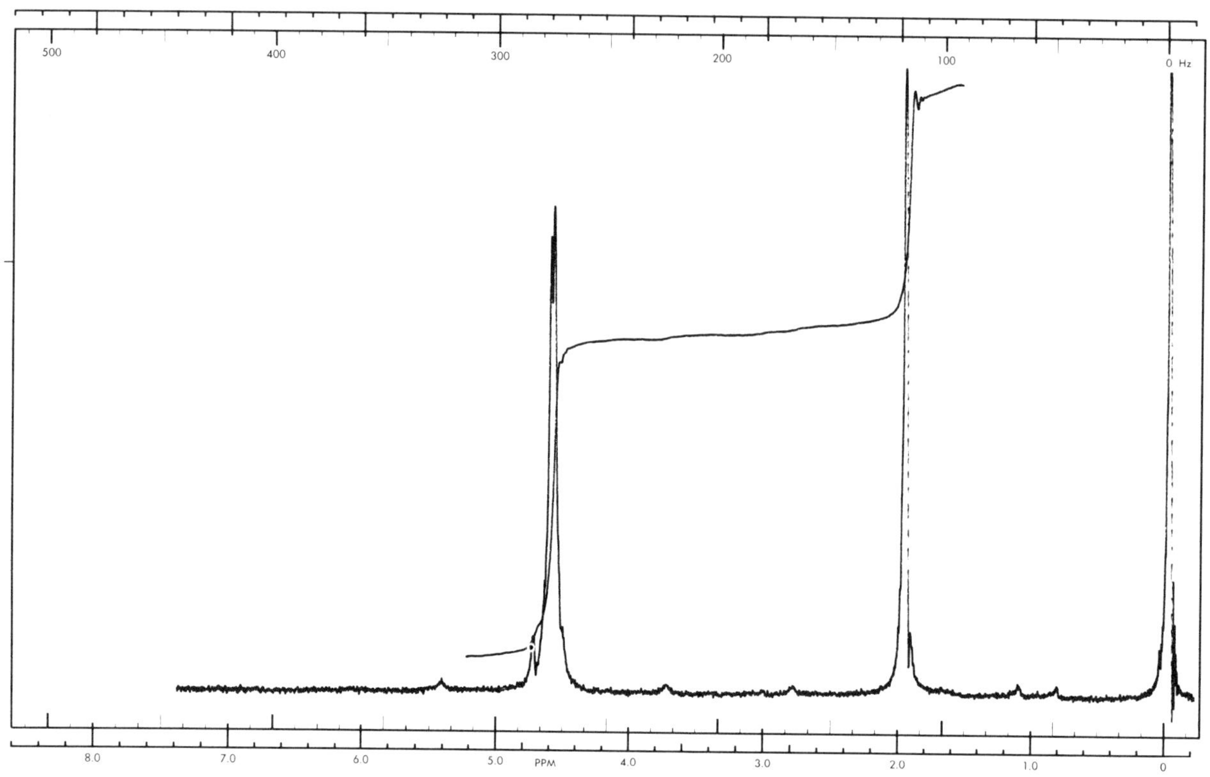

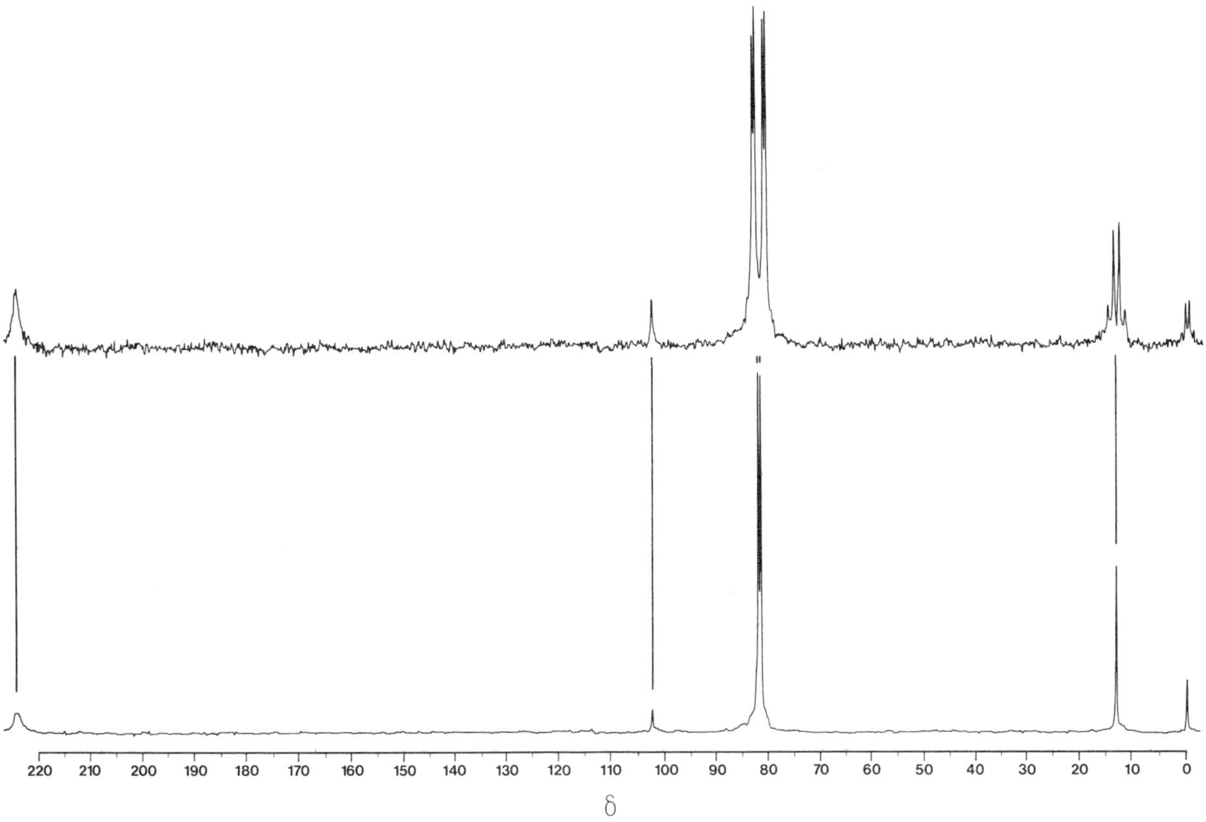
220
210
200
190
180
170
160
150
140
130
120
110
100
90
80
70
60
50
40
30
20
10
0
δ

Section 2

More Complex Carbon-13 NMR Spectra

Problem 2-1

2-1

$C_8H_6N_2$

MS: m130(s), 103, 76

IR: -

UV: Strong absorption

PMR: Upper scan offset 200 Hz

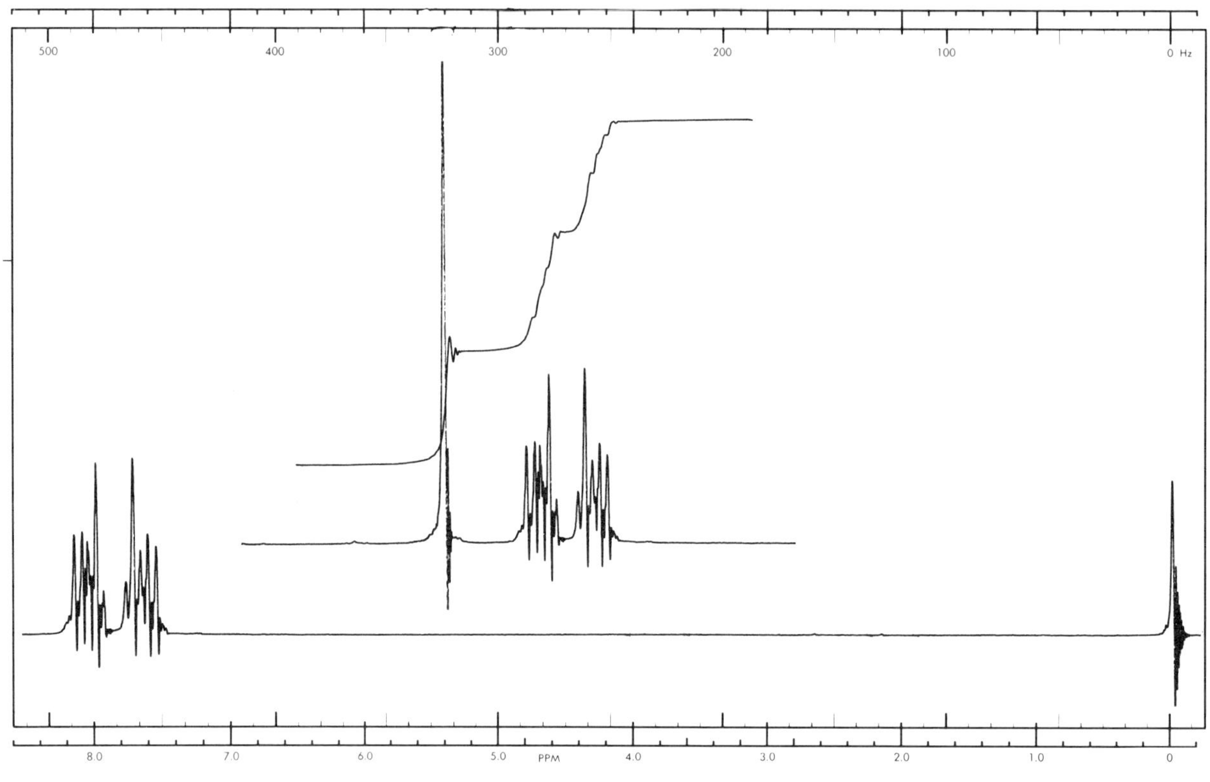

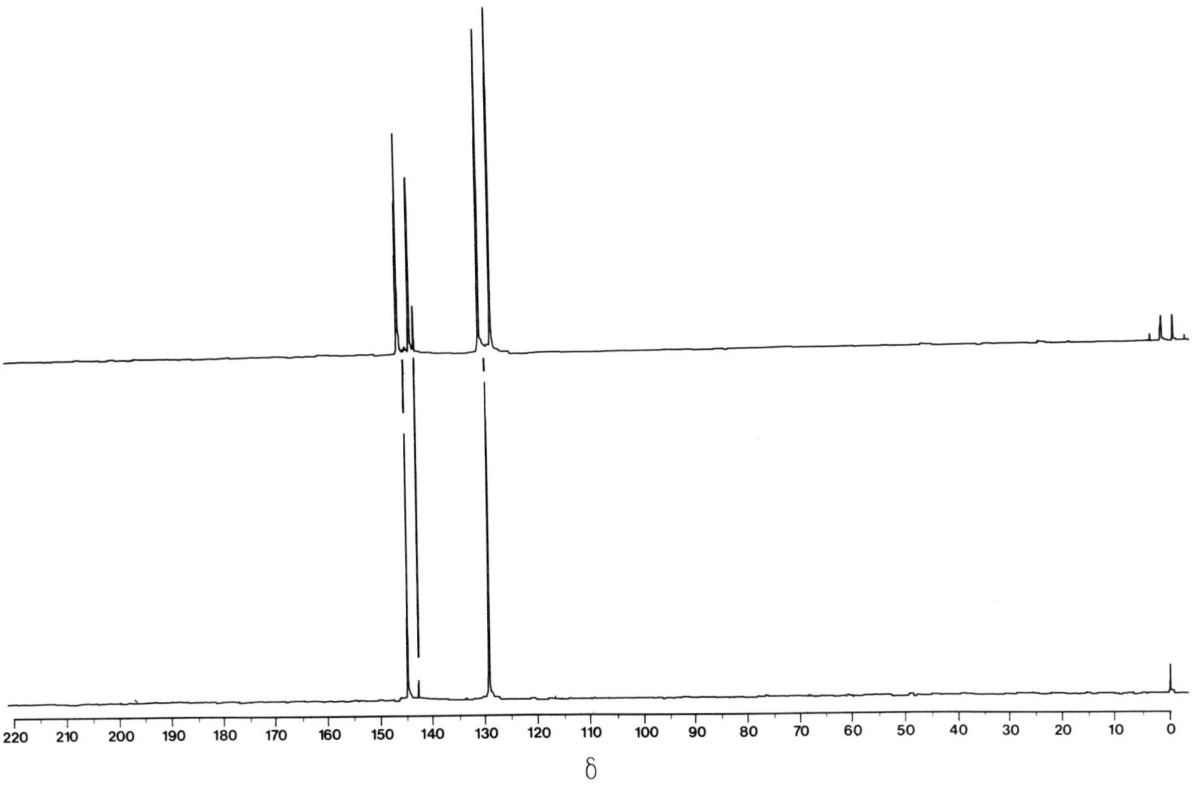
220
210
200
190
180
170
160
150
140
130
120
110
100
90
80
70
60
50
40
30
20
10
0
δ

Problem 2-2

2-2

$C_{14}H_{10}$

MS: m178(s), 101, 77

IR: -

UV: Strong absorption

CMR: 5 sec delay

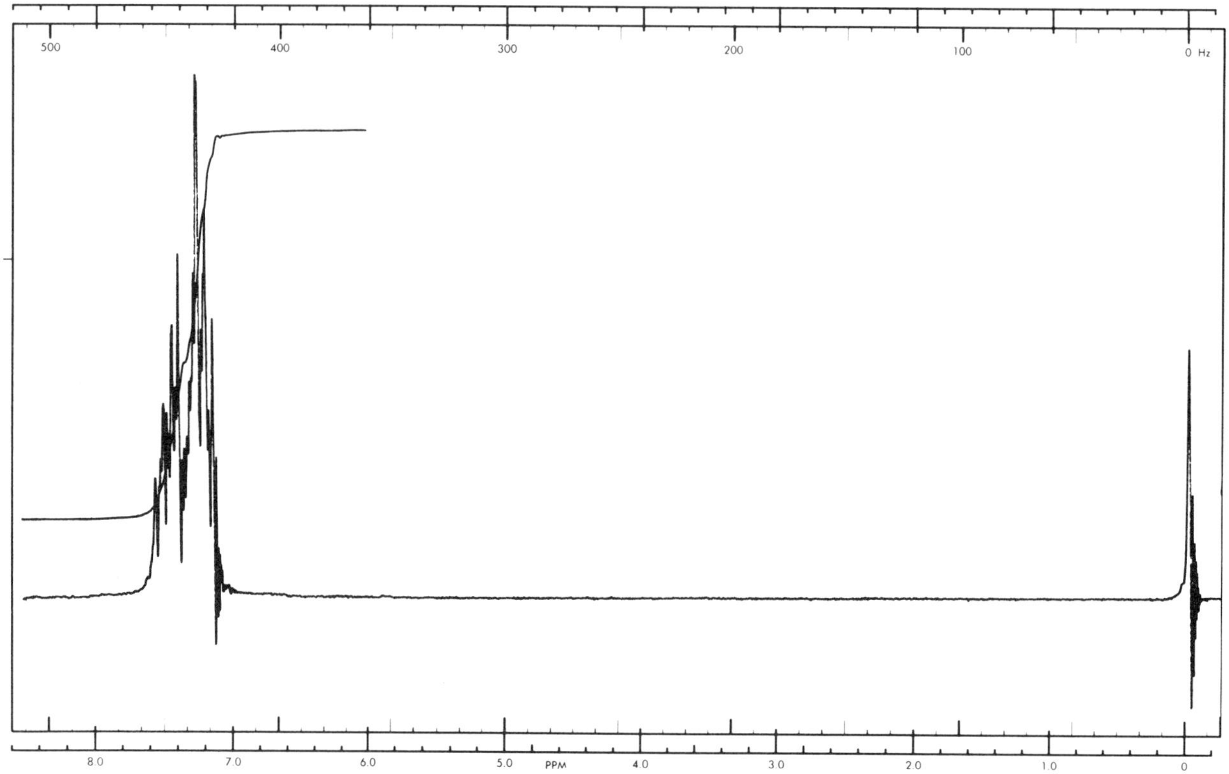

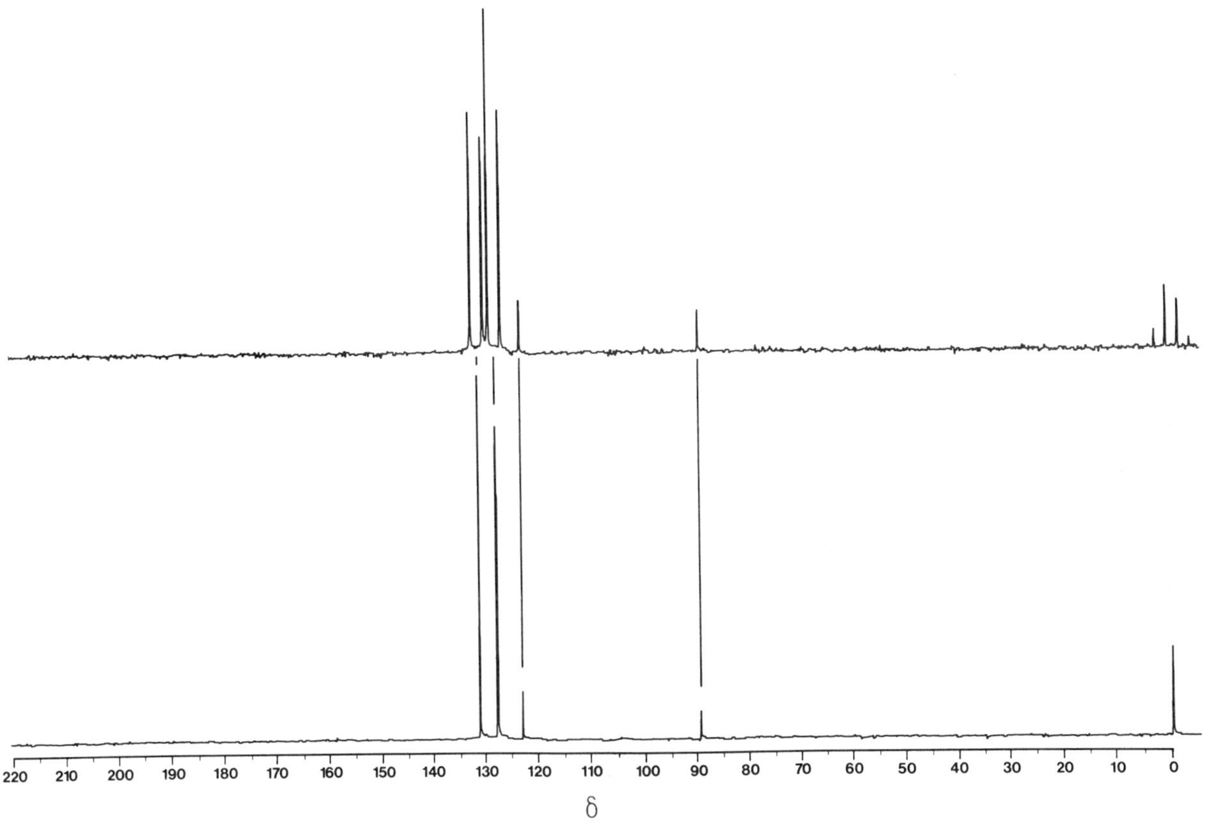
220
210
200
190
180
170
160
150
140
130
120
110
100
90
80
70
60
50
40
30
20
10
0
δ

Problem 2-3

2-3

$C_{12}H_{10}O$

MS: m170(s), 169, 142, 141, 77

IR: 3350 cm^{-1}

UV: Strong absorption

PMR and CMR: d_6-Me_2CO + TMS

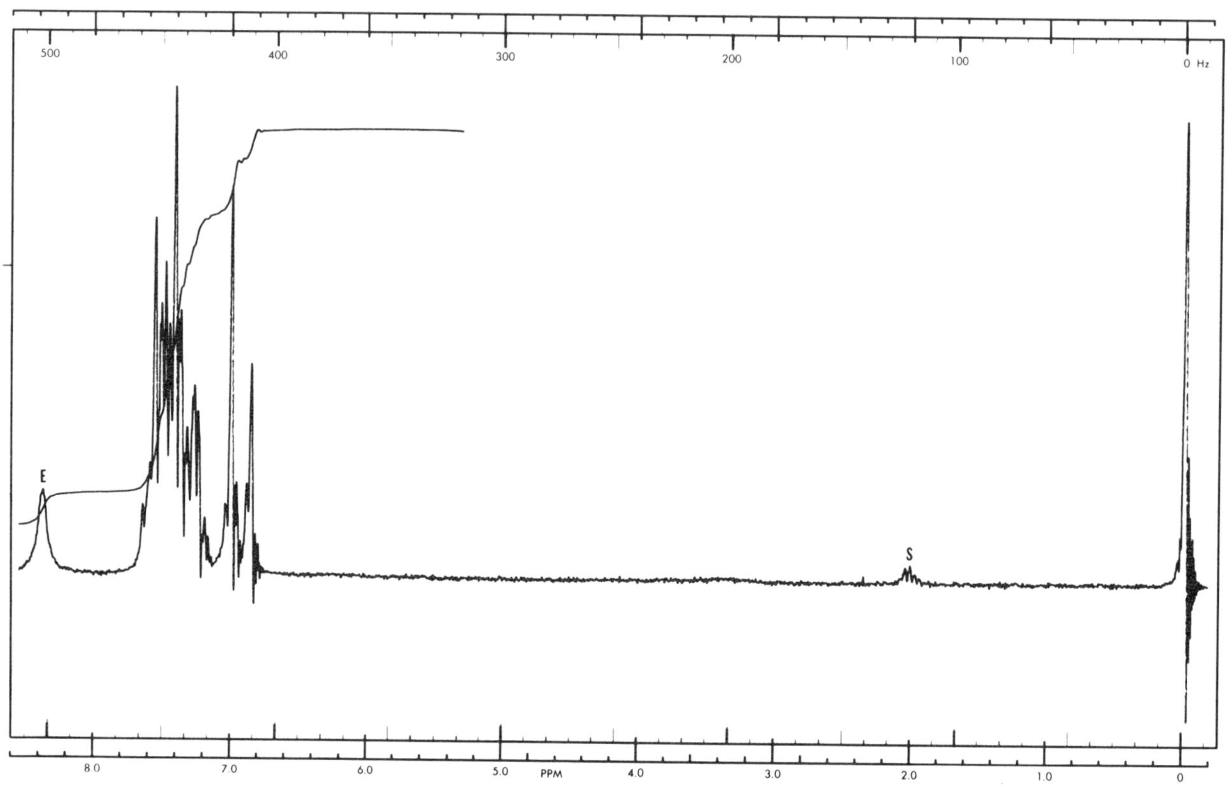

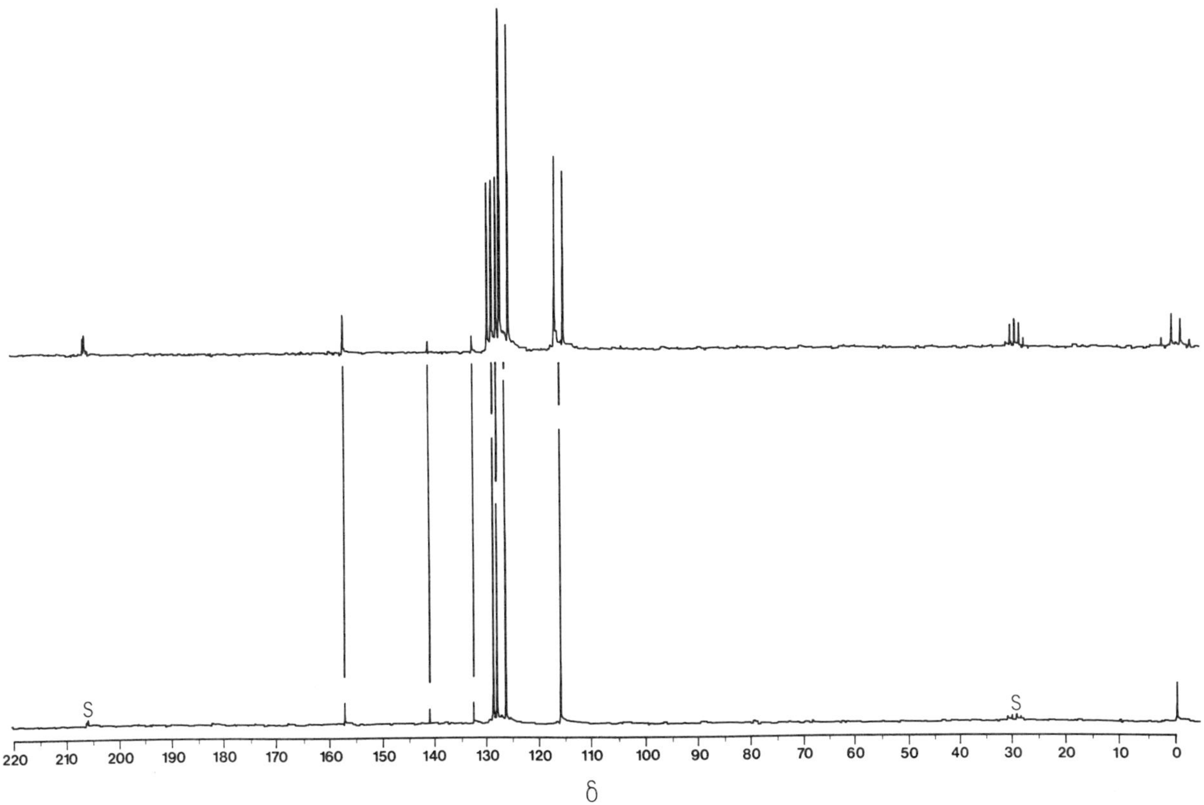
S
S
220
210
200
190
180
170
160
150
140
130
120
110
100
90
80
70
60
50
40
30
20
10
0
δ

Problem 2-4

2-4

$C_{17}H_{14}O$

MS: m234, 131, 103

IR: 1655, 1630 cm^{-1}

UV: Strong absorption

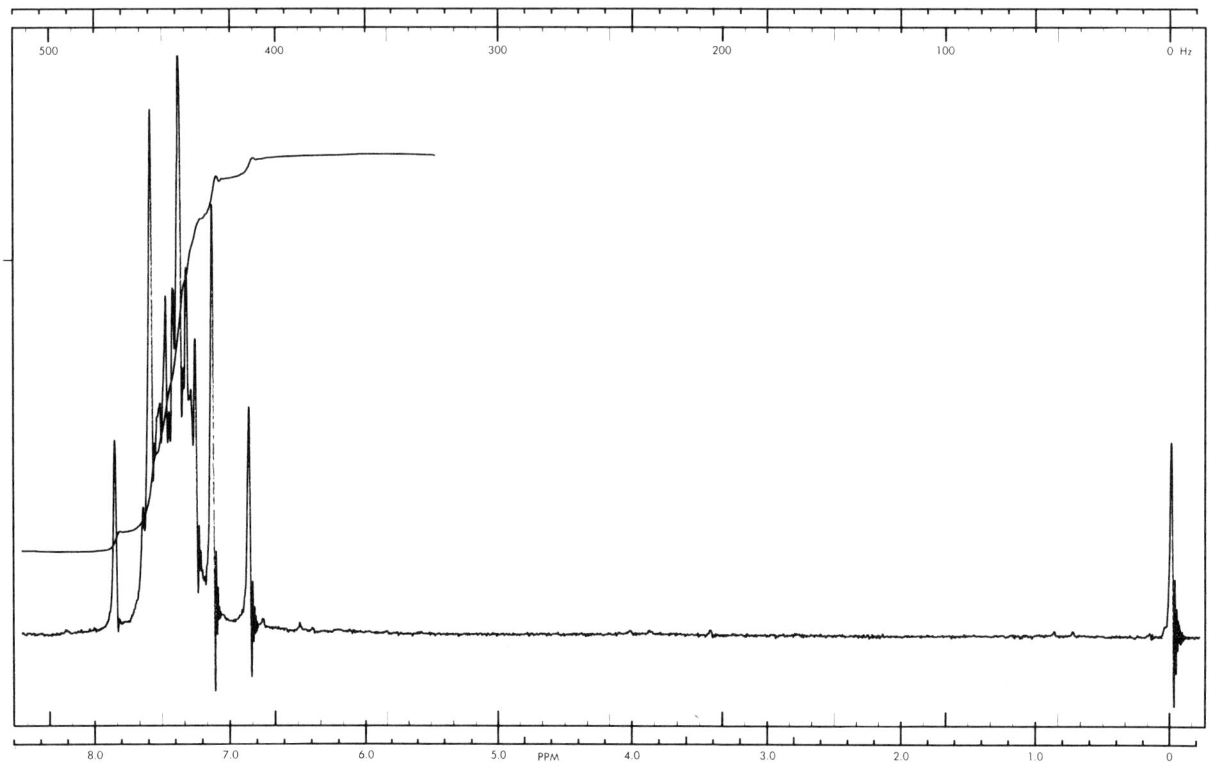

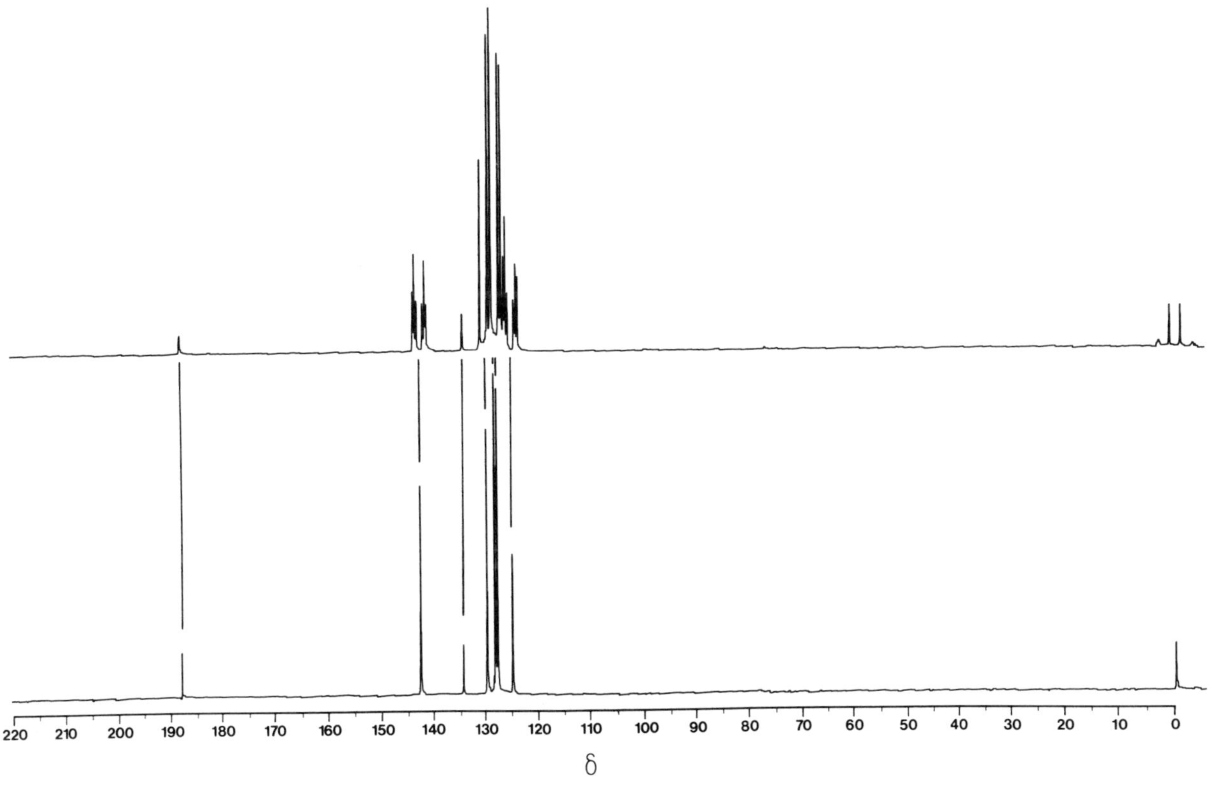
220
210
200
190
180
170
160
150
140
130
120
110
100
90
80
70
60
50
40
30
20
10
0
δ

Problem 2-5

2-5

$C_7H_{10}NO^+Cl^-$

MS: m159(w), 107, 91, 77

IR: 2900 cm^{-1} (broad)

UV: Strong absorption

PMR: d_6-DMSO + TMS; upper scan offset 500 Hz

CMR: d_6-DMSO

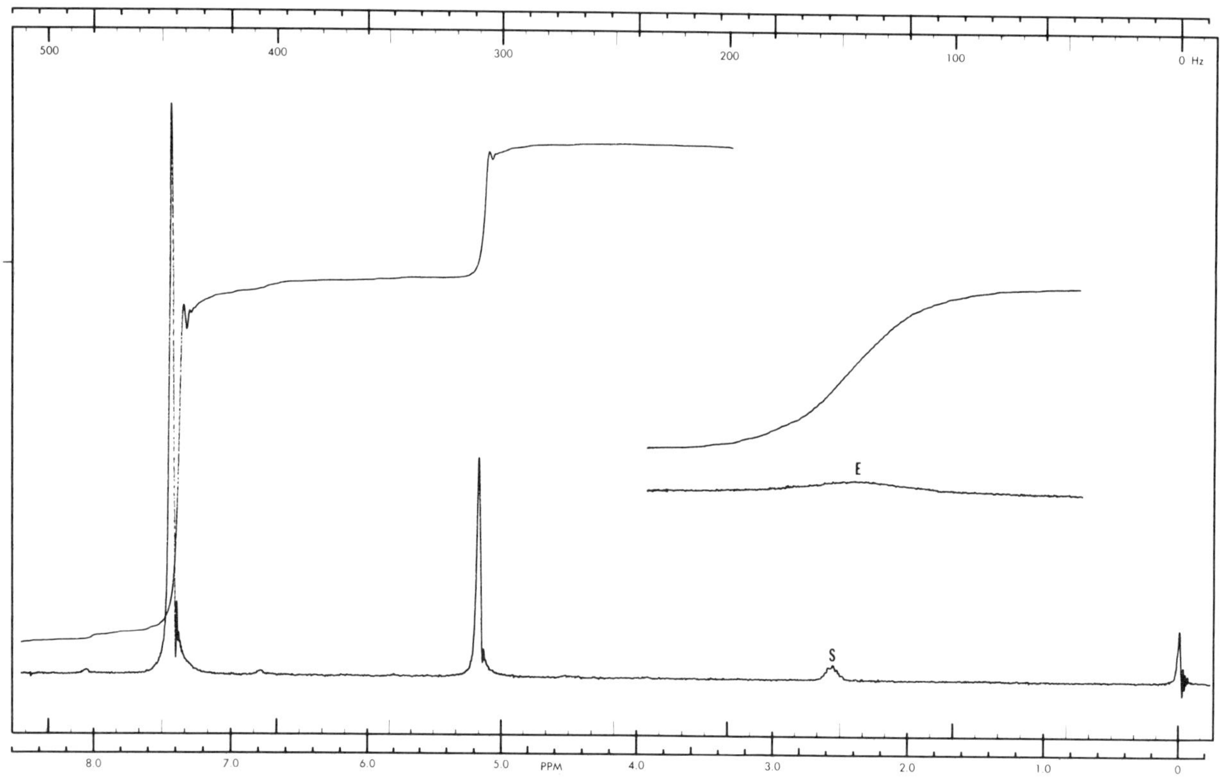

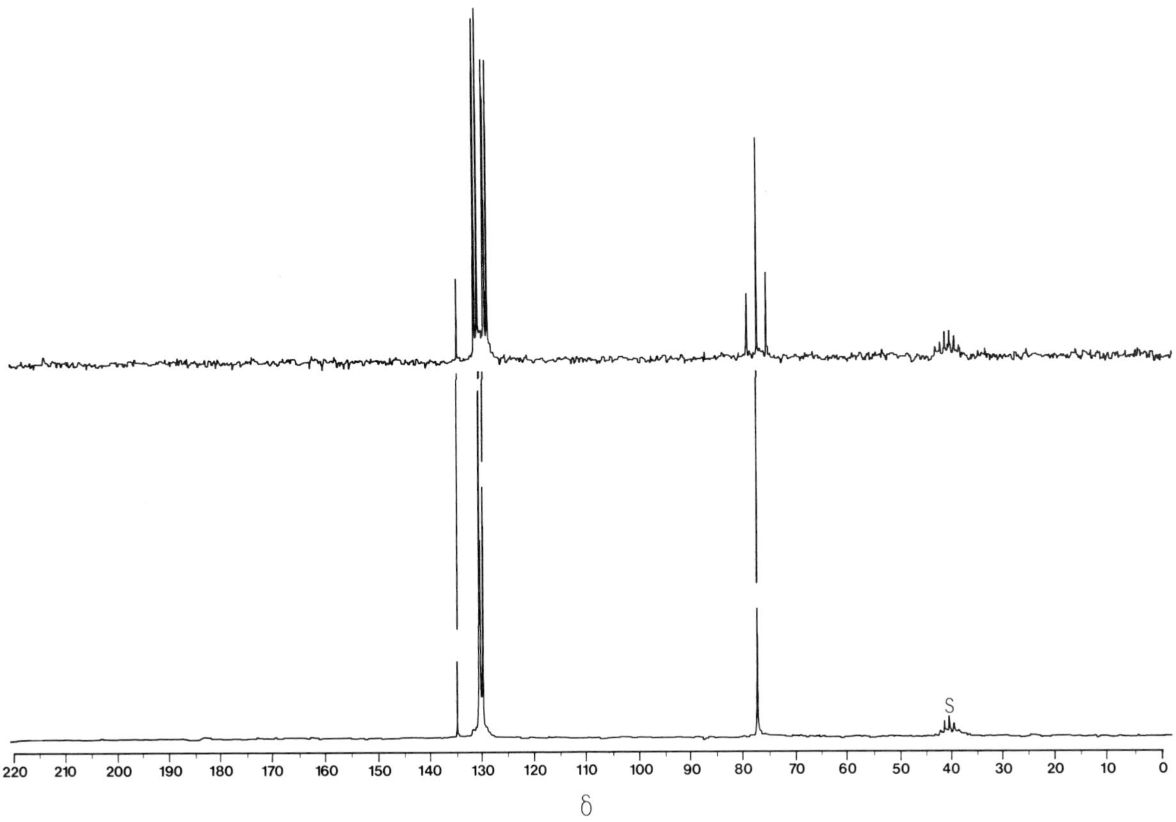
S
220 210 200 190 180 170 160 150 140 130 120 110 100 90 80 70 60 50 40 30 20 10 0
δ

Problem 2-6

2-6

$C_{16}H_{18}O_4$

MS: m274(s), 259, 137

IR: -

UV: Strong absorption

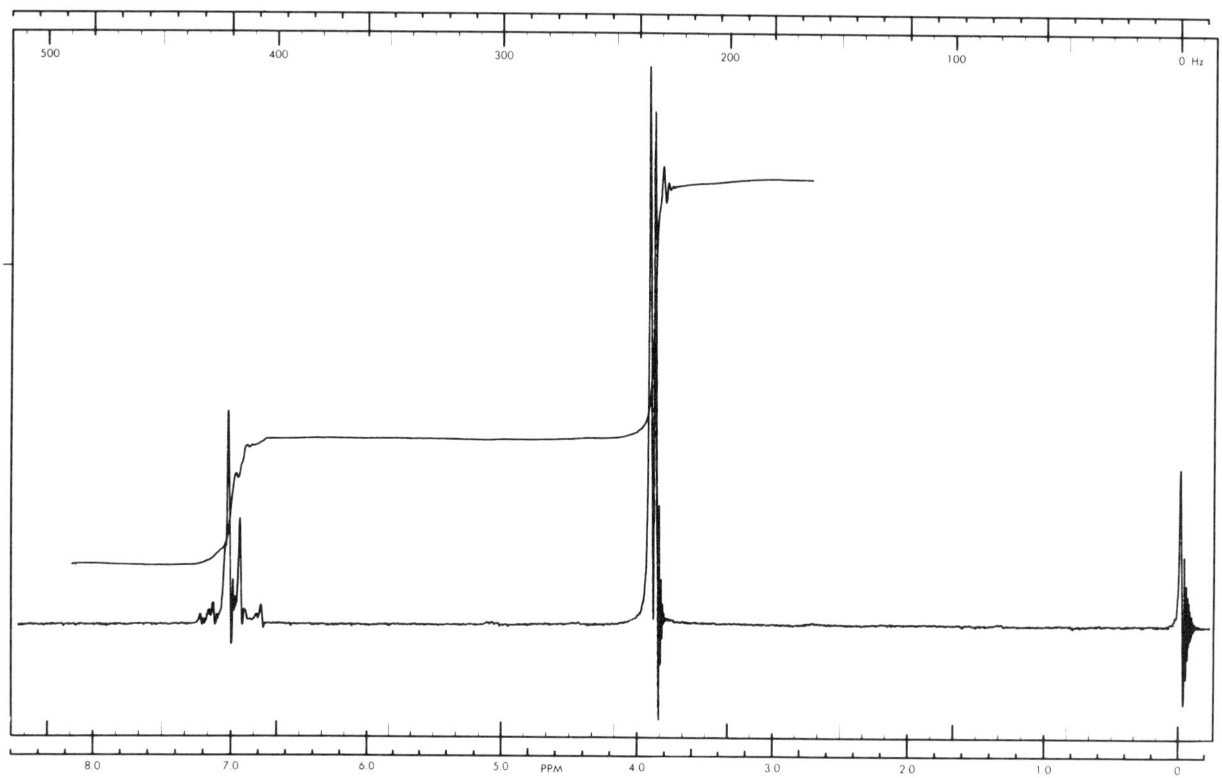

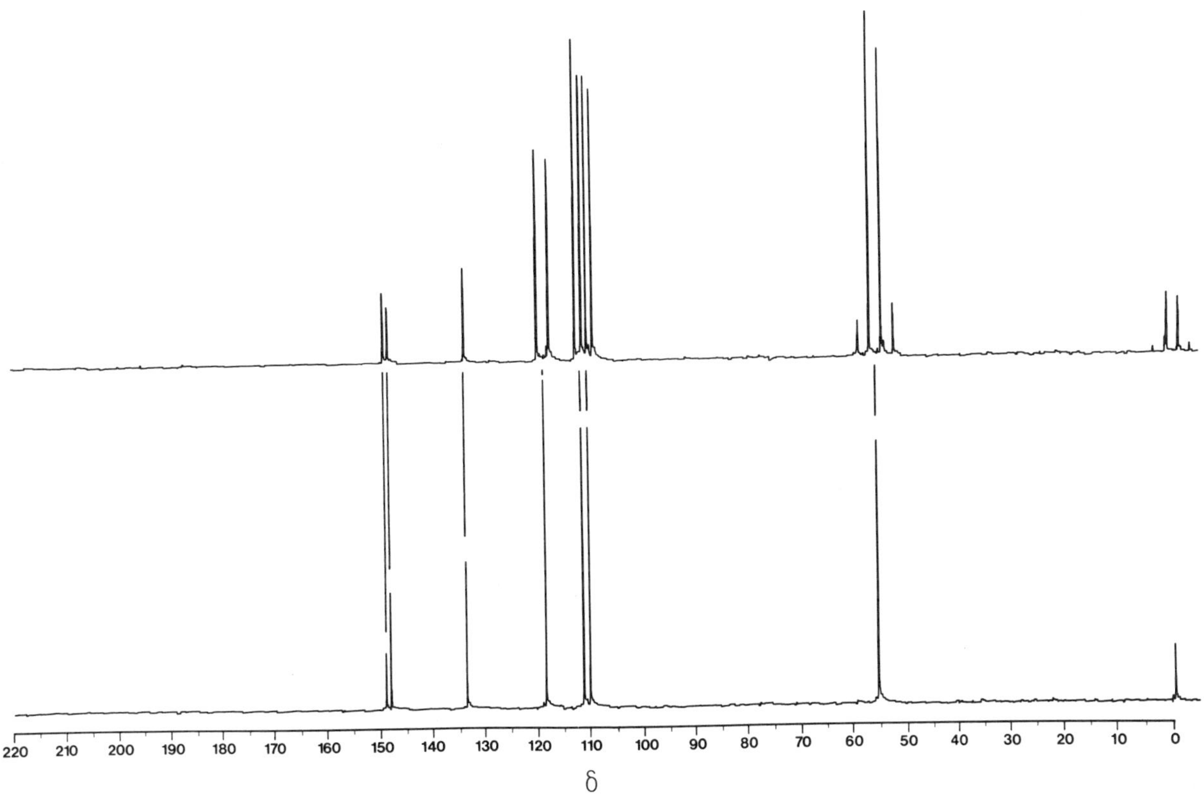
220
210
200
190
180
170
160
150
140
130
120
110
100
90
80
70
60
50
40
30
20
10
0
δ

Problem 2-7

2-7

$C_{13}H_{13}NO_2$

MS: m215, 146, 69

IR: 1780, 1712, 1660 cm^{-1}

UV: Strong absorption

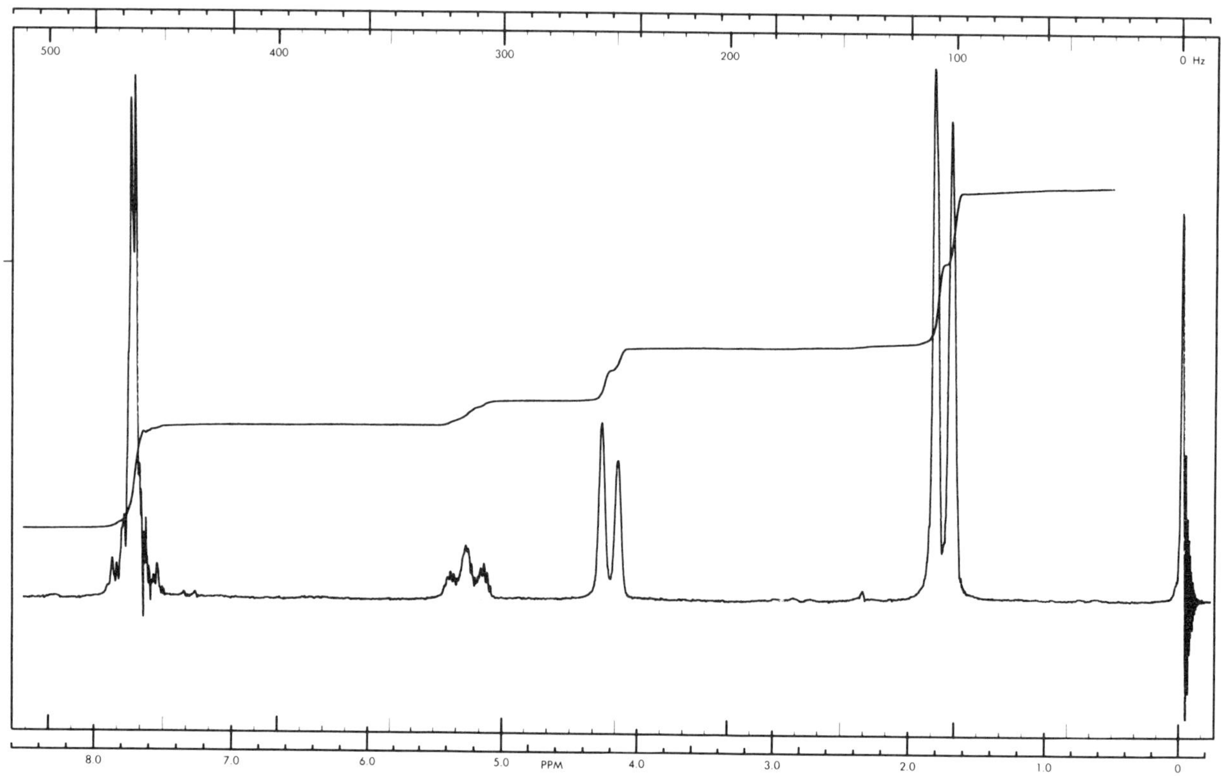

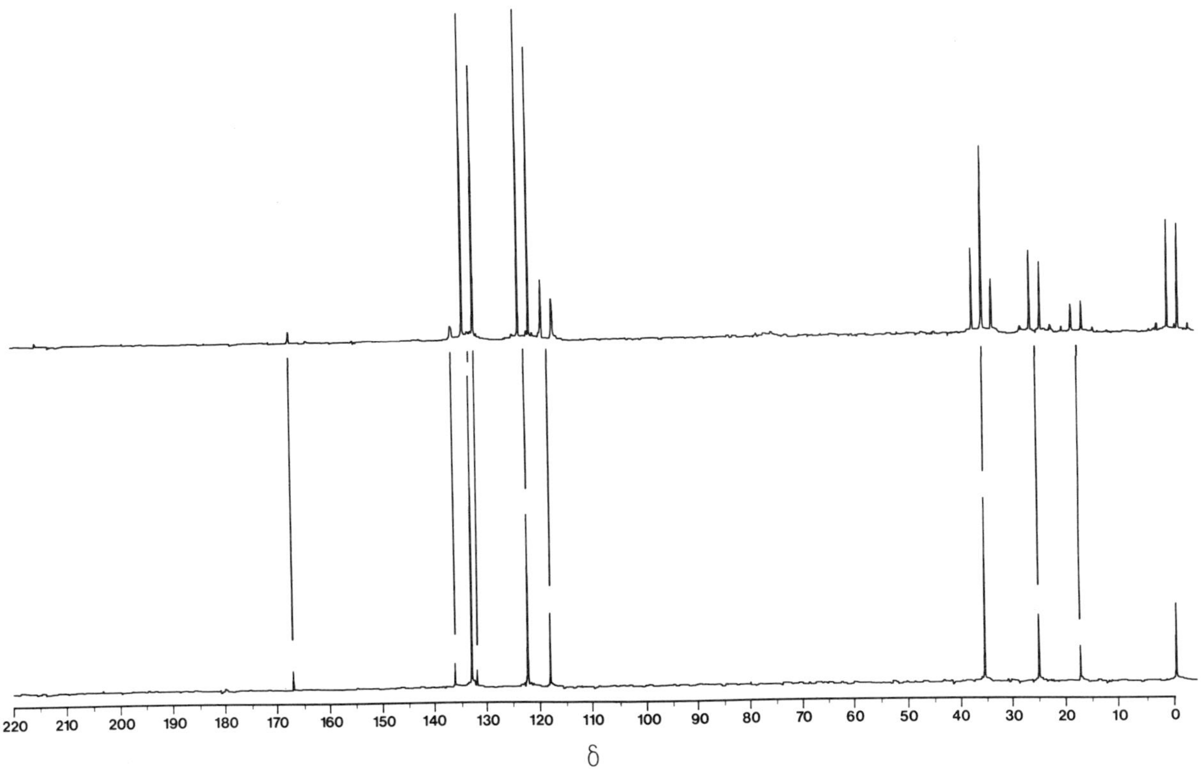
220
210
200
190
180
170
160
150
140
130
120
110
100
90
80
70
60
50
40
30
20
10
0
δ

Problem 2-8

2-8

$C_{11}H_{11}NO_5$

MS: m237, 207, 191, 150, 60

IR: 2900 (broad), 1700 cm^{-1}

UV: Strong absorption

PMR: d_5-Pyridine + TMS; lower scan offset 410 Hz, upper scan offset 200 Hz

CMR: d_6-DMSO + dioxane

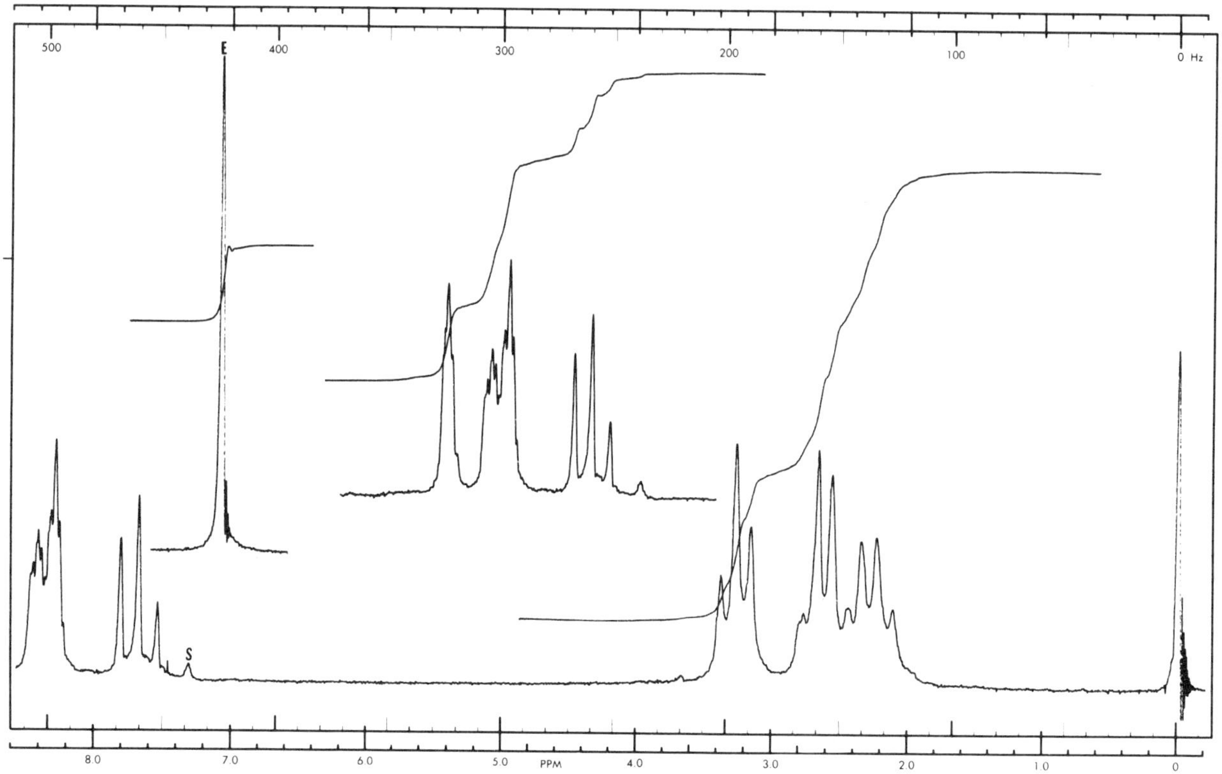

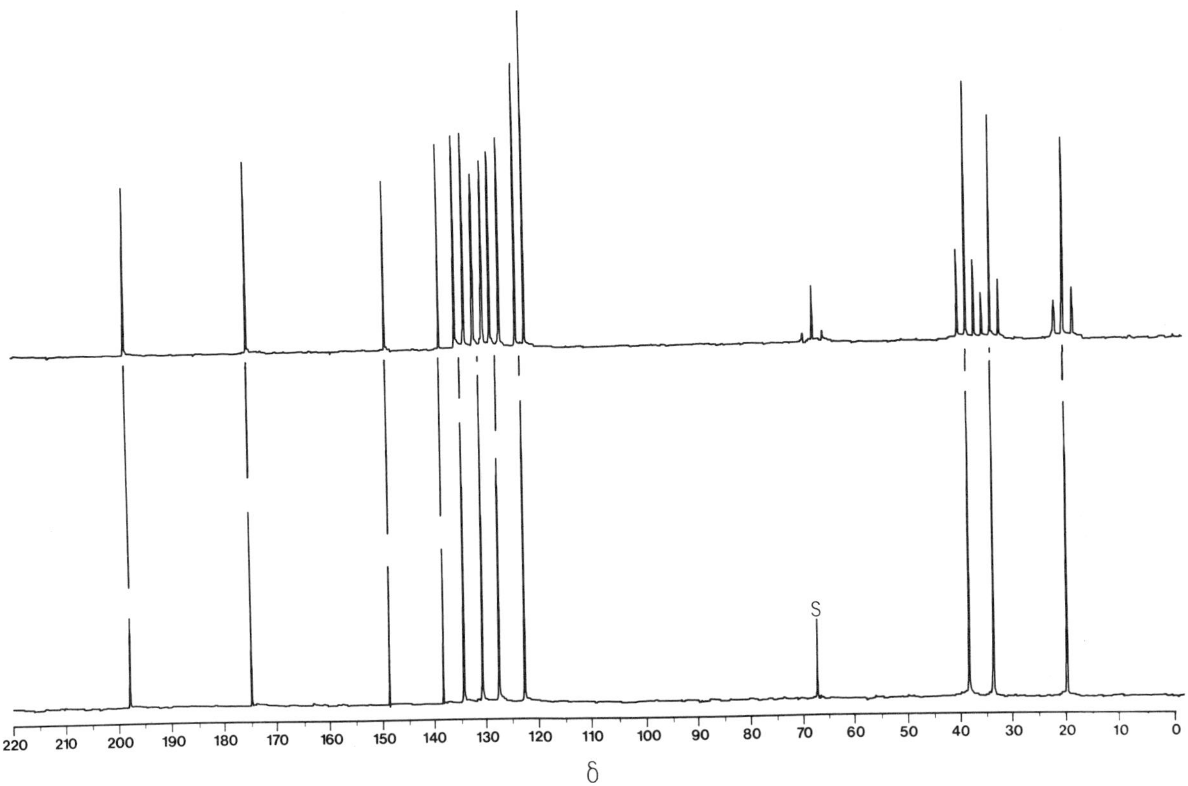
S
220
210
200
190
180
170
160
150
140
130
120
110
100
90
80
70
60
50
40
30
20
10
0
δ

Problem 2-9

2-9

$C_5H_9NO_2$

MS: m115(w), 70

IR: 3200-2200 (broad), 1630, 1570 cm^{-1}

UV: none

PMR: D_2O + DSS

CMR: D_2O + dioxane; 5 sec delay

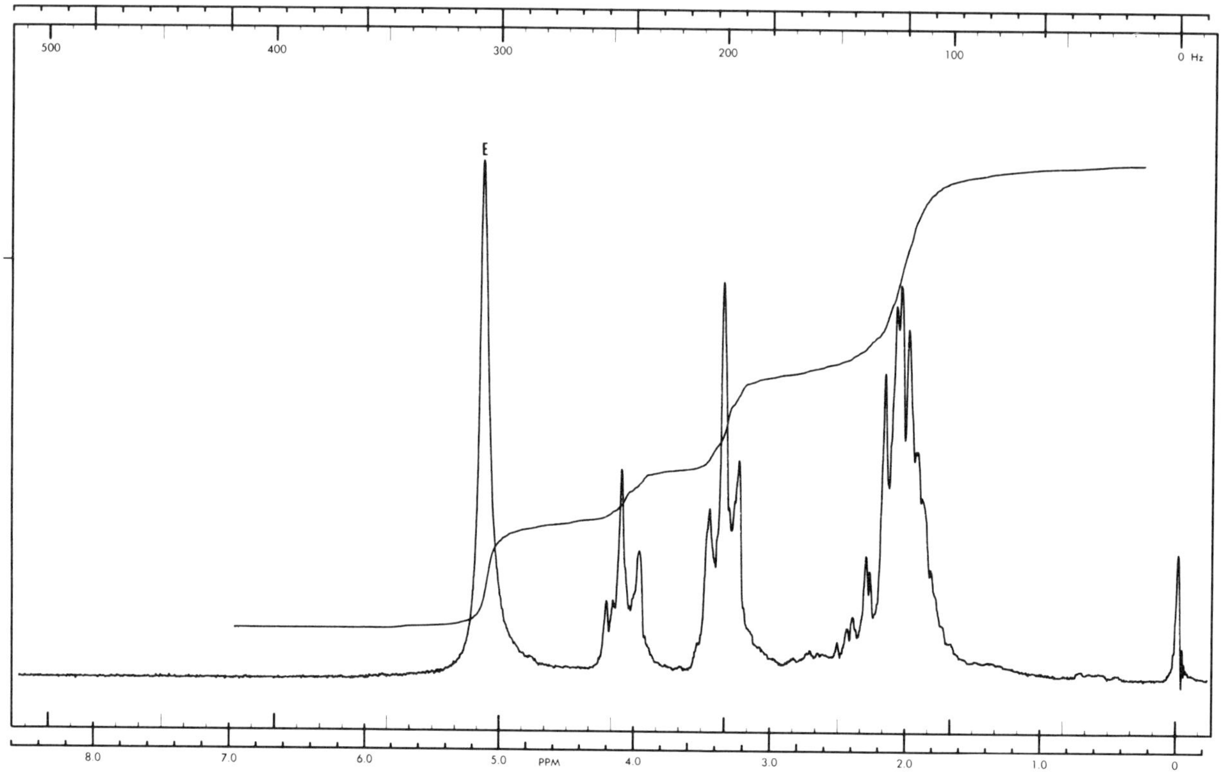

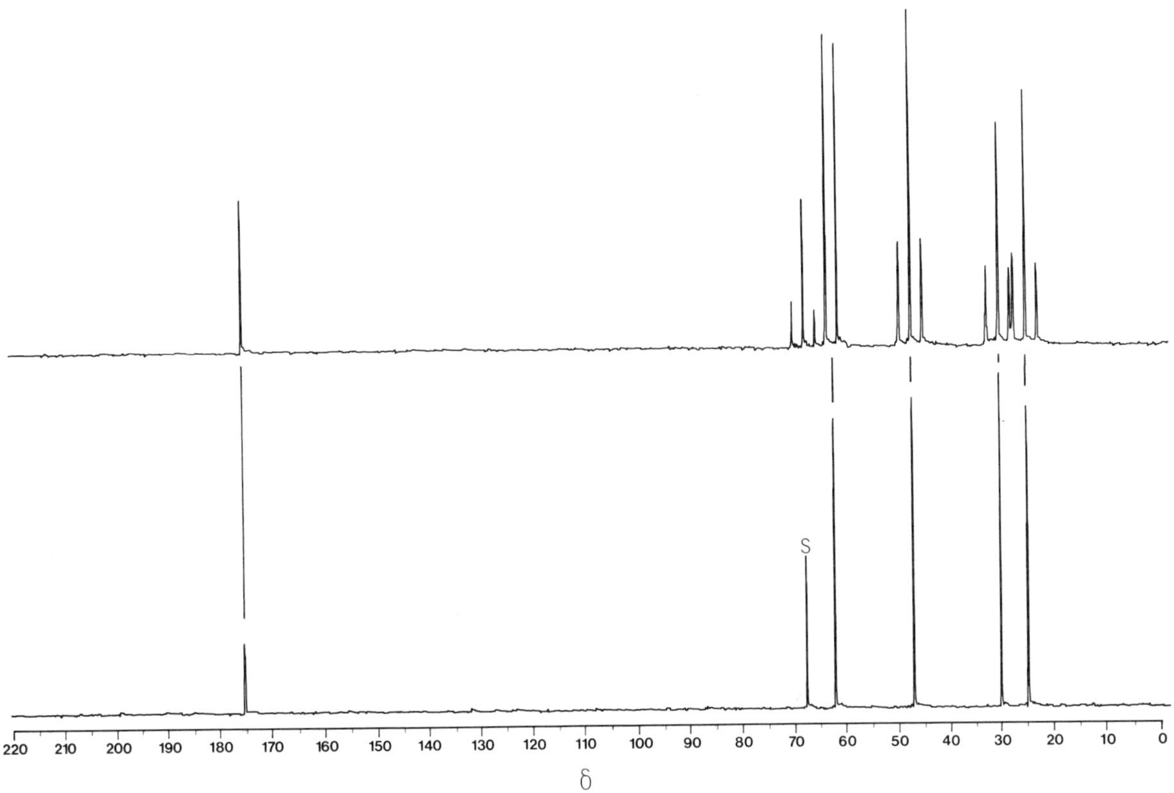
S
220
210
200
190
180
170
160
150
140
130
120
110
100
90
80
70
60
50
40
30
20
10
0
δ

Problem 2-10

2-10

$C_{24}H_{38}O_4$

MS: m390, 261, 112

IR: 1715 cm^{-1}

UV: Strong absorption

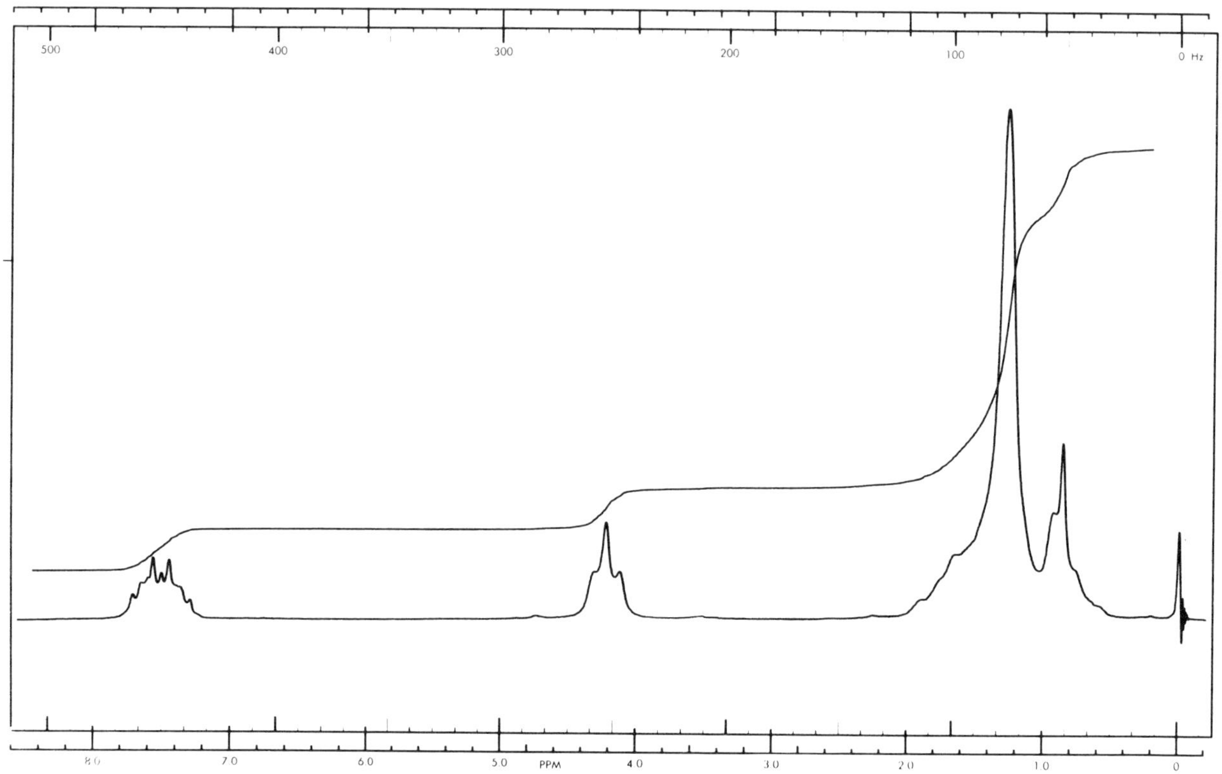

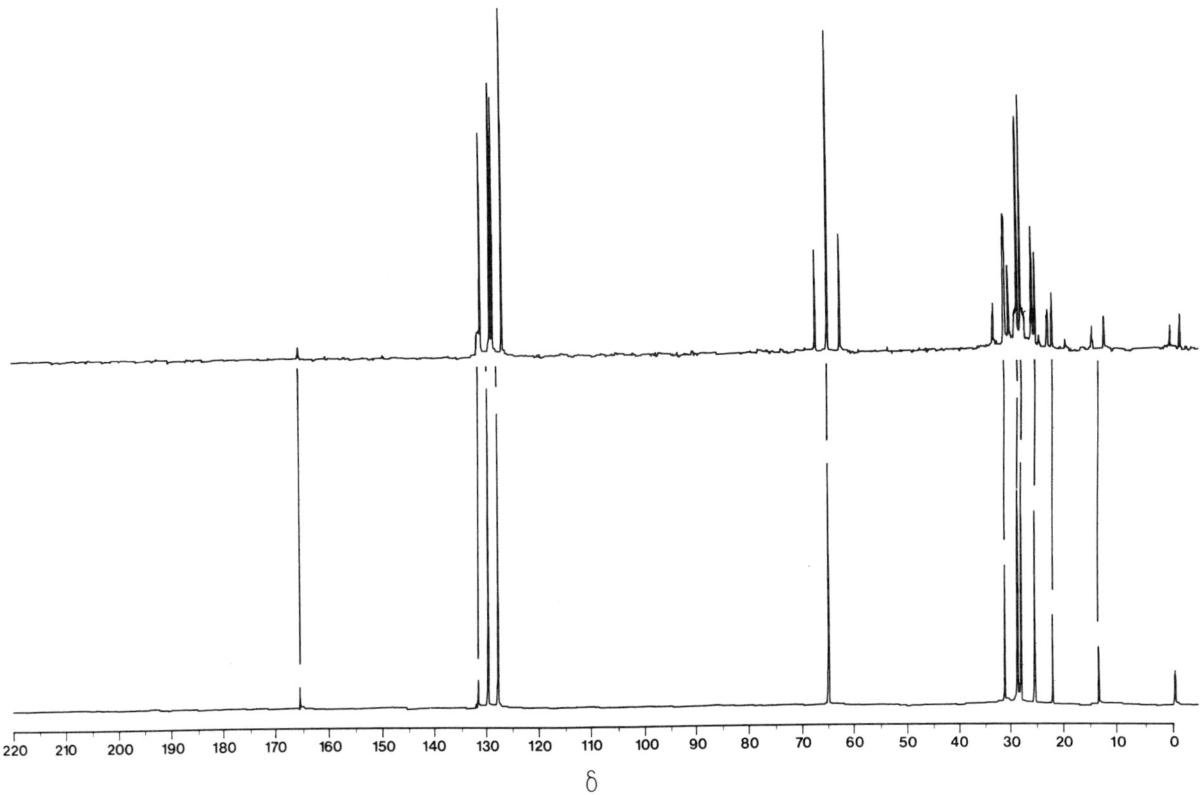
220
210
200
190
180
170
160
150
140
130
120
110
100
90
80
70
60
50
40
30
20
10
0
δ

Problem 2-11

2-11
$C_{13}H_9NS$

MS: m211, 134, 108, 103, 77

IR:

UV: Strong absorption

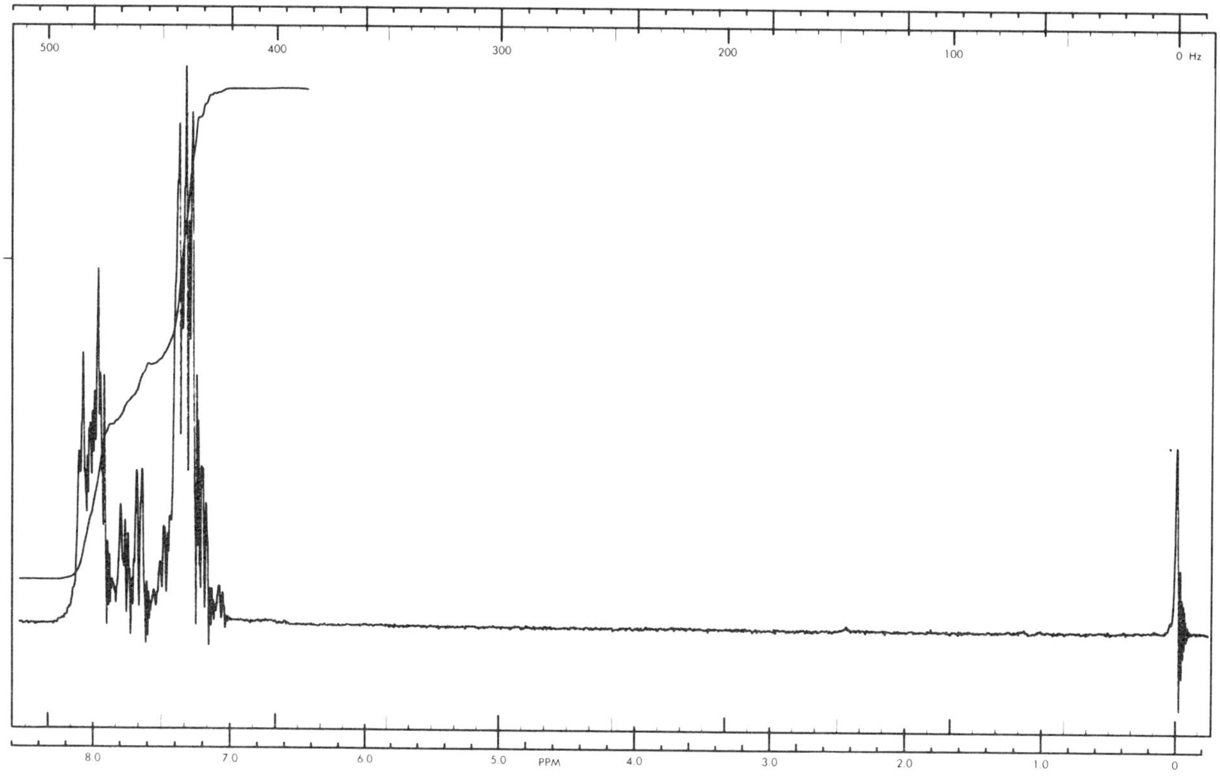

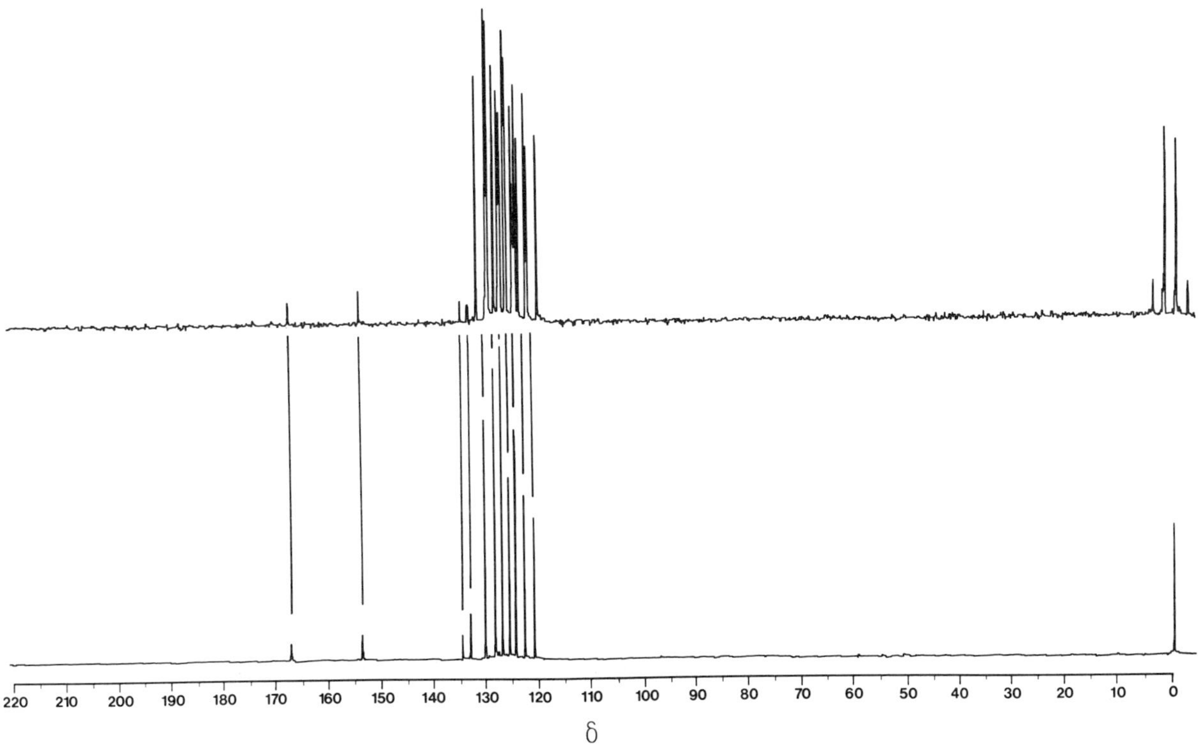
220 210 200 190 180 170 160 150 140 130 120 110 100 90 80 70 60 50 40 30 20 10 0
δ

Problem 2-12

2-12
$C_{12}H_{12}N_2$

MS: m184(s), 183, 92, 91, 77
IR: 3400, 3350 cm^{-1}
UV: Strong absorption

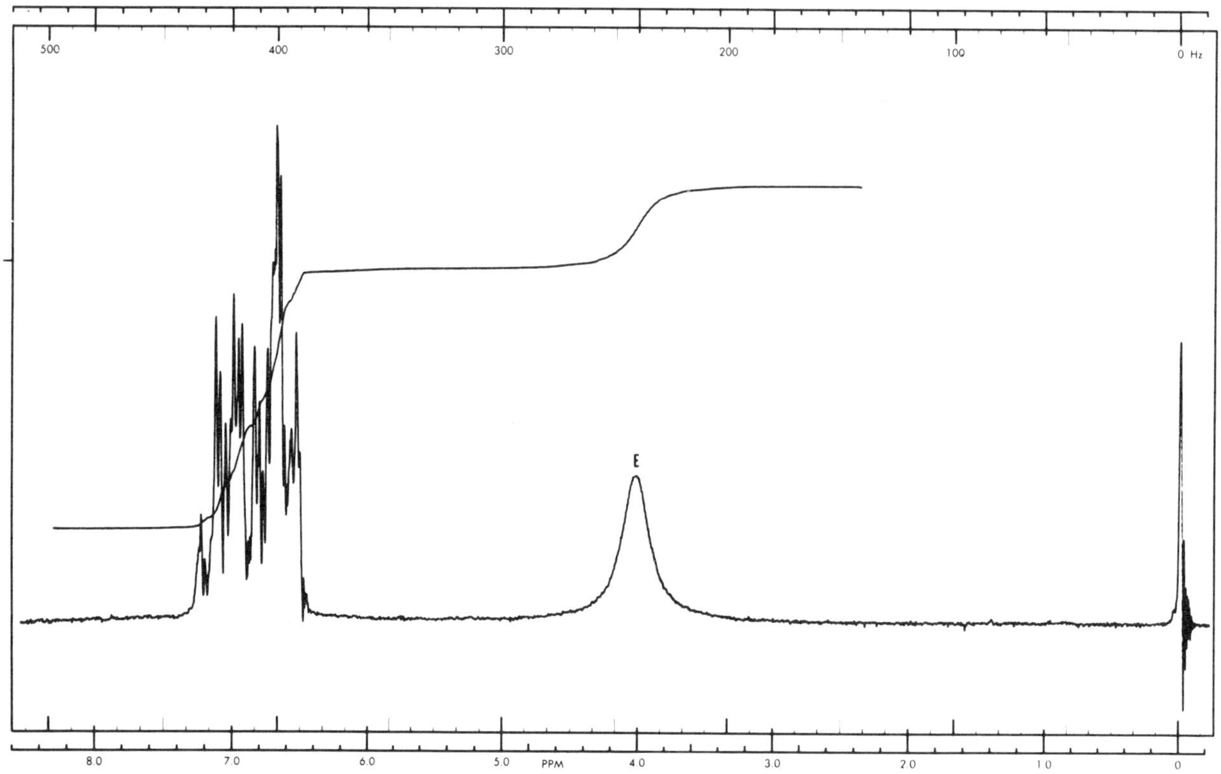

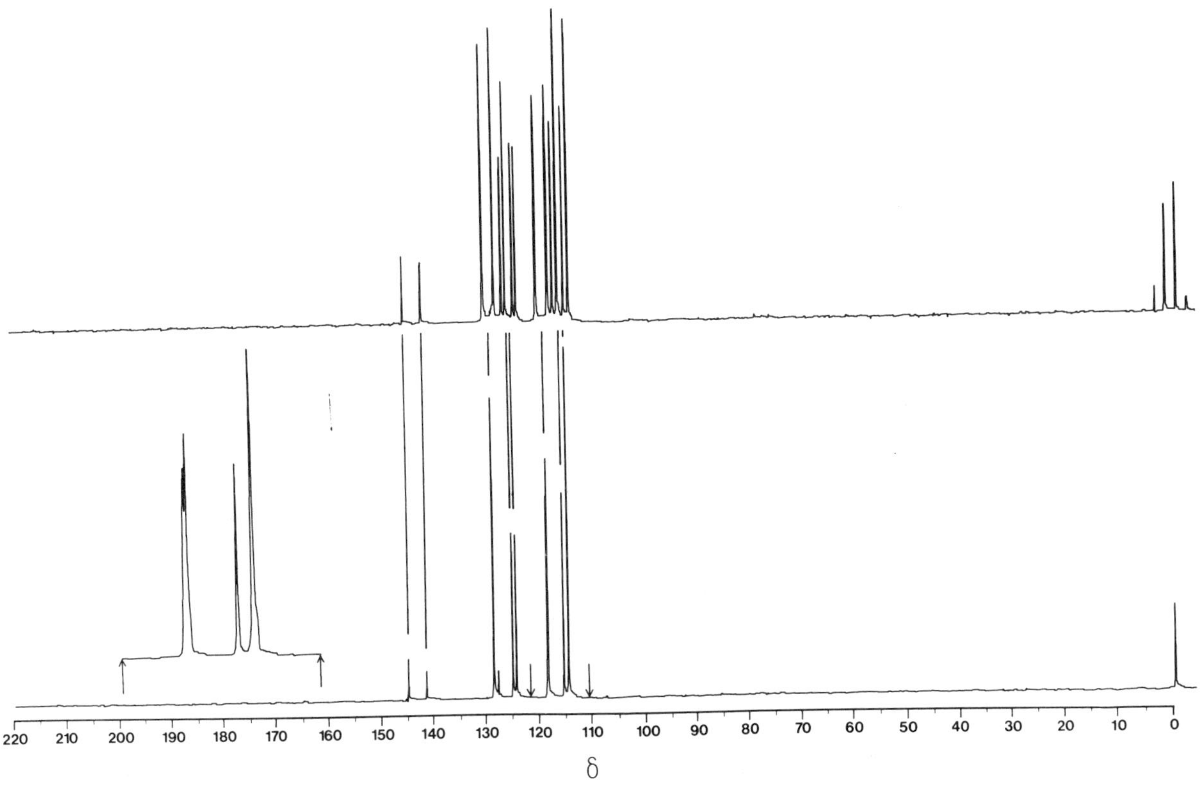

220 210 200 190 180 170 160 150 140 130 120 110 100 90 80 70 60 50 40 30 20 10 0
δ

Problem 2-13

2-13

$C_{10}H_8O_2$

MS: m160(s), 159, 132, 131

IR: 3300 cm^{-1}

UV: Strong absorption

PMR: d_6-Me_2CO + TMS; upper scan offset 200 Hz

CMR: d_6-Me_2CO + TMS; 20 sec delay

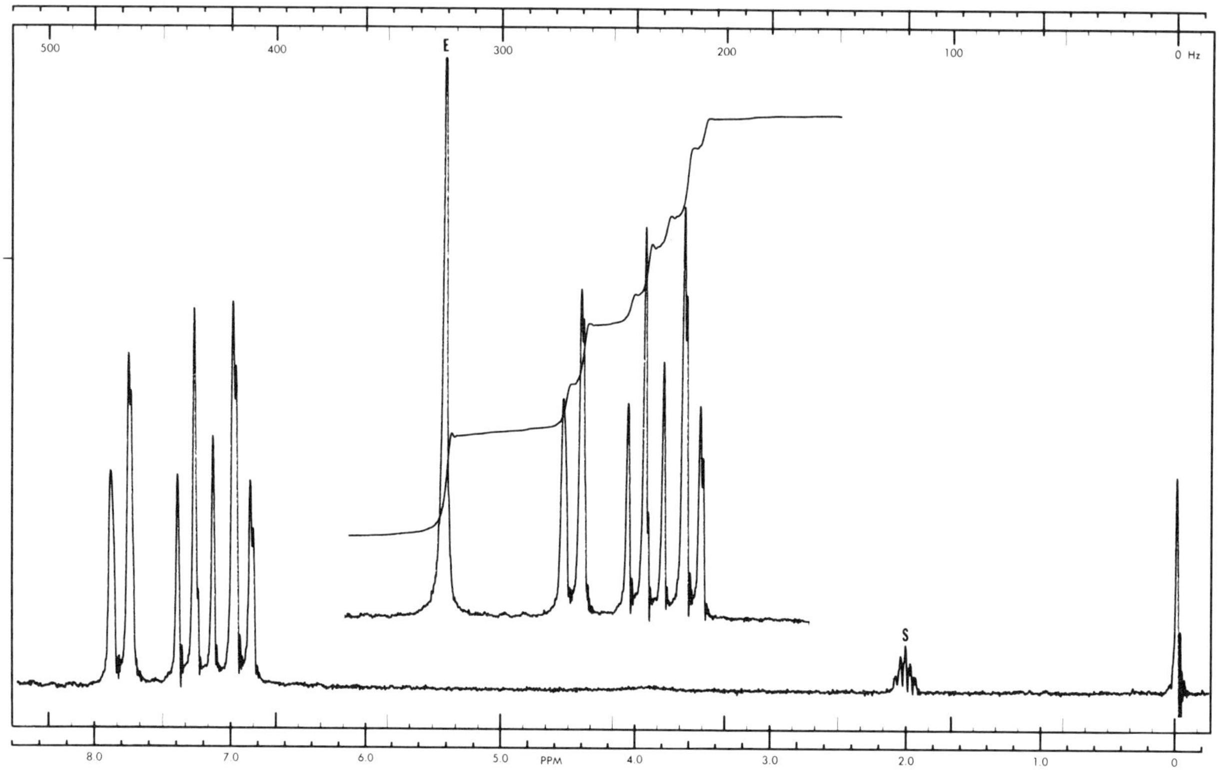

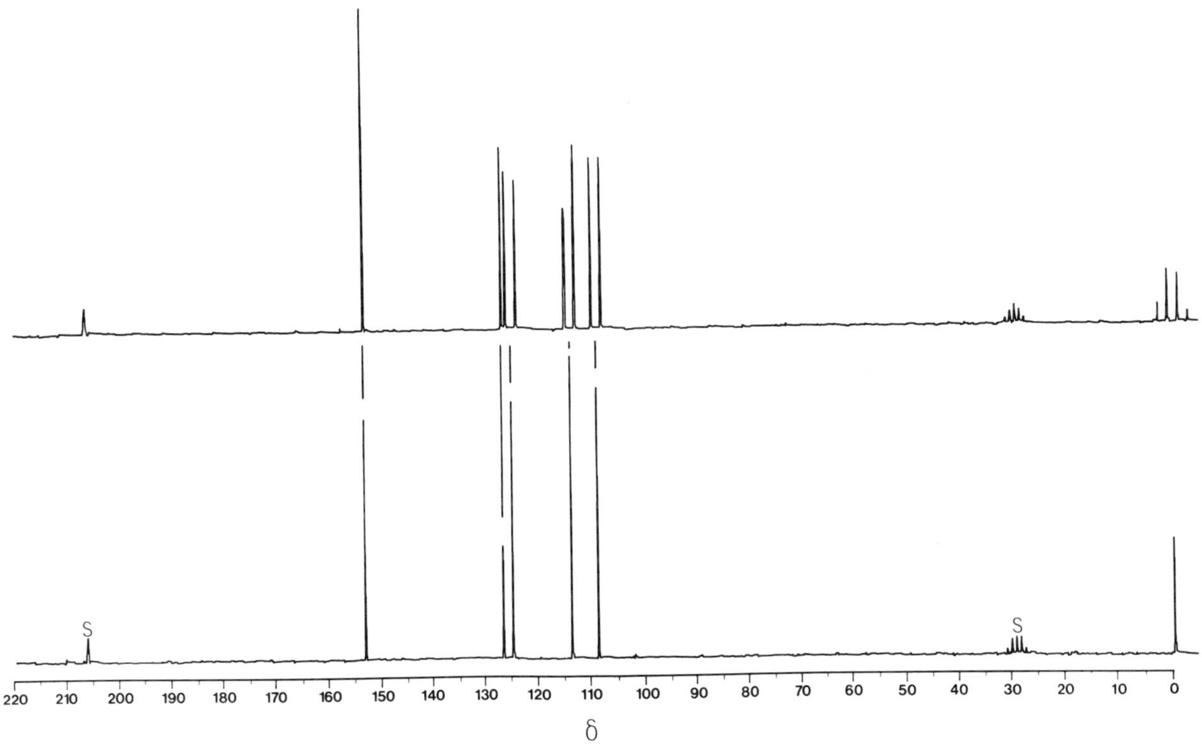
S
S
220
210
200
190
180
170
160
150
140
130
120
110
100
90
80
70
60
50
40
30
20
10
0
δ

Problem 2-14

2-14

$C_{10}H_8O_2$

MS: m160(s), 159, 132, 131

IR: 3350 cm^{-1}

UV: Strong absorption

PMR and CMR: d_6-Me_2CO + TMS; 10 sec delay

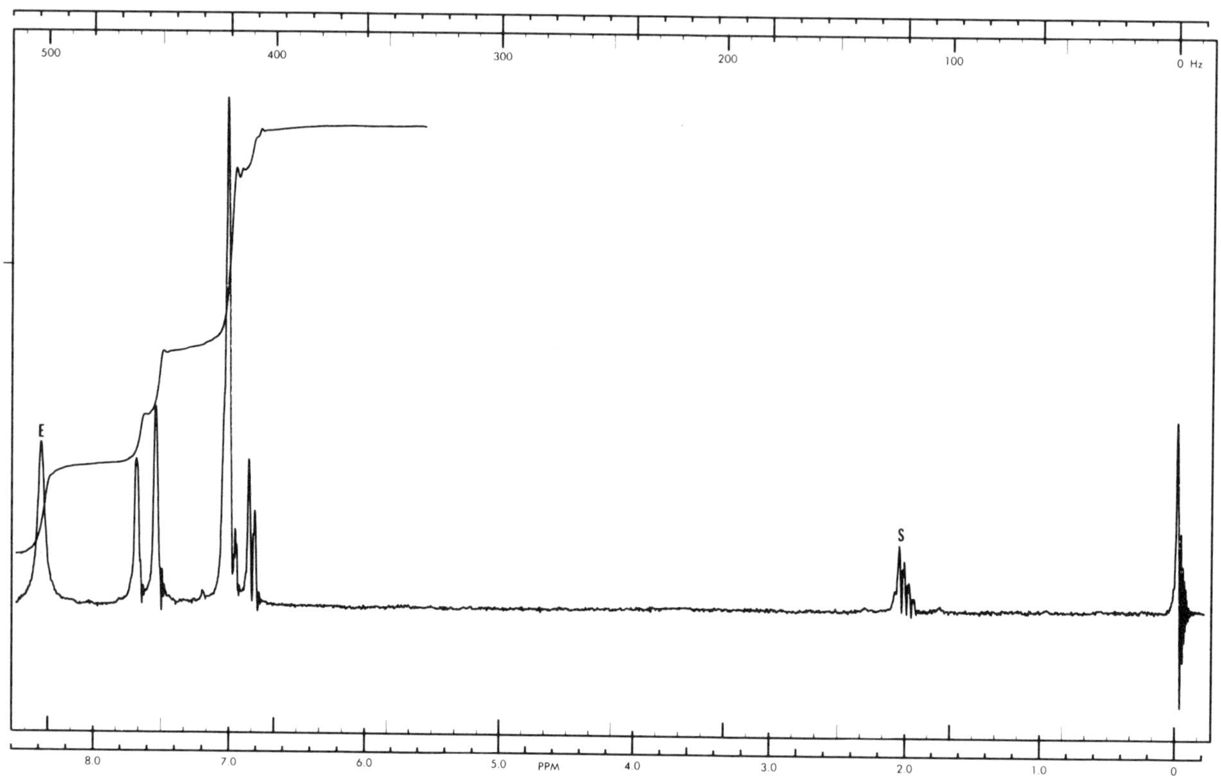

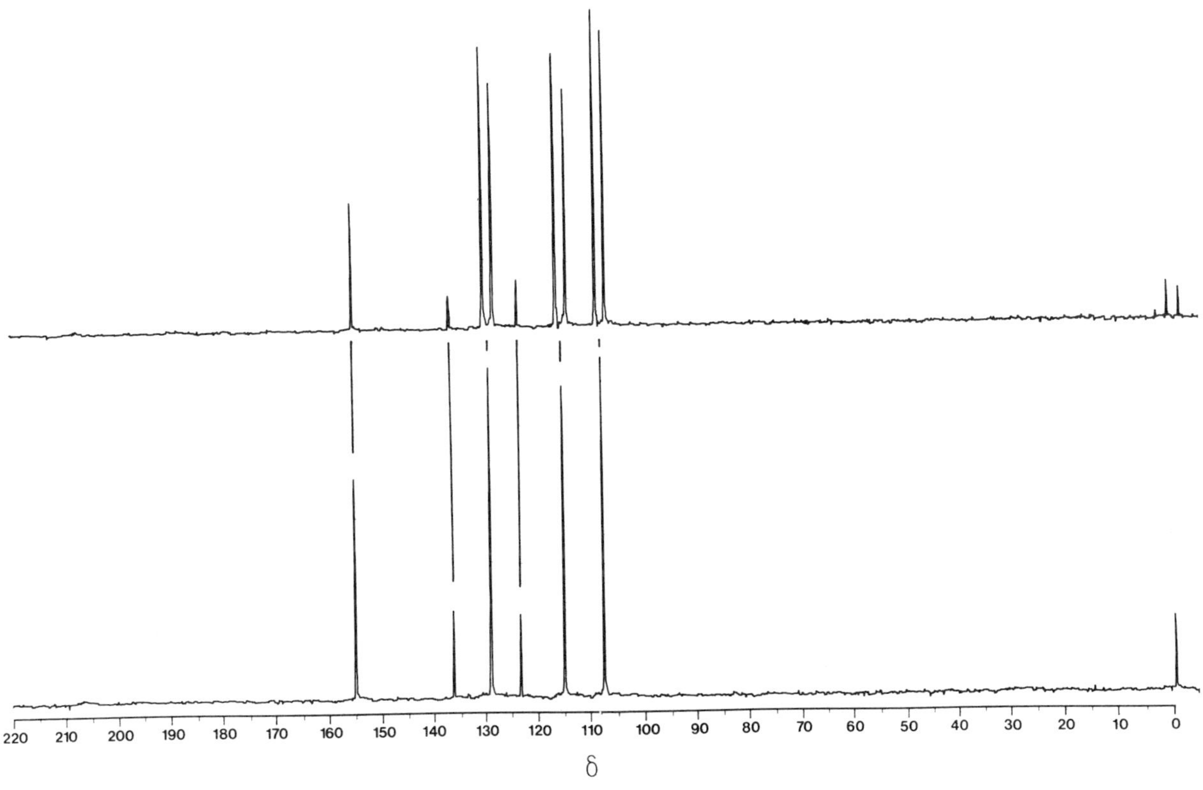
220
210
200
190
180
170
160
150
140
130
120
110
100
90
80
70
60
50
40
30
20
10
0
δ

Problem 2-15

2-15

$C_{10}H_{16}O_3$

MS: m184, 156, 139, 111

IR: 1740, 1710 cm^{-1}

UV: 285 (ε 14)

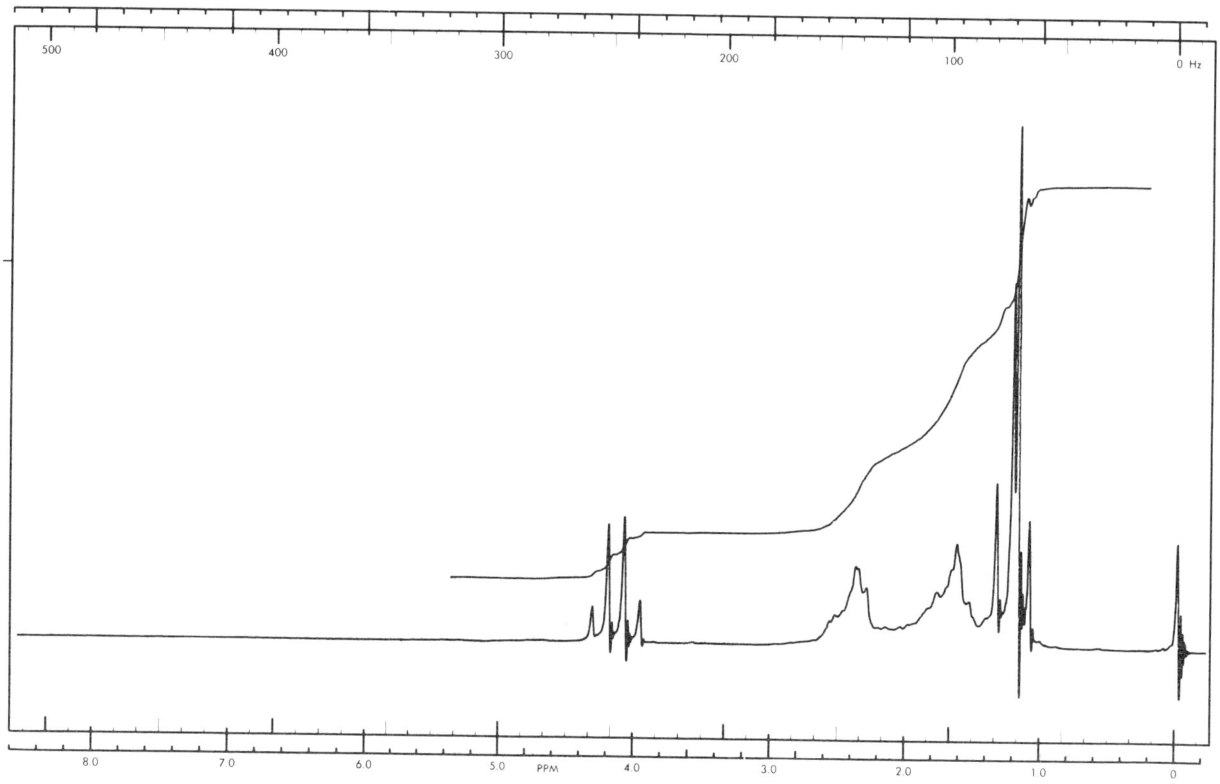

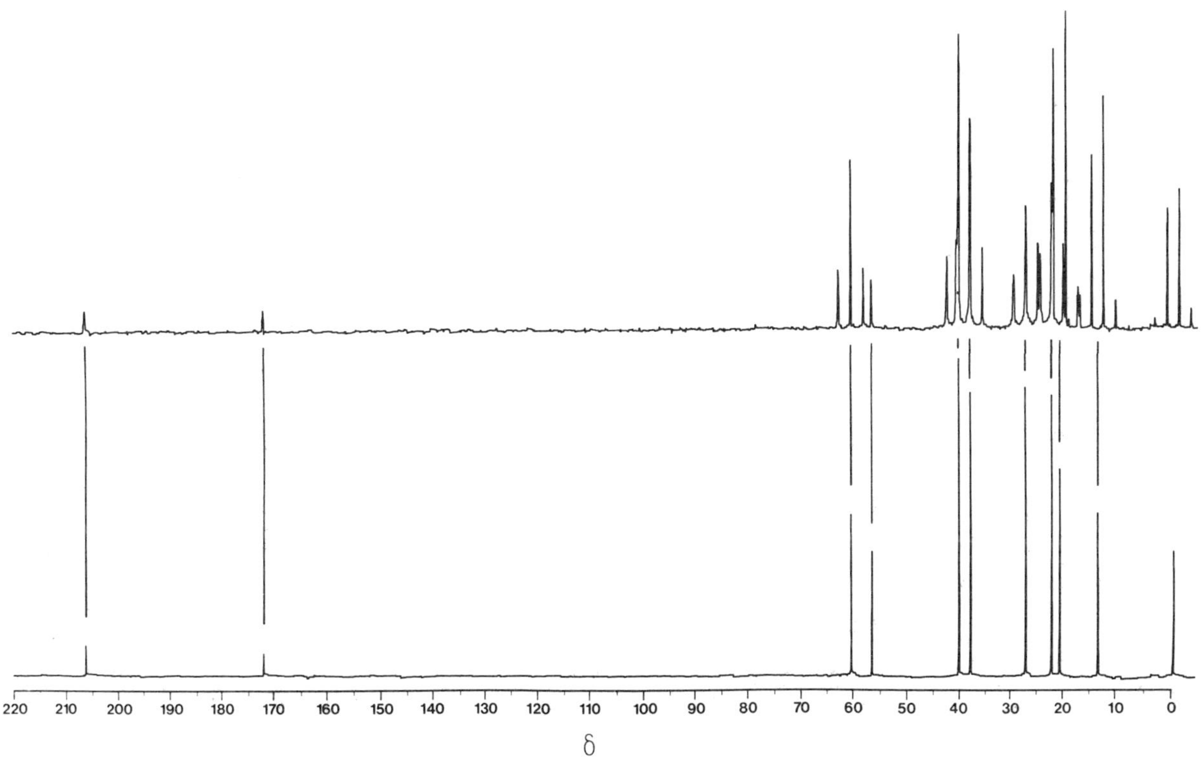
220
210
200
190
180
170
160
150
140
130
120
110
100
90
80
70
60
50
40
30
20
10
0
δ

Problem 2-16

2-16

$C_{10}H_8Br_2O_4$

MS: m350, 321, 319, 293, 291

IR: 1740 cm^{-1}

UV: Strong absorption

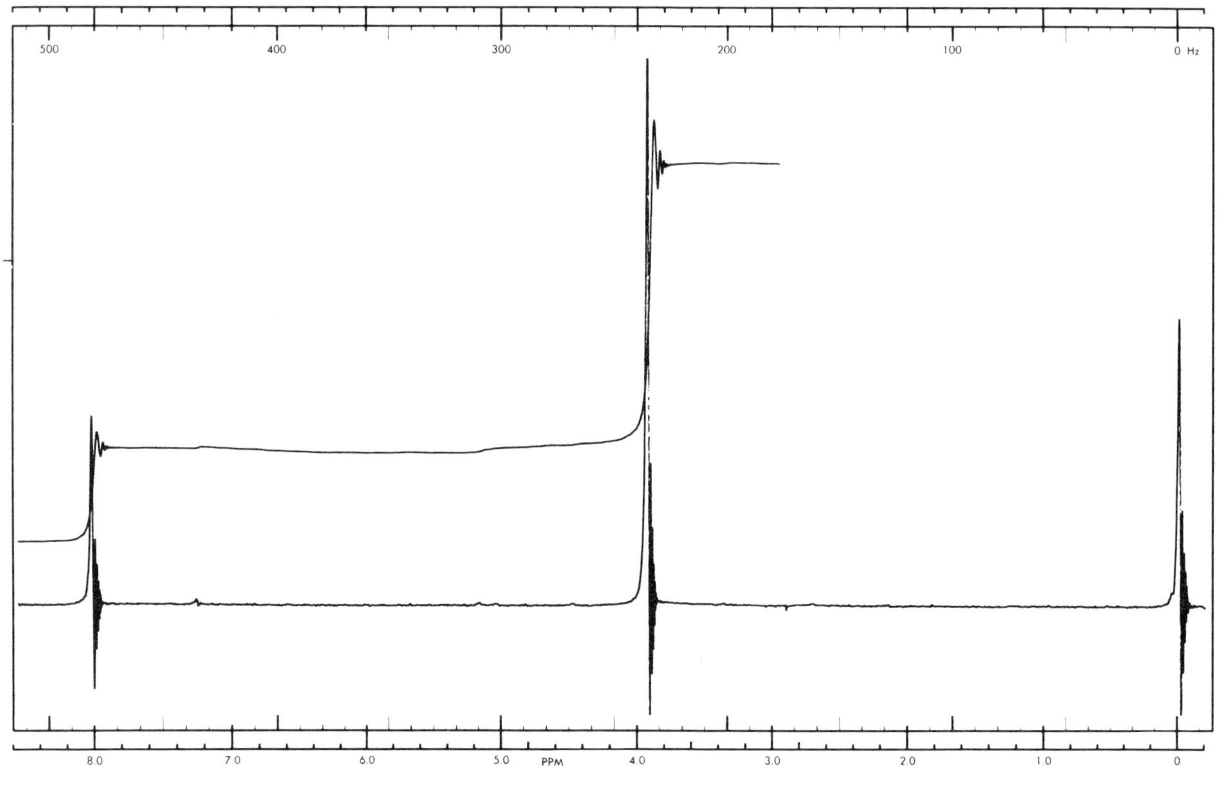

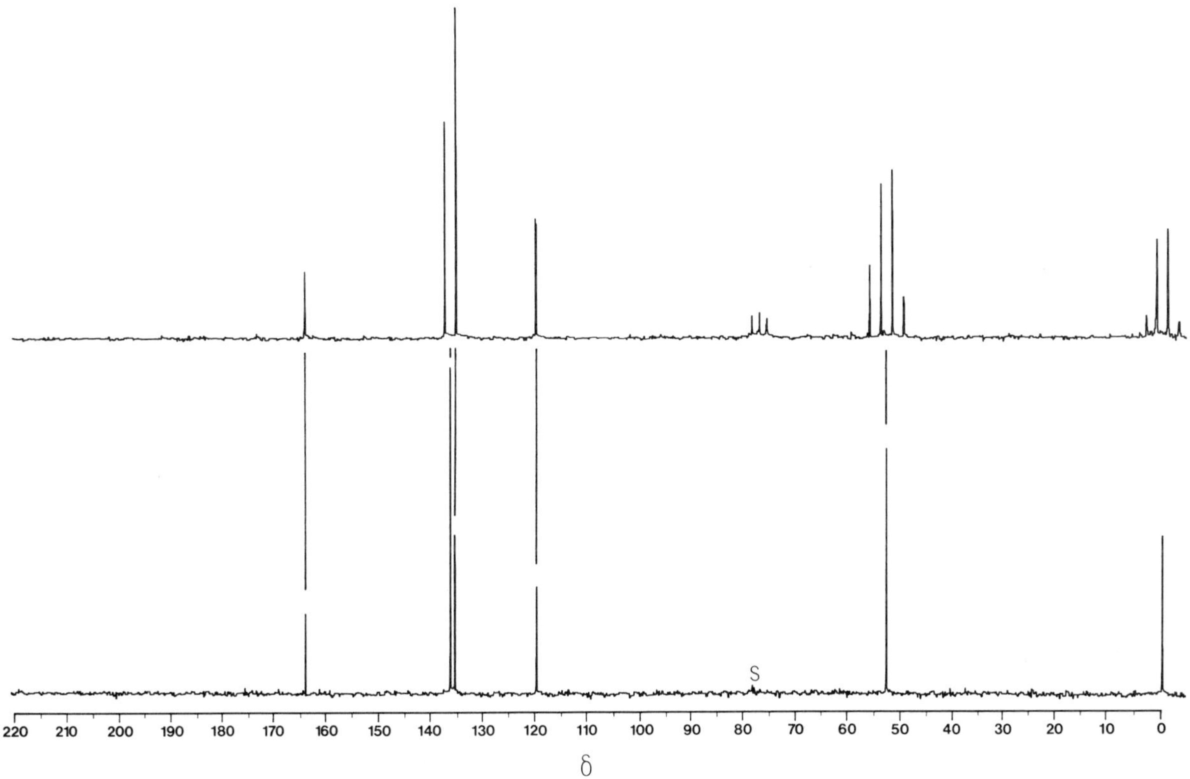
S
220 210 200 190 180 170 160 150 140 130 120 110 100 90 80 70 60 50 40 30 20 10 0
δ

Problem 2-17

2-17

$C_9H_6N_2O_2$

MS: m174(s), 146, 118

IR: 2220 cm^{-1} (broad)

UV: Strong absorption

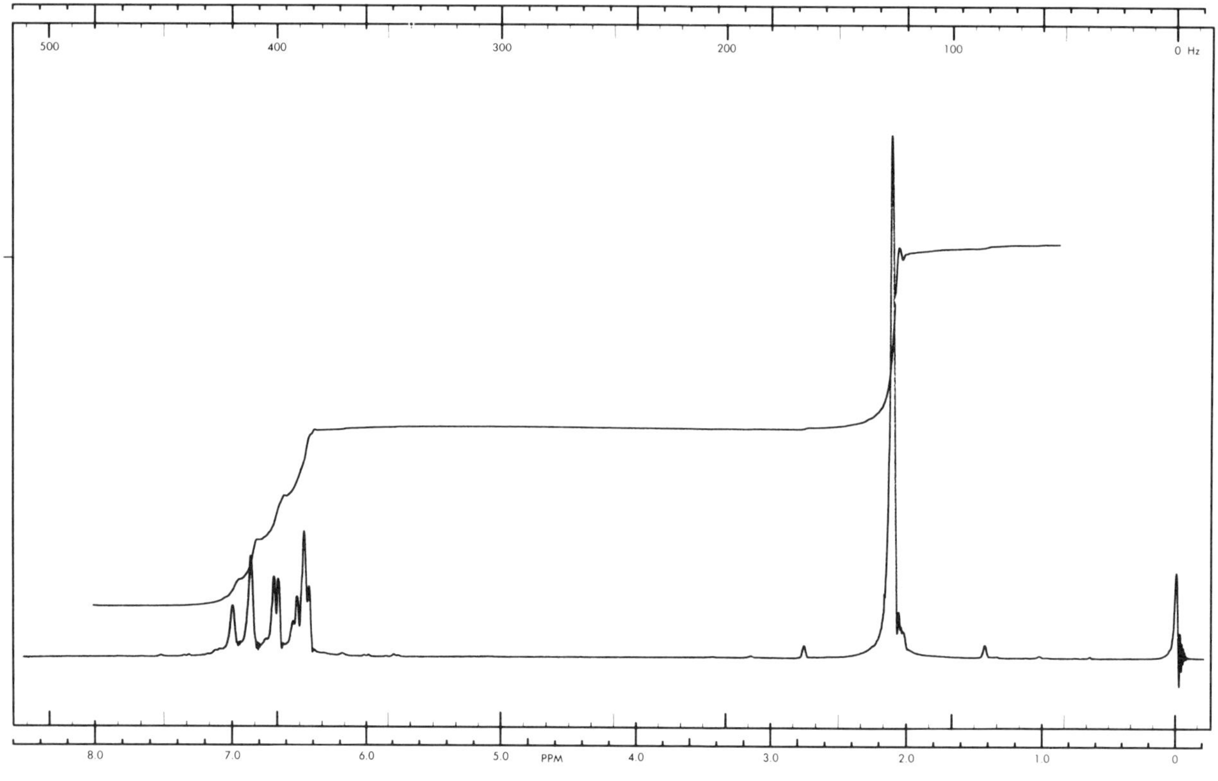

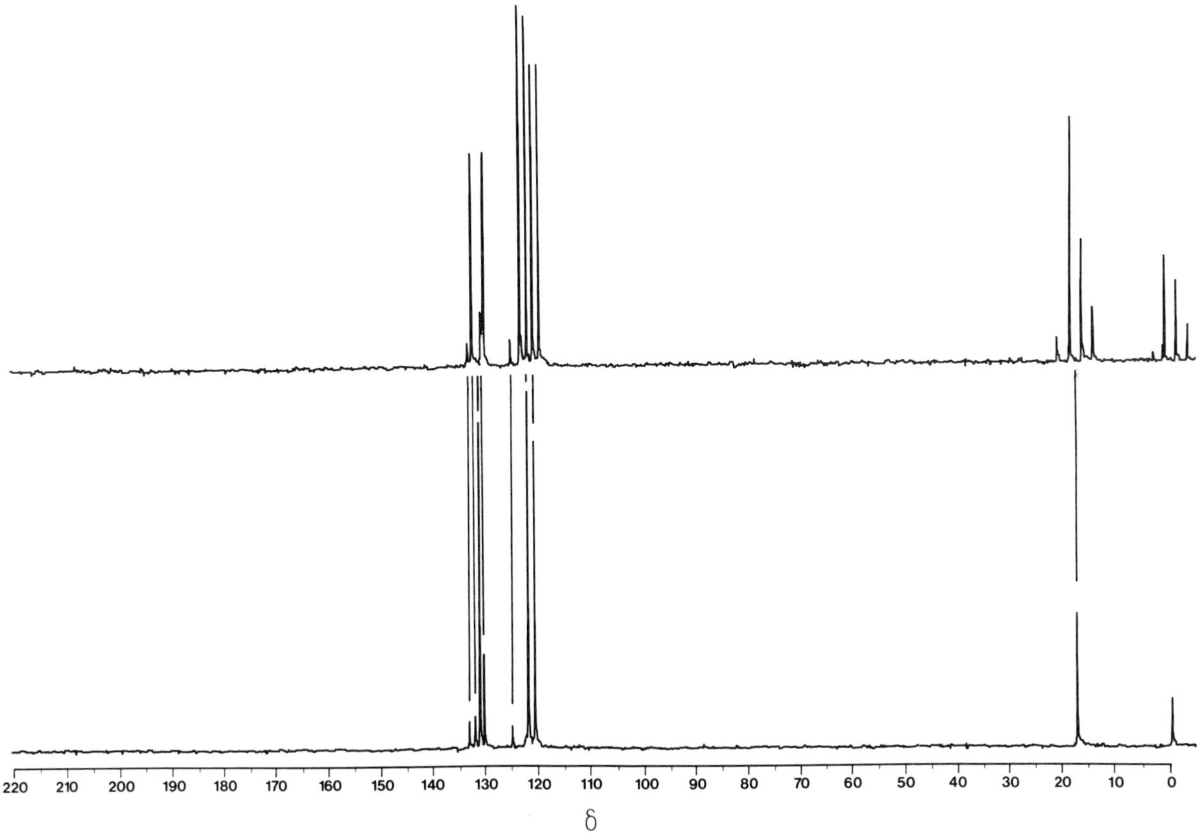
220
210
200
190
180
170
160
150
140
130
120
110
100
90
80
70
60
50
40
30
20
10
0
δ

Problem 2-18

2-18

$C_8H_{12}O_2$

MS: m140, 109, 81, 54

IR: 1735, 1665 cm^{-1}

UV: none

CMR: 20 sec delay

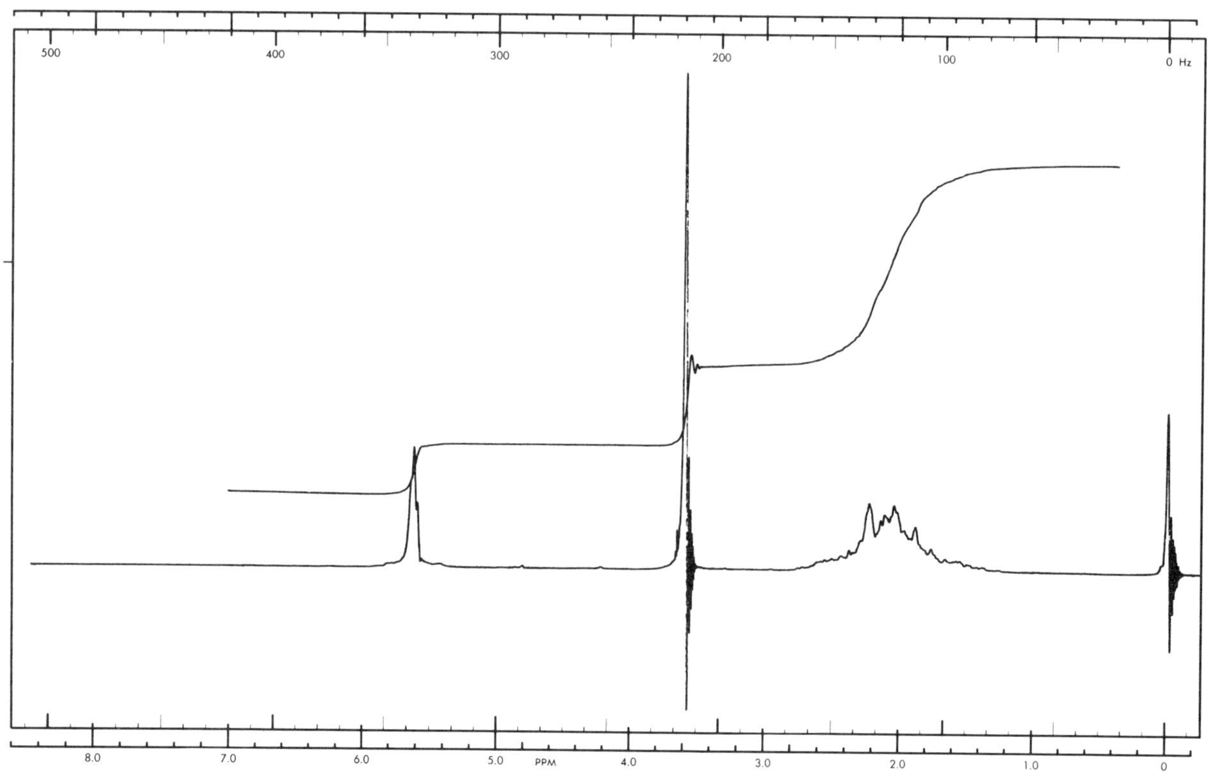

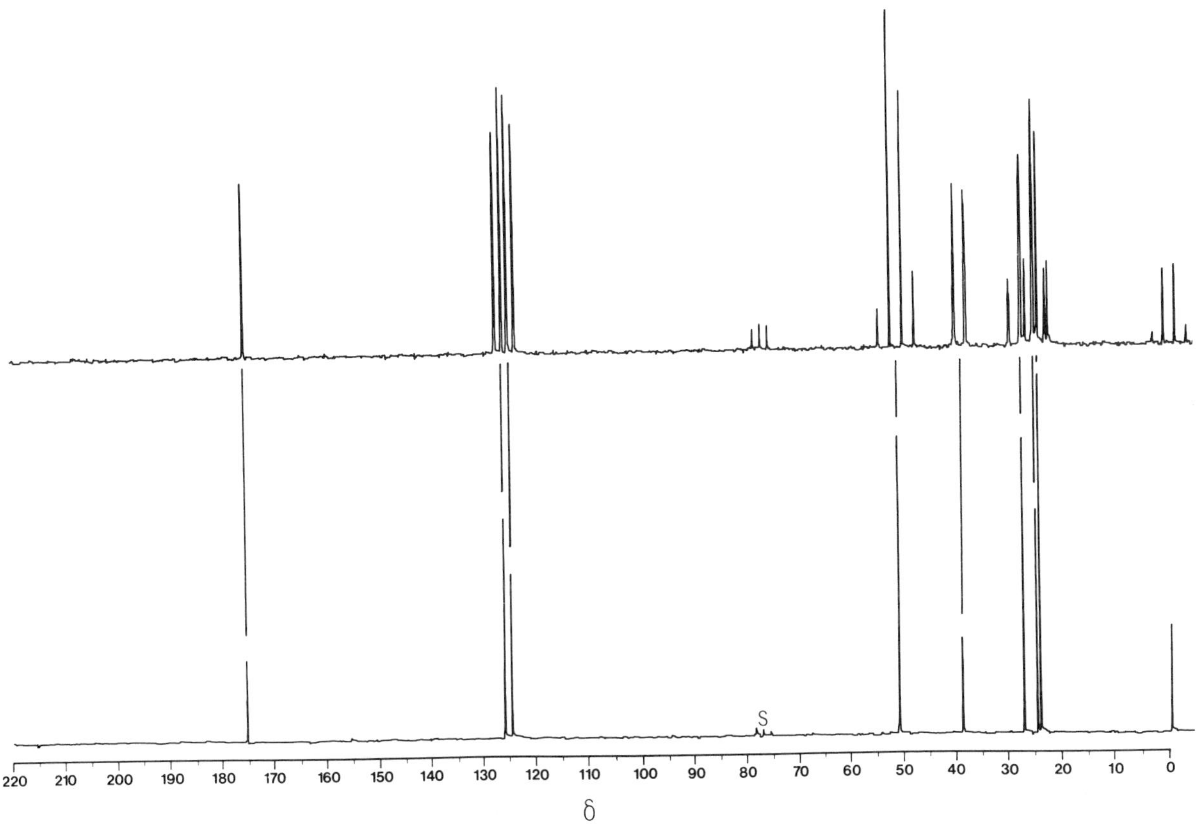
S
220
210
200
190
180
170
160
150
140
130
120
110
100
90
80
70
60
50
40
30
20
10
0
δ

Problem 2-19

2-19

$C_{14}H_{22}O_4$

MS: m254(w), 211

IR: 1755, 1643 cm^{-1}

UV: none

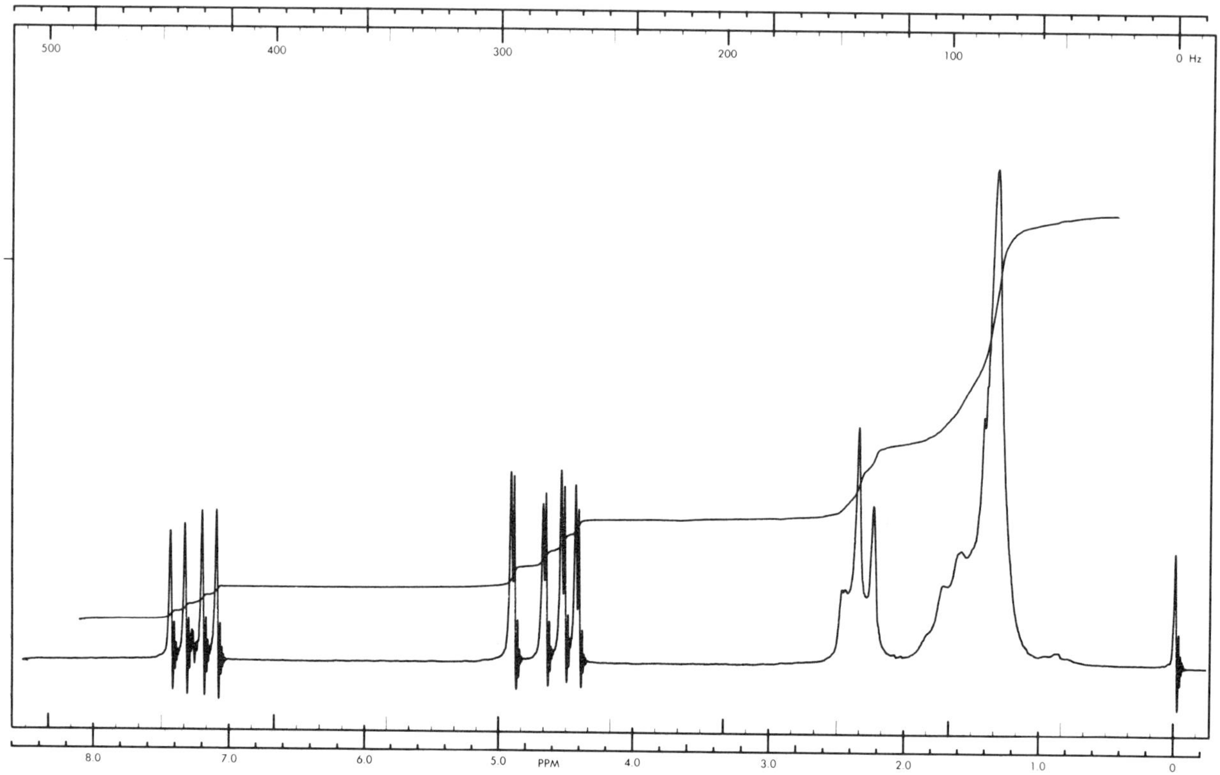

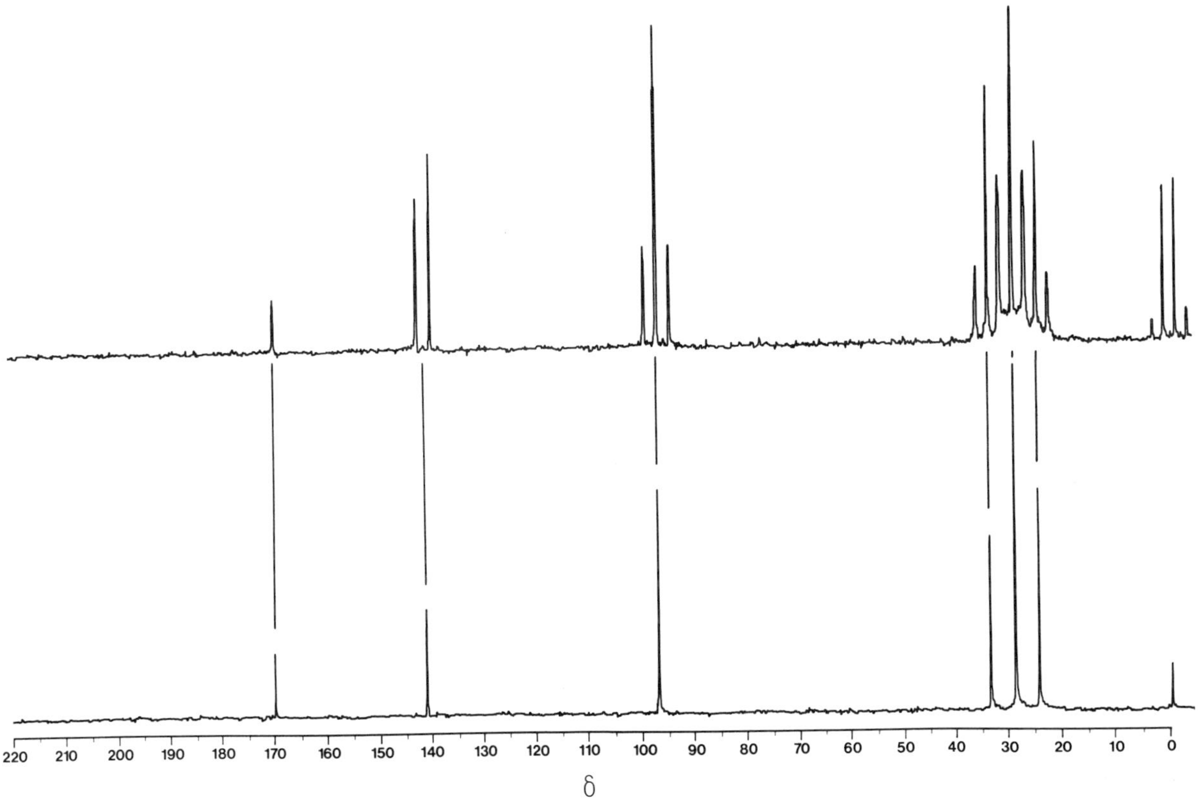
220
210
200
190
180
170
160
150
140
130
120
110
100
90
80
70
60
50
40
30
20
10
0
δ

Problem 2-20

2-20

$C_{10}H_{18}O$

MS: m154, 139, 96

IR: -

UV: none

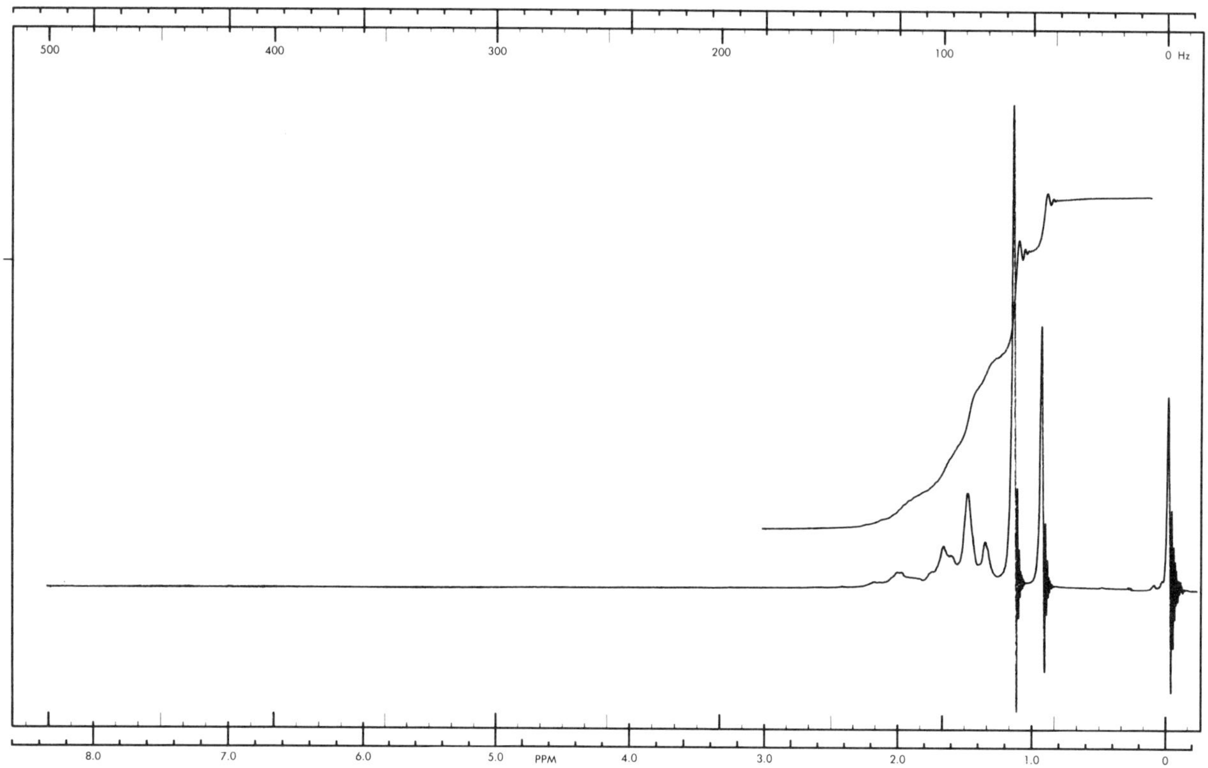

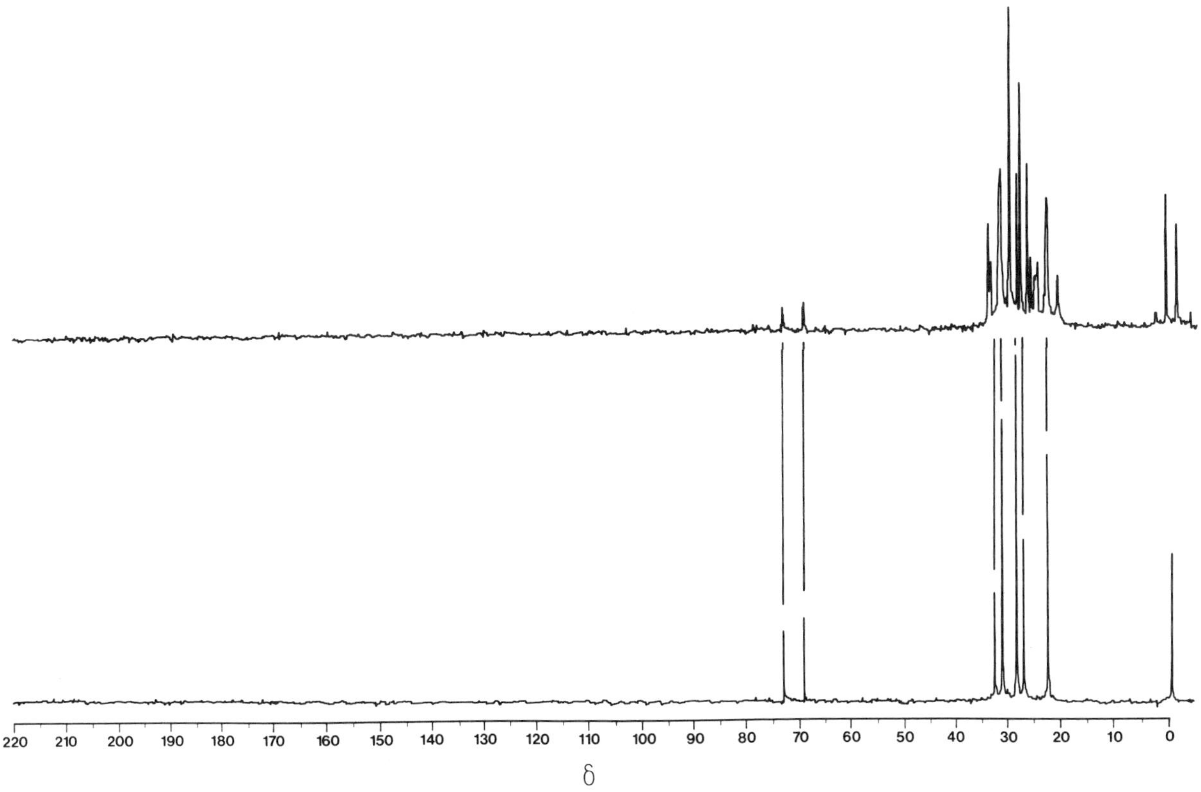
220
210
200
190
180
170
160
150
140
130
120
110
100
90
80
70
60
50
40
30
20
10
0
δ

Problem 2-21

2-21

$C_{10}H_{18}O$

MS: m154(w), 137, 136

IR: 3300, 1665 cm^{-1}

UV: none

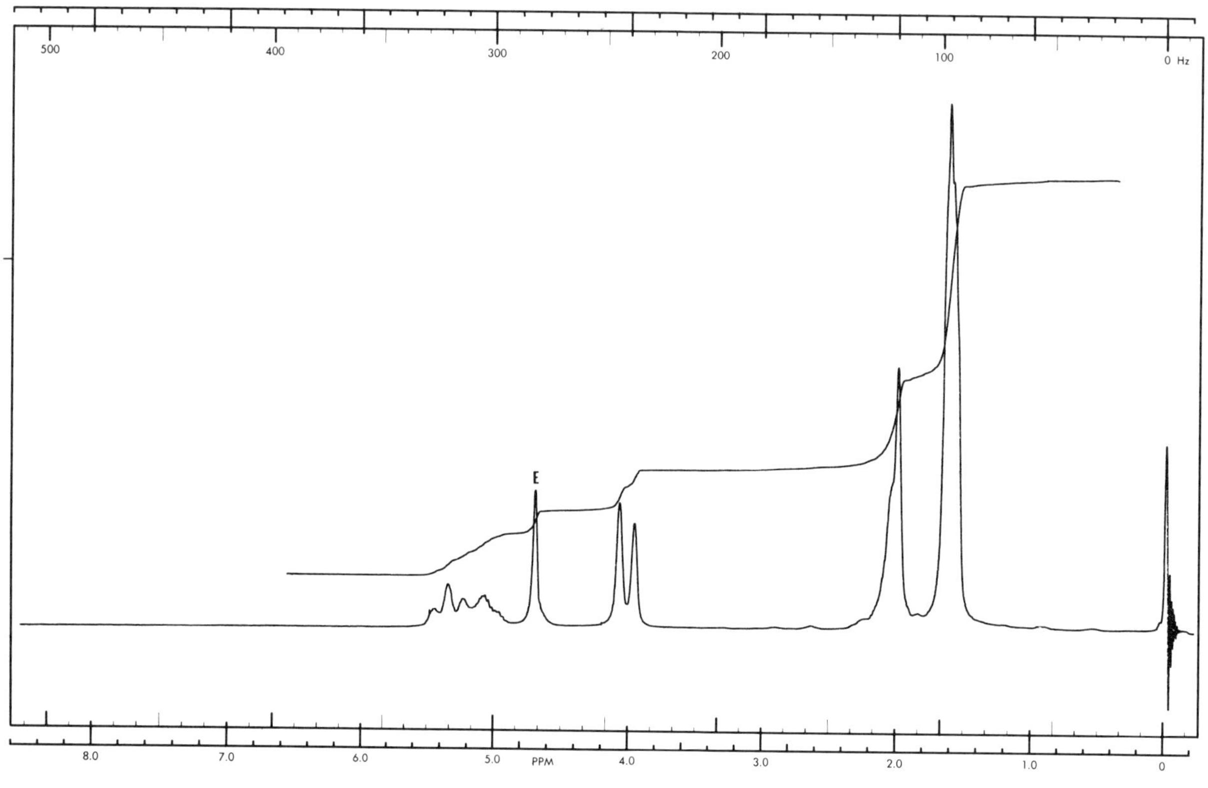

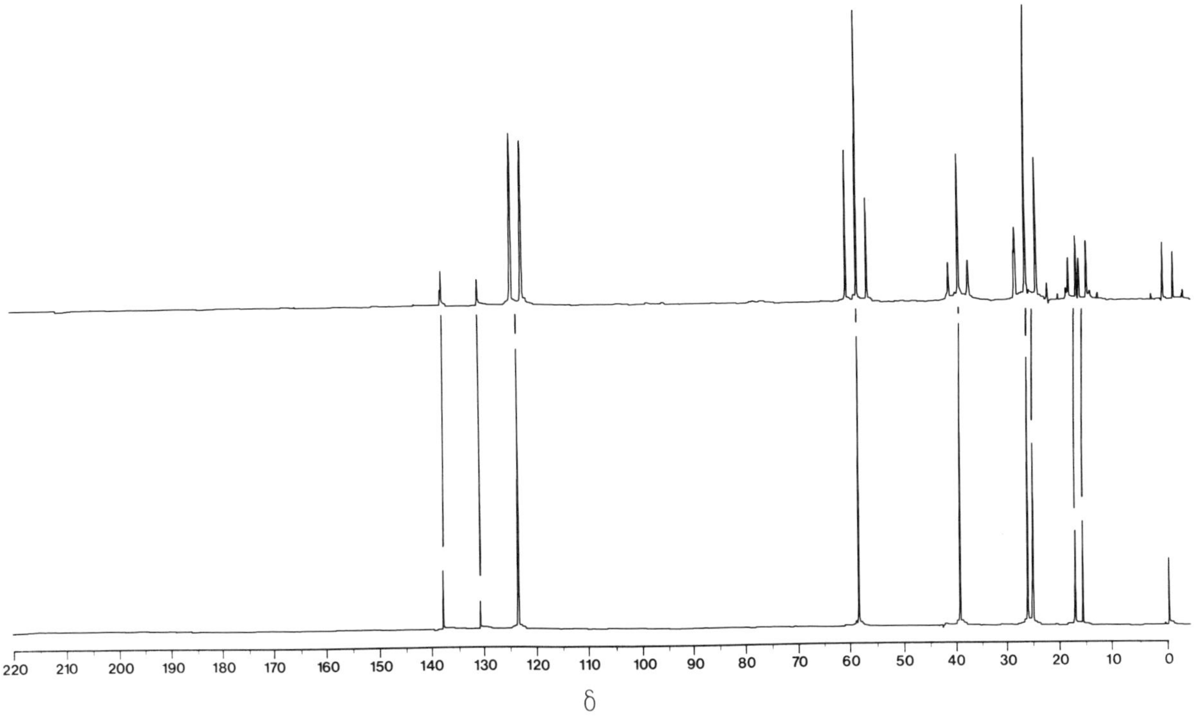
220
210
200
190
180
170
160
150
140
130
120
110
100
90
80
70
60
50
40
30
20
10
0
δ

Problem 2-22

2-22 $C_{10}H_{18}O$

MS: m154(w), 136
IR: 3400, 1650 cm^{-1}
UV: none

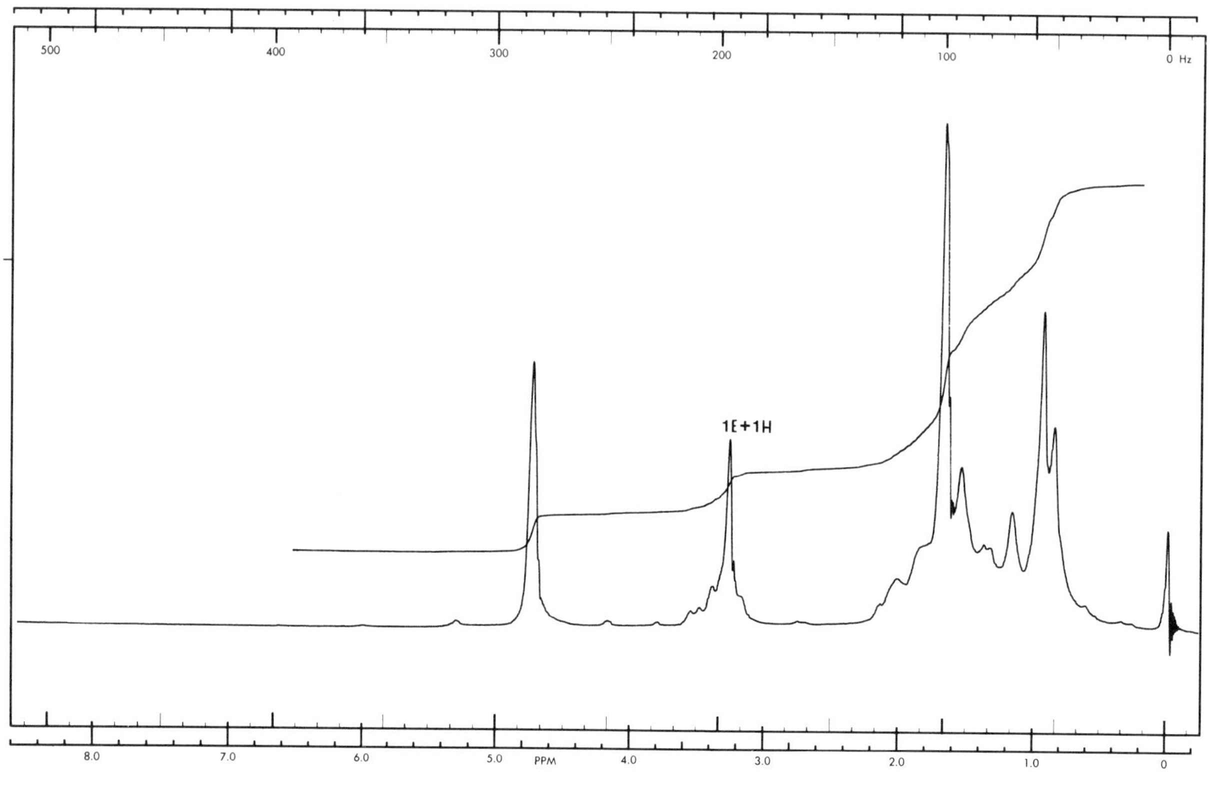

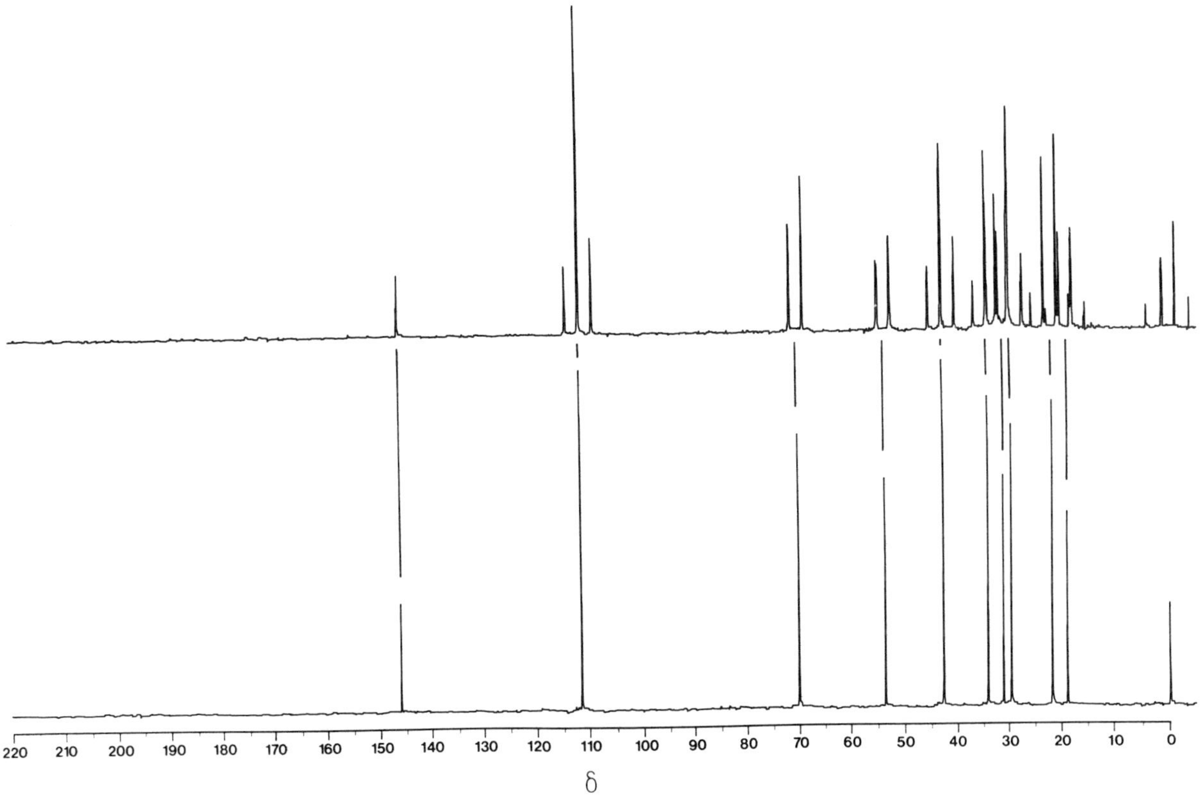
220 210 200 190 180 170 160 150 140 130 120 110 100 90 80 70 60 50 40 30 20 10 0
δ

Problem 2-23

2-23

$C_{10}H_{18}O$

MS: m154, 139, 136

IR: 3400 cm^{-1}

UV: none

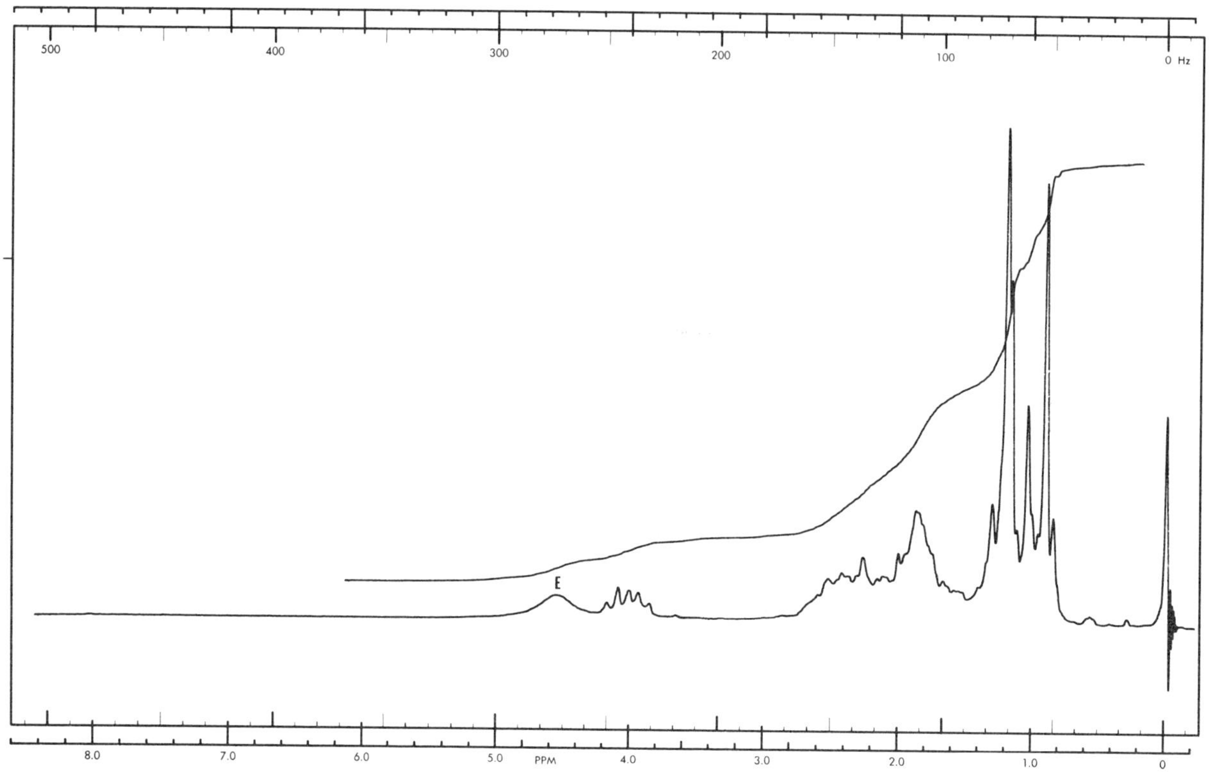

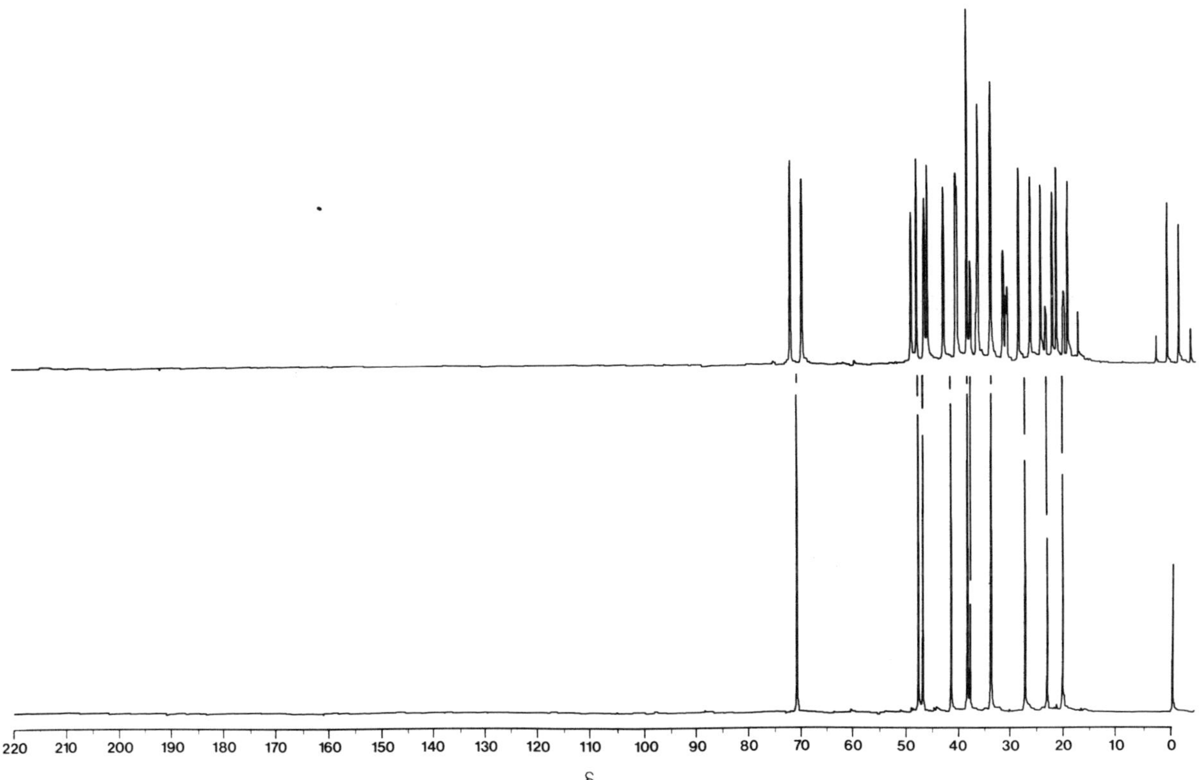
220
210
200
190
180
170
160
150
140
130
120
110
100
90
80
70
60
50
40
30
20
10
0
δ

Problem 2-24

2-24

$C_{11}H_{20}O_{2}$

MS: m184(w), 60, 41

IR: 2900 (broad), 1700, 1640 cm^{-1}

UV: none

PMR: Upper scan offset 300 Hz

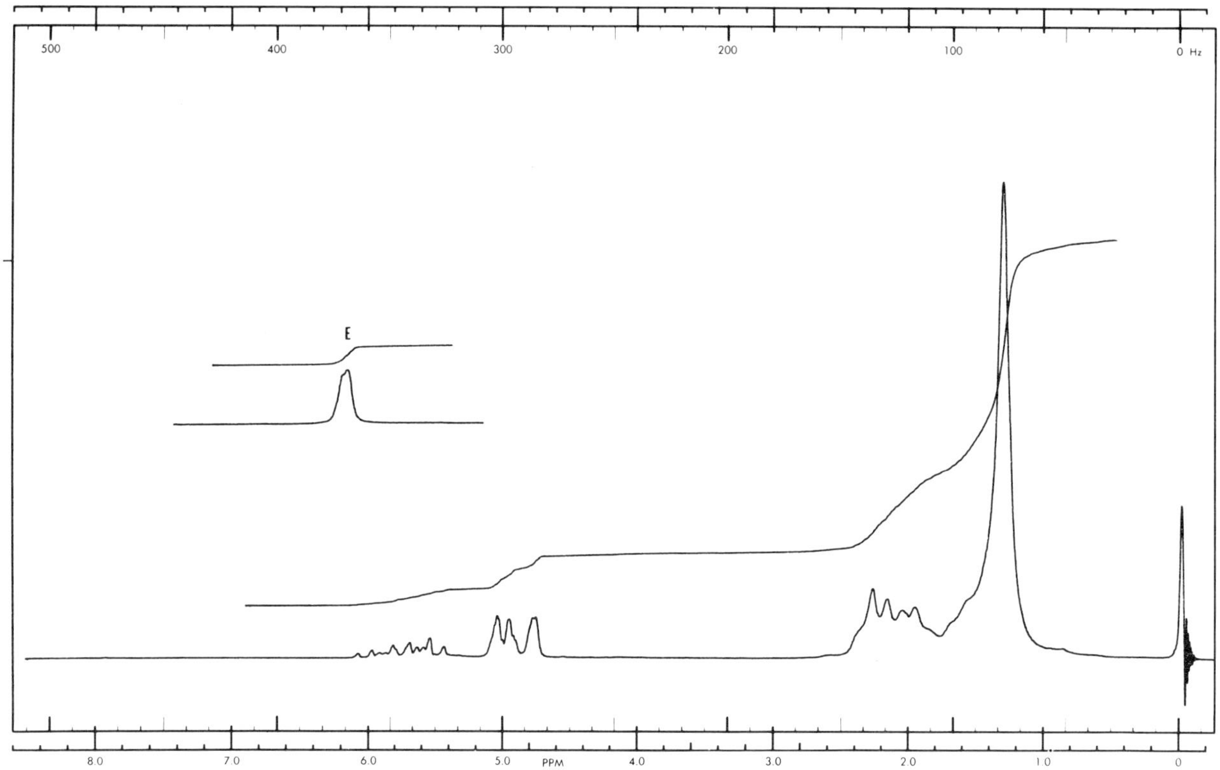

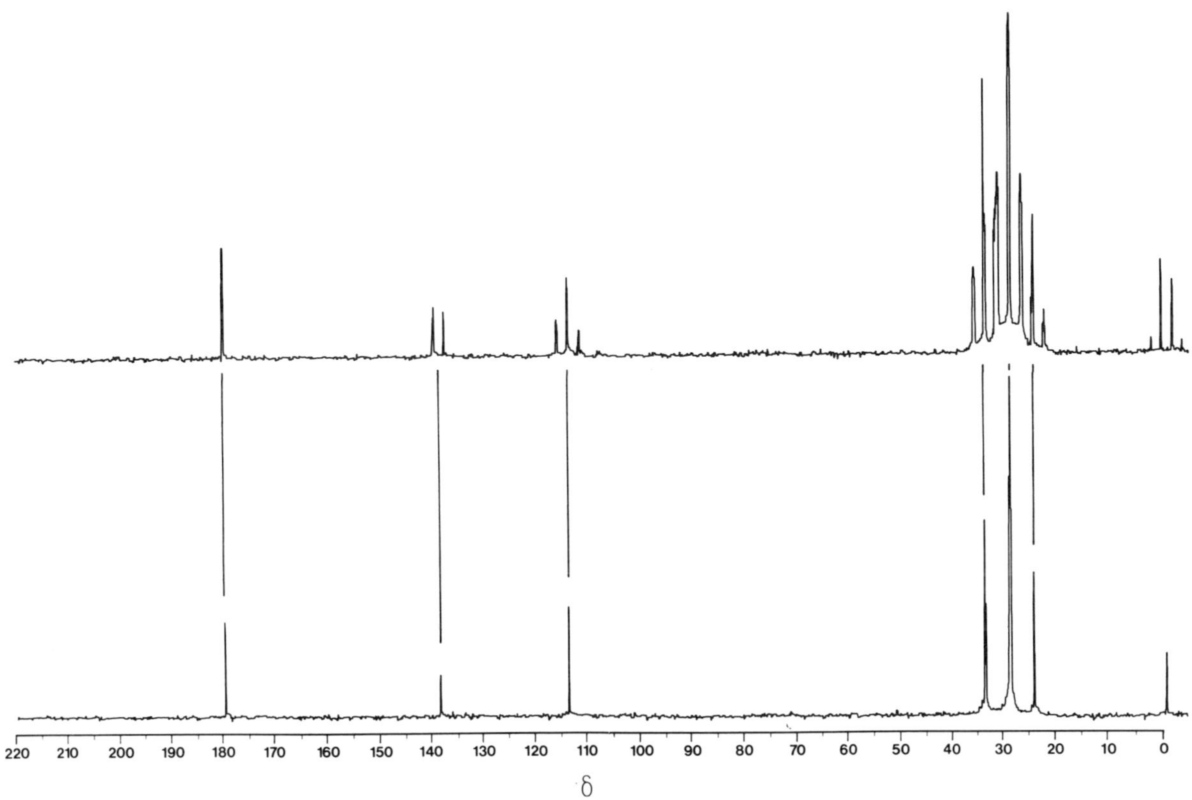
220
210
200
190
180
170
160
150
140
130
120
110
100
90
80
70
60
50
40
30
20
10
0
δ

Problem 2-25

2-25

$C_5H_9N_3$

MS: m111, 112, 110, 95, 81

IR: 3300, 2700 (broad), 1600 cm^{-1}

UV: Strong absorption

PMR: D_2O + DSS; upper scan offset 100 Hz

CMR: D_2O + dioxane

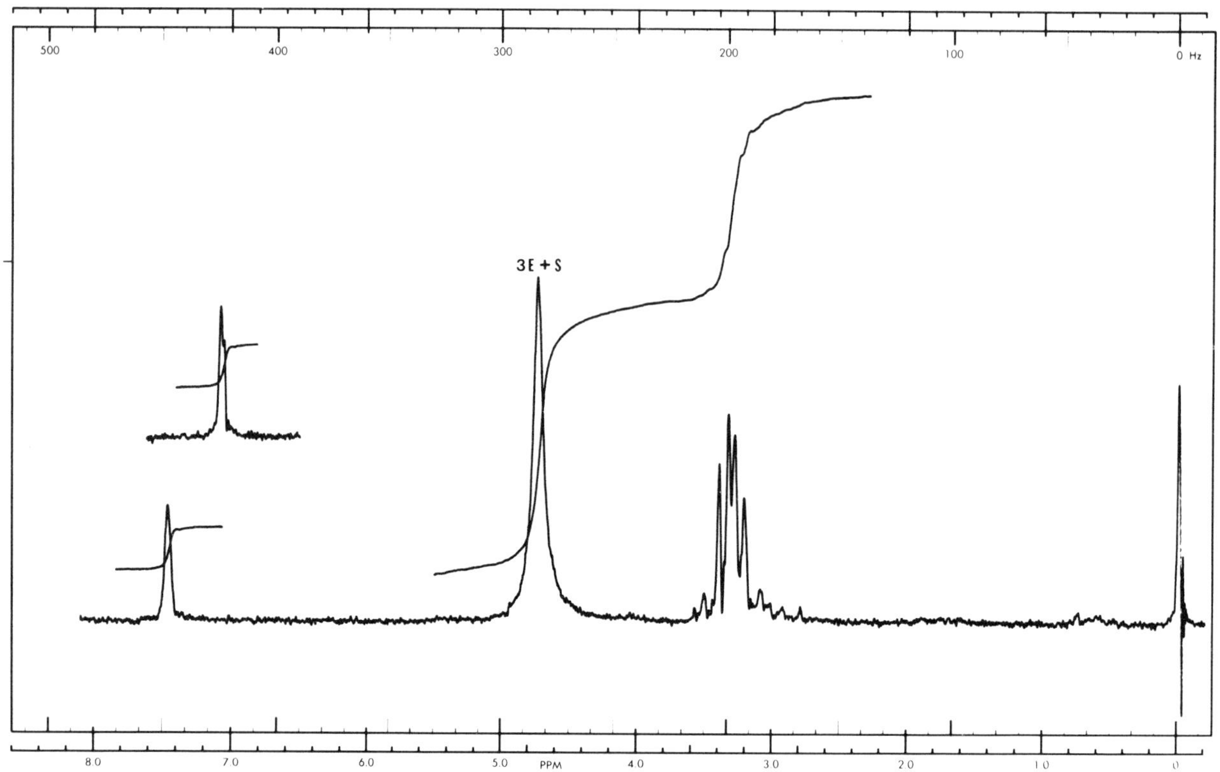

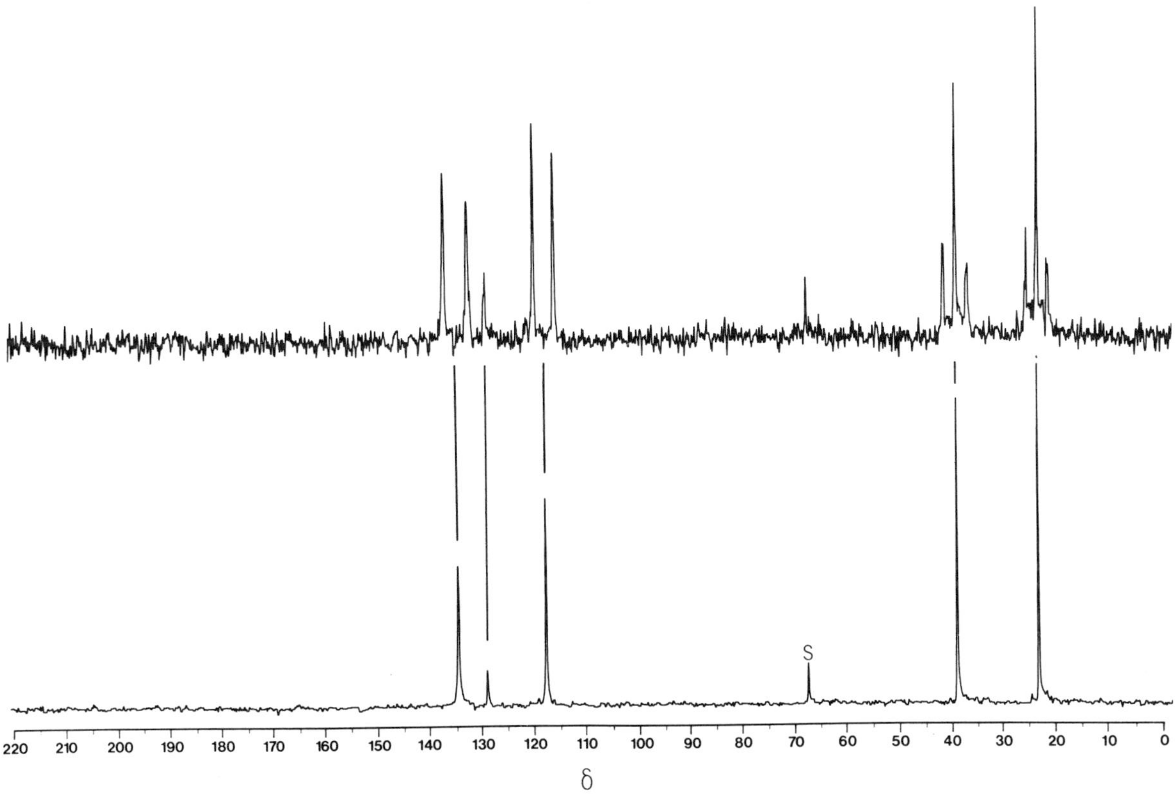
S
220
210
200
190
180
170
160
150
140
130
120
110
100
90
80
70
60
50
40
30
20
10
0
δ

Problem 2-26

2-26

$C_{12}H_{11}N_2O_3Na$

MS: Free acid m232, 231, 204, 146, 118, 77

IR: 3200, 1700, 1600 cm^{-1}

UV: 239 (ε 4620)

PMR: D_2O + DSS

CMR: D_2O + dioxane

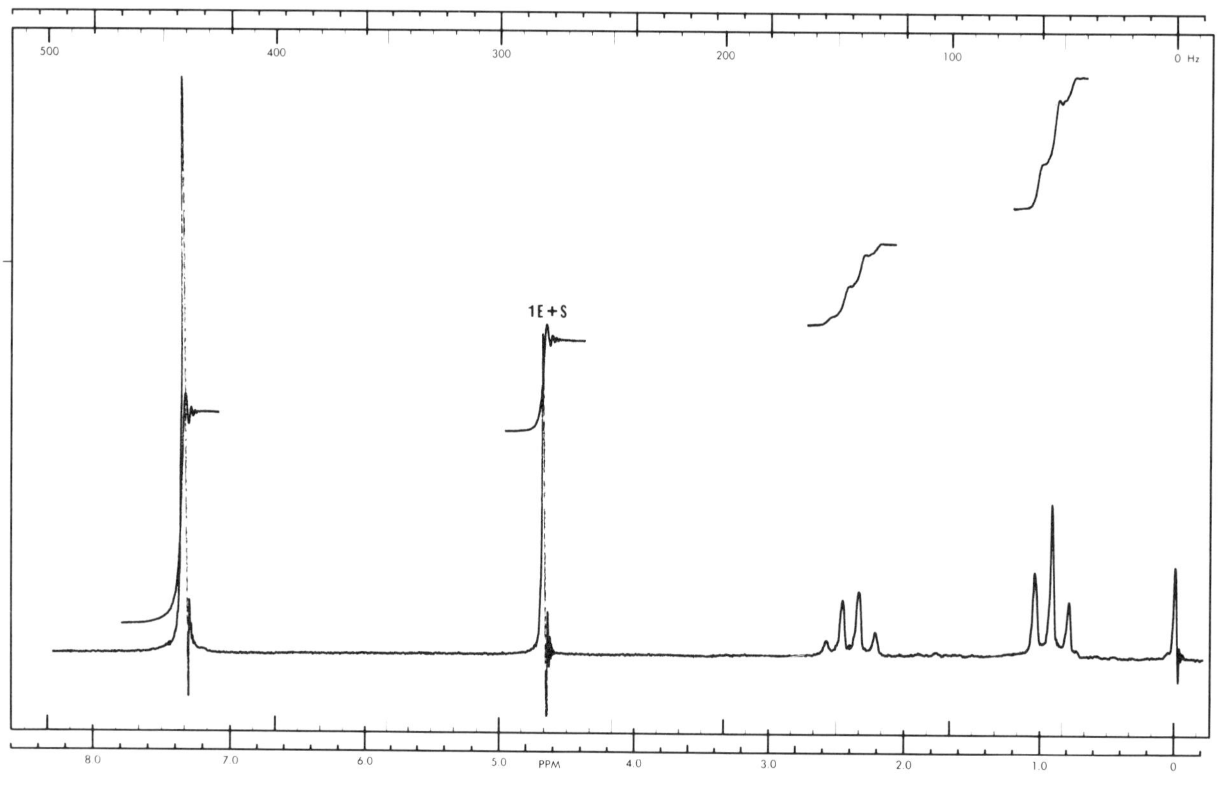

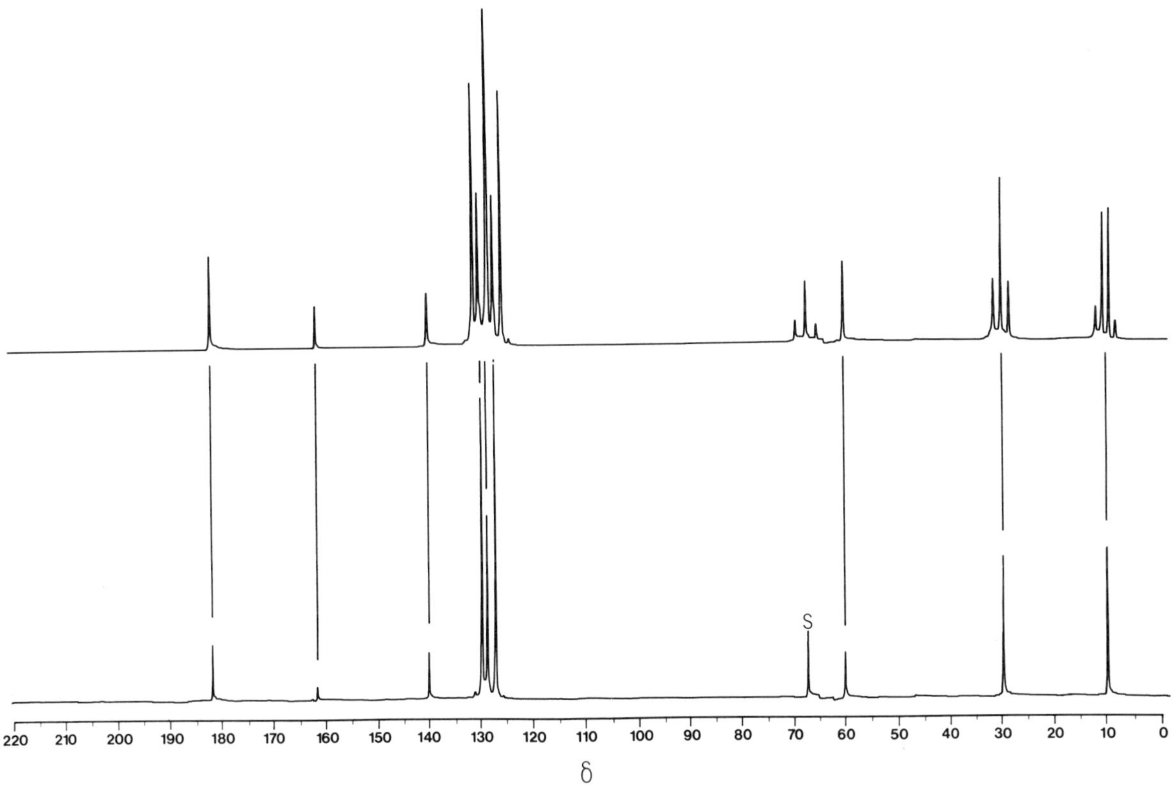
S
220 210 200 190 180 170 160 150 140 130 120 110 100 90 80 70 60 50 40 30 20 10 0
δ

Problem 2-27

2-27

$C_9H_{14}NO_3^+Cl^-$

MS: Free base m183(w), 165, 139, 44

IR: 4000-2400 (broad) cm^{-1}

UV: 282 (ε 4630)

PMR: D_2O + DSS

CMR: D_2O + dioxane

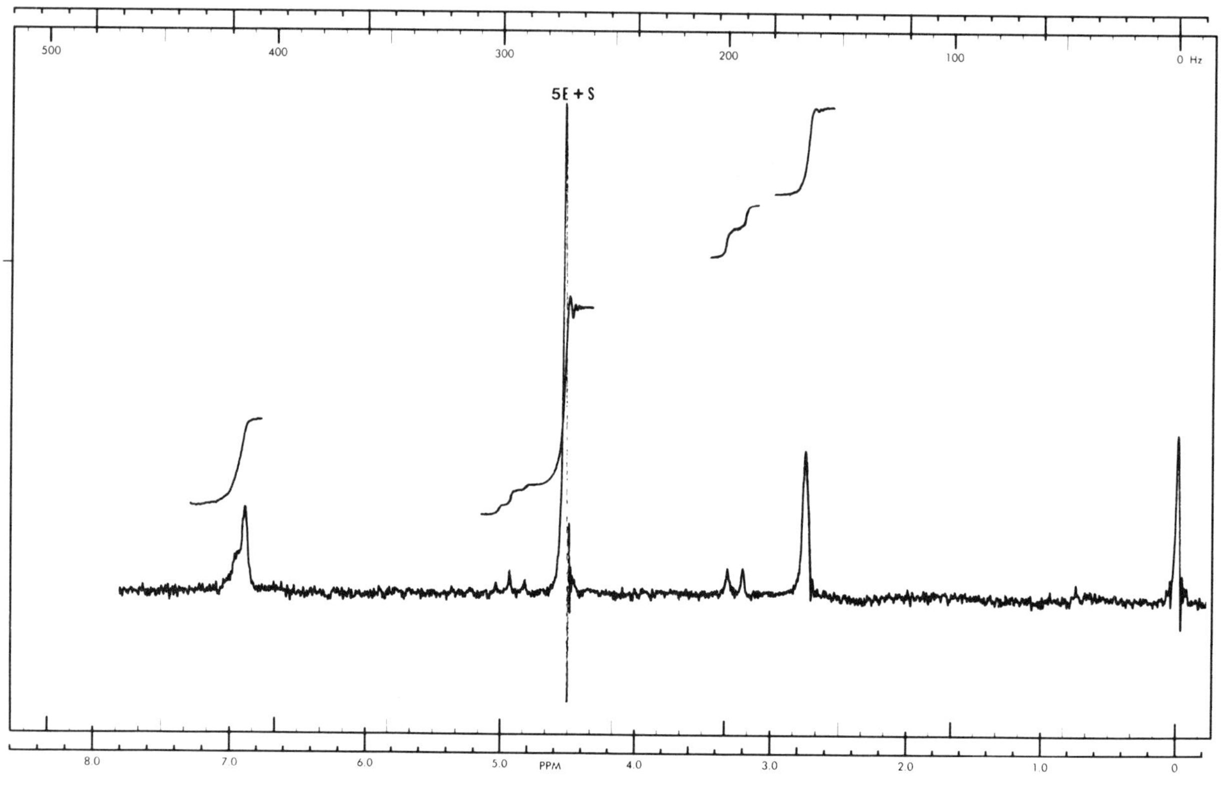

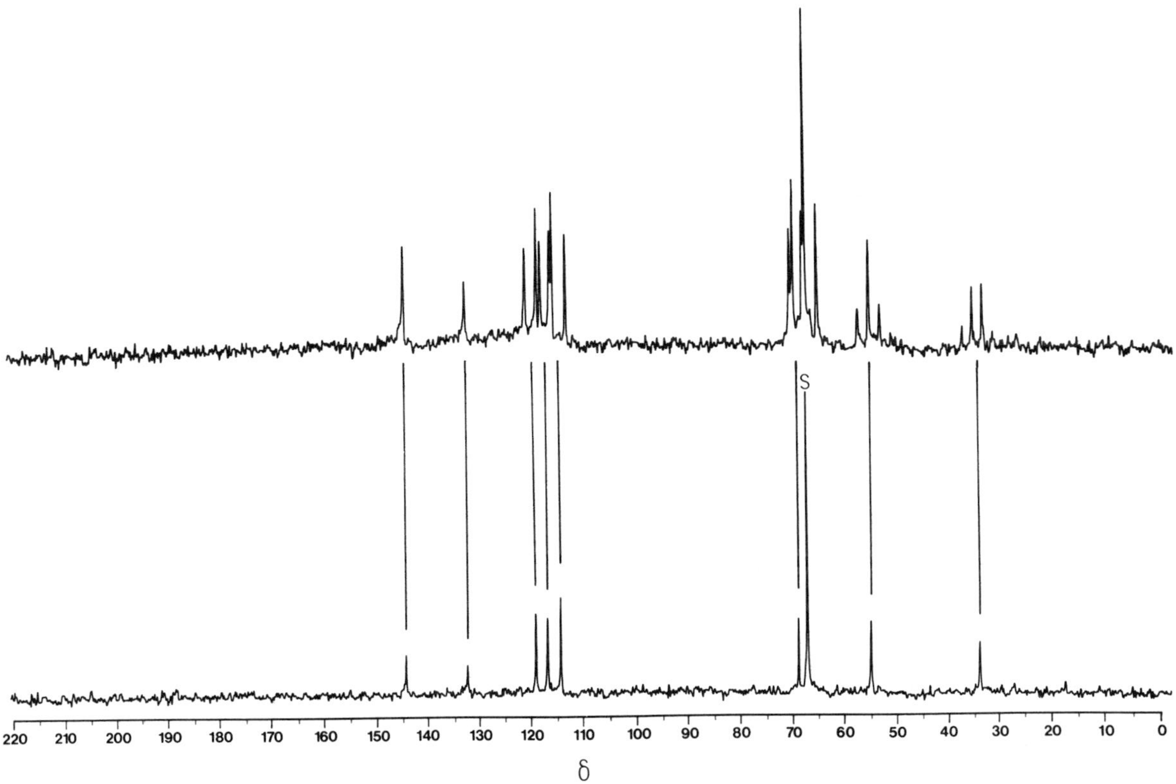
S
220
210
200
190
180
170
160
150
140
130
120
110
100
90
80
70
60
50
40
30
20
10
0
δ

Problem 2-28

2-28

$C_6H_8O_6$

MS: m176(w), 158, 145, 115

IR: 3100 (broad), 1760, 1670 cm^{-1}

UV: 247 (ε 7120)

PMR: D_2O + DSS

CMR: D_2O + dioxane

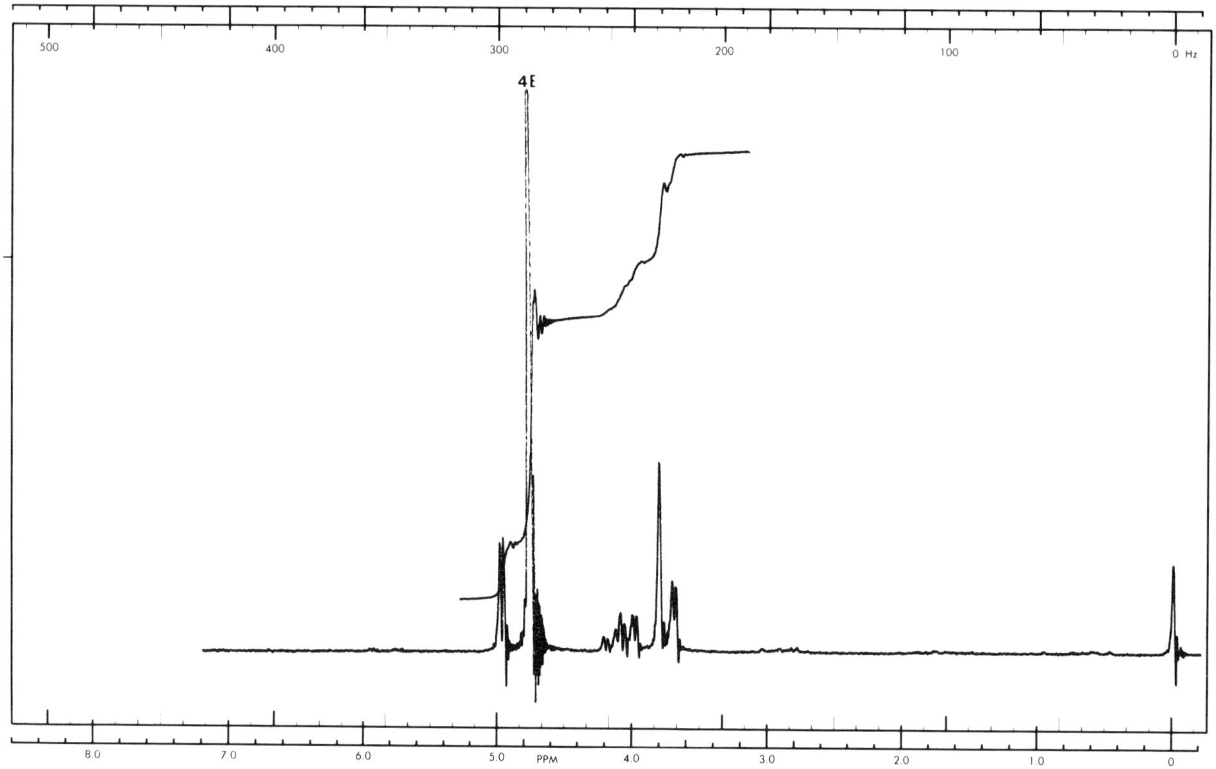

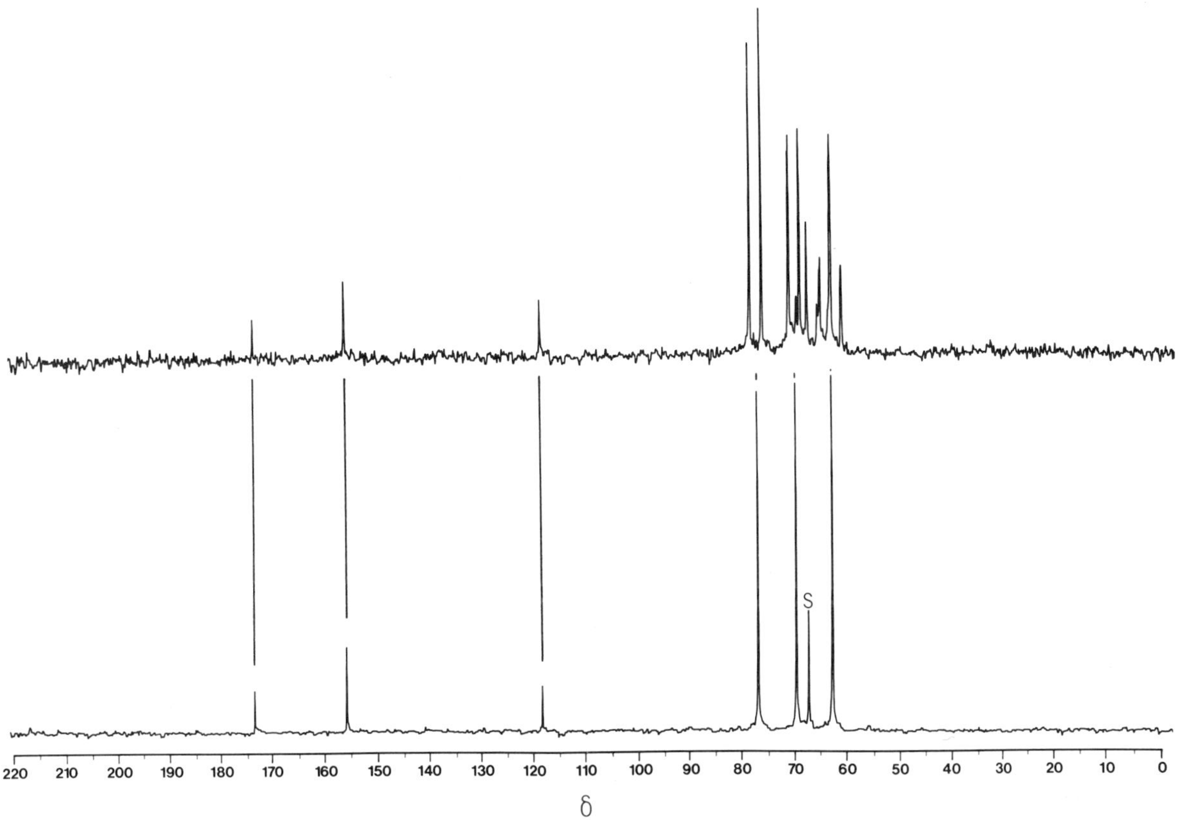
S
220
210
200
190
180
170
160
150
140
130
120
110
100
90
80
70
60
50
40
30
20
10
0
δ

Problem 2-29

2-29

$C_4H_3FN_2O_2$

MS: m130(s), 129, 111

IR: 3100 (broad), 1730, 1690 cm^{-1}

UV: 265 (ε 7100)

PMR: D_2O + DSS

CMR: D_2O + dioxane

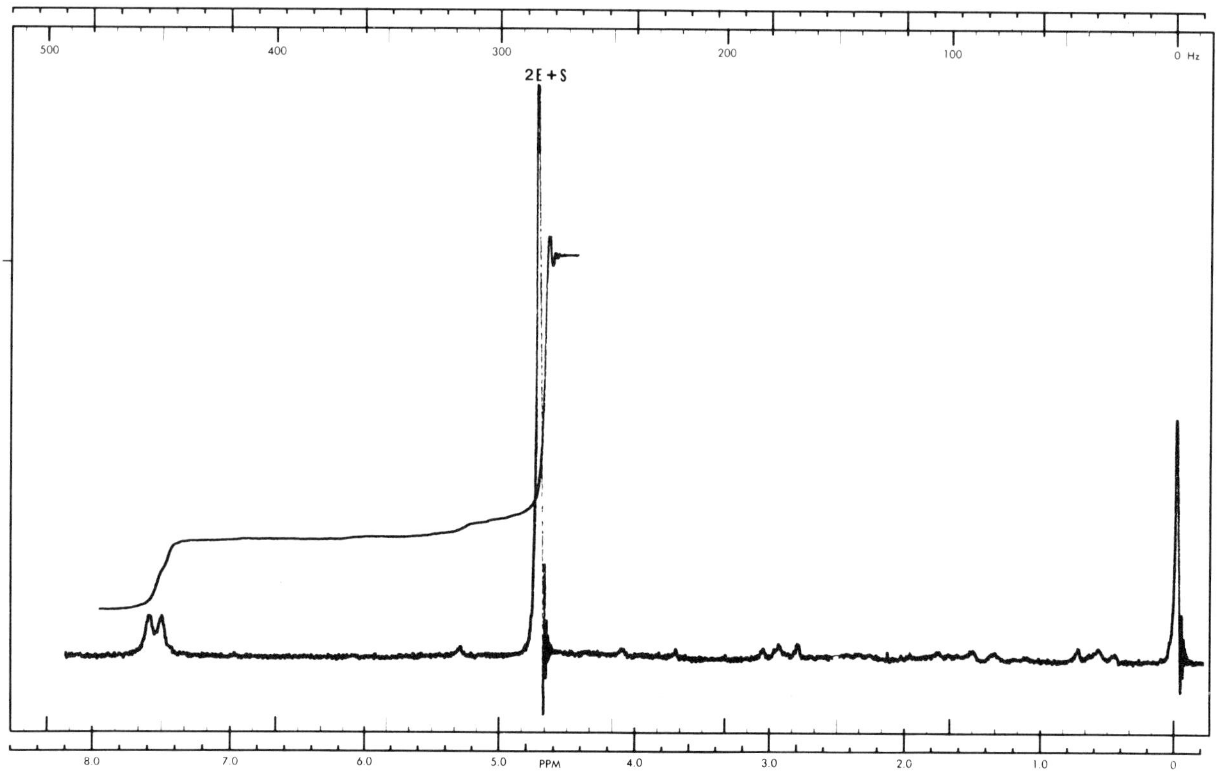

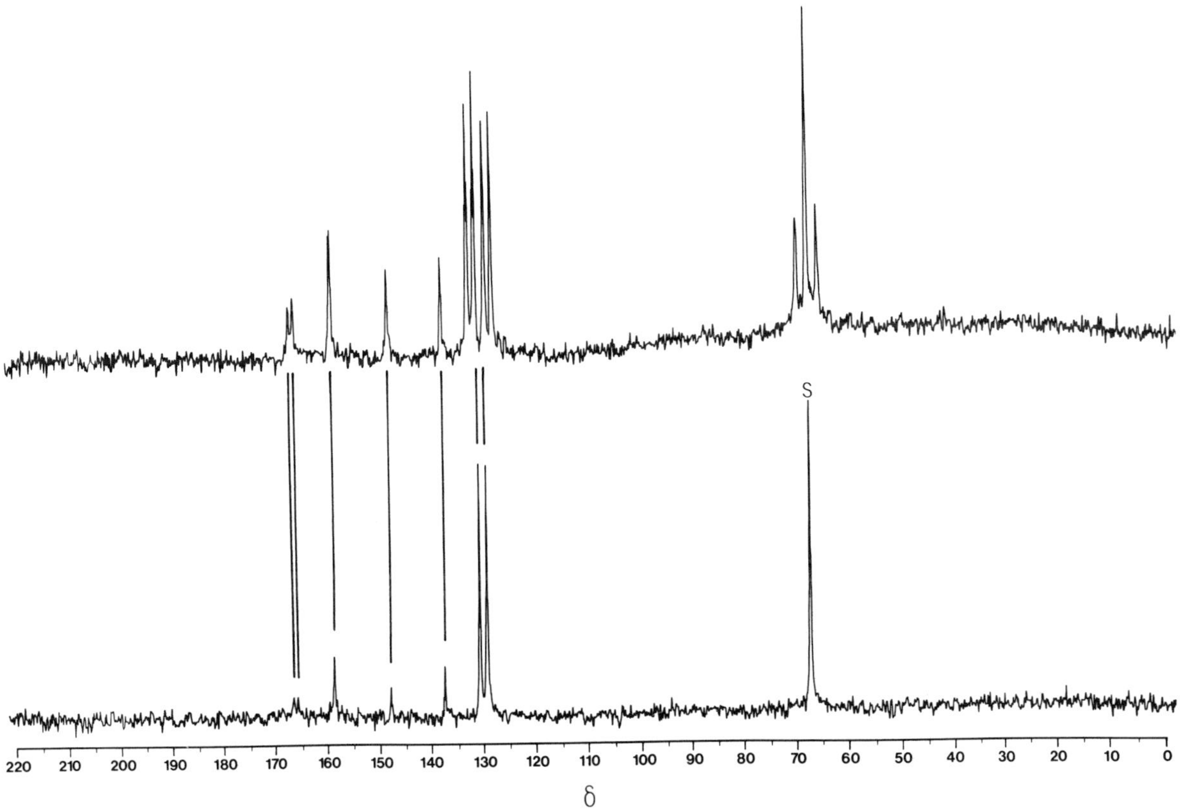
S
220 210 200 190 180 170 160 150 140 130 120 110 100 90 80 70 60 50 40 30 20 10 0
δ

Problem 2-30

2-30	MS:	Free base m255, 197, 167, 77, 58
$C_{17}H_{22}NO^{+}Cl^{-}$	IR:	2600 (broad) cm^{-1}
	UV:	258 (ε 452)
	PMR:	D_2O + DSS
	CMR:	D_2O + dioxane

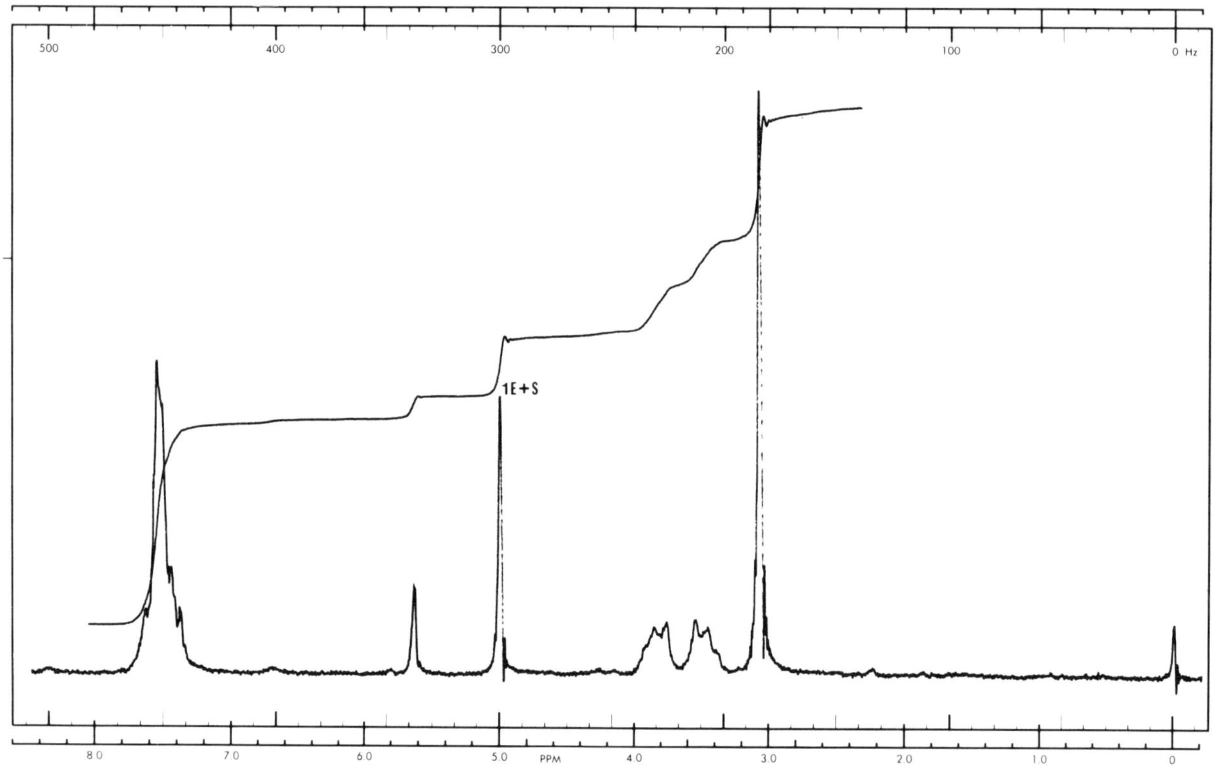

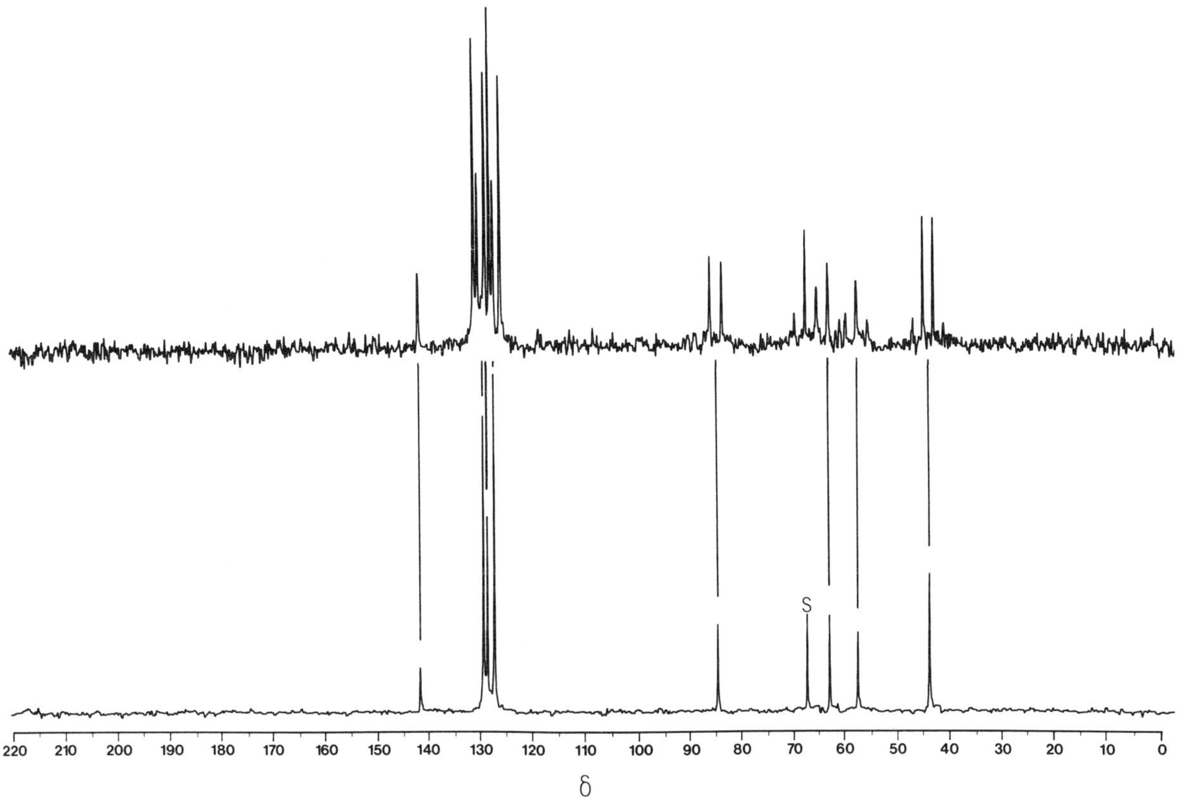
S
220
210
200
190
180
170
160
150
140
130
120
110
100
90
80
70
60
50
40
30
20
10
0
δ

Problem 2-31

2-31	MS:	m174(w), 129, 74, 73
$C_6H_{14}N_4O_2$	IR:	3300-2300 (broad), 1680, 1630 cm^{-1}
	UV:	End absorption
	PMR:	D_2O + DSS
	CMR:	D_2O + dioxane

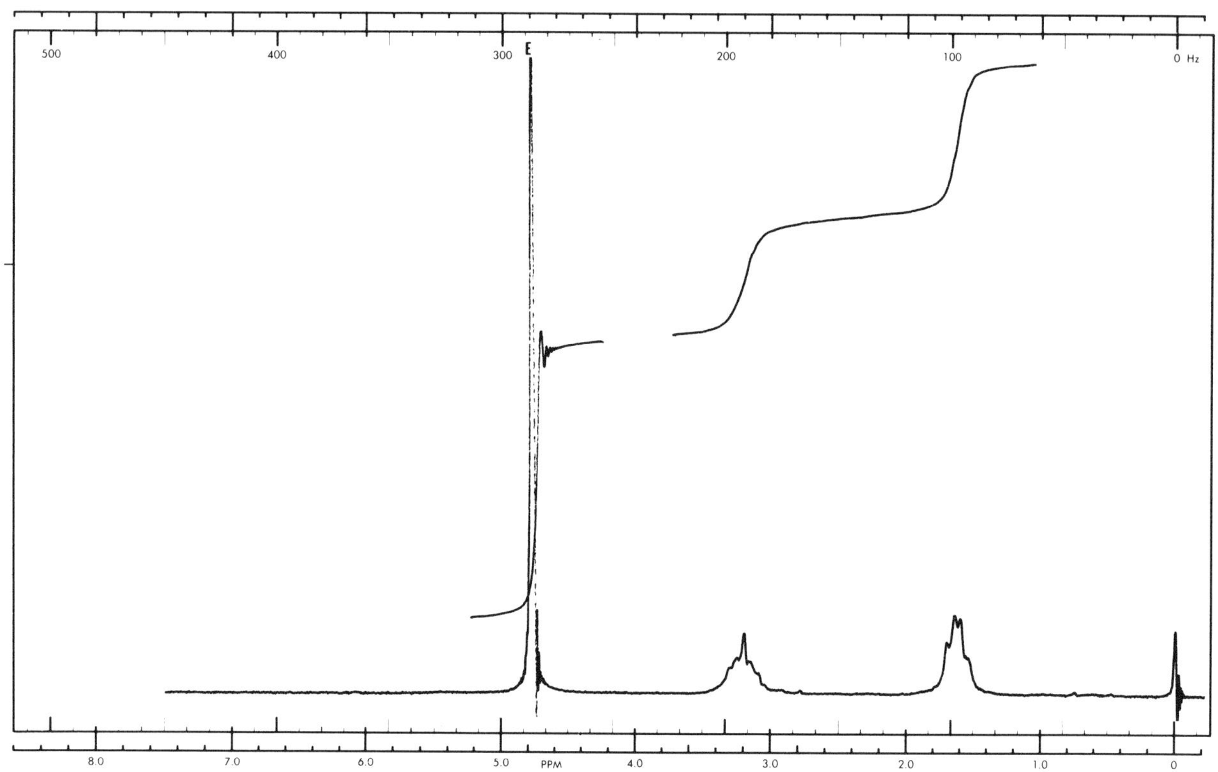

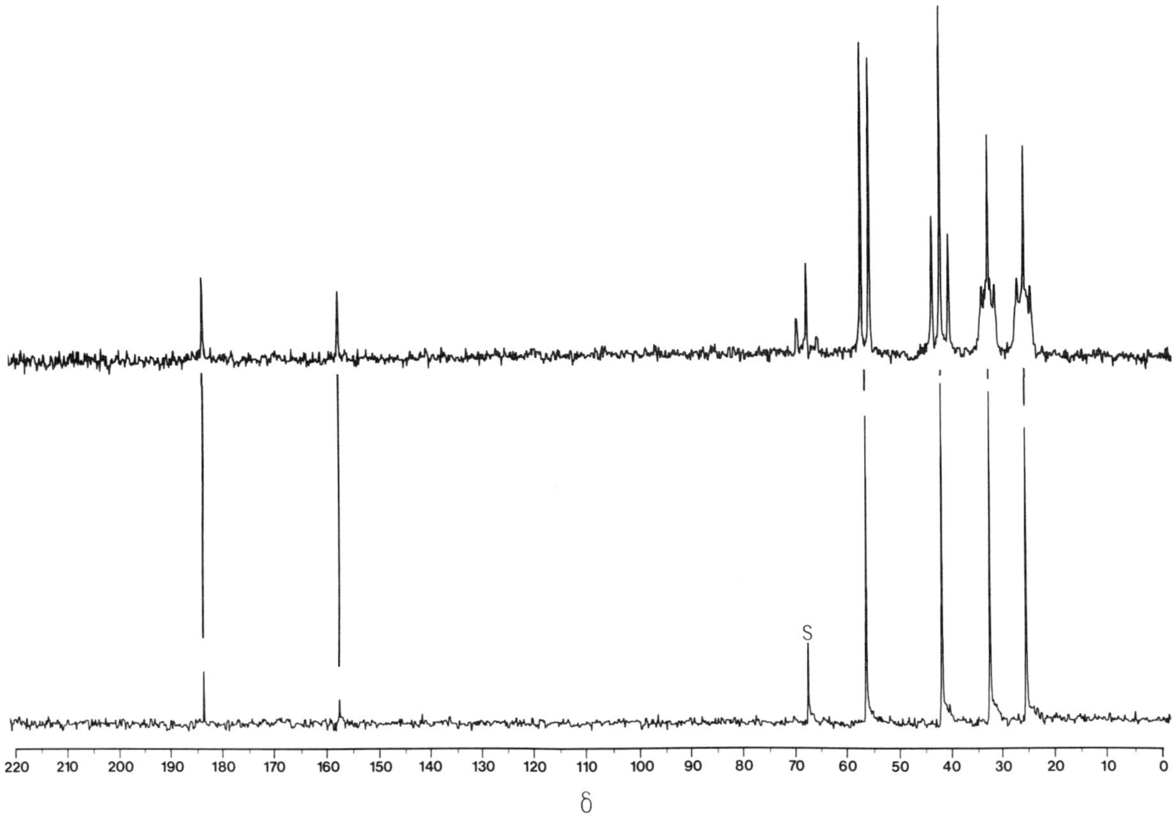
S
220
210
200
190
180
170
160
150
140
130
120
110
100
90
80
70
60
50
40
30
20
10
0
δ

Problem 2-32

2-32	MS:	m156(w), 141, 138
$C_{10}H_{20}O$	IR:	3250 cm^{-1}
	UV:	none

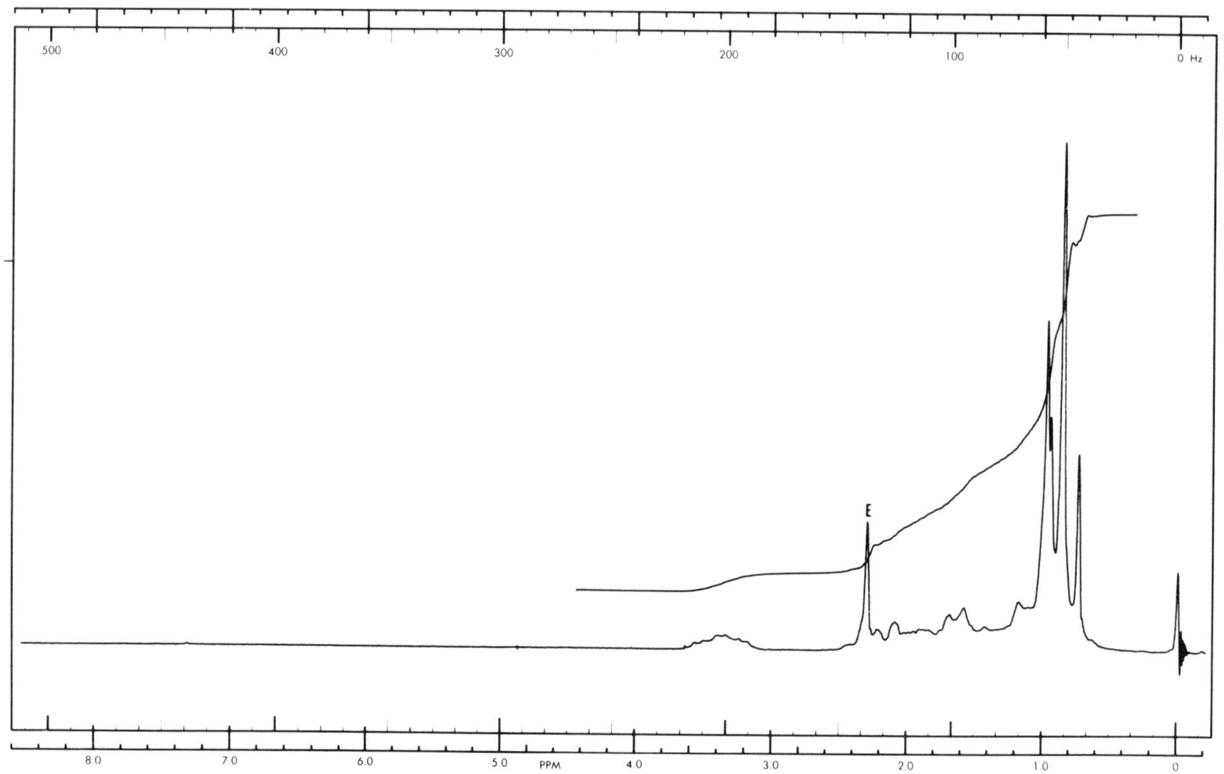

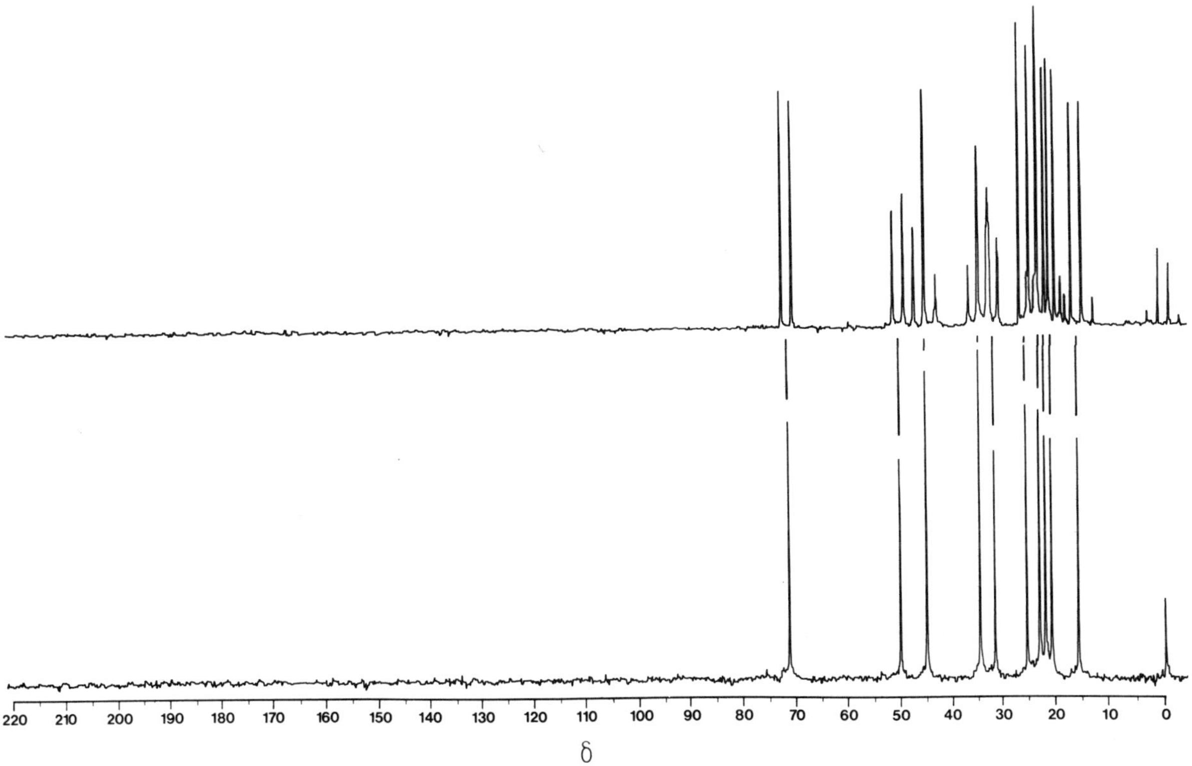
220
210
200
190
180
170
160
150
140
130
120
110
100
90
80
70
60
50
40
30
20
10
0
δ

Problem 2-33

2-33 MS: m106, 66, 40

C_8H_{10} IR: 1650 cm^{-1}

UV: none

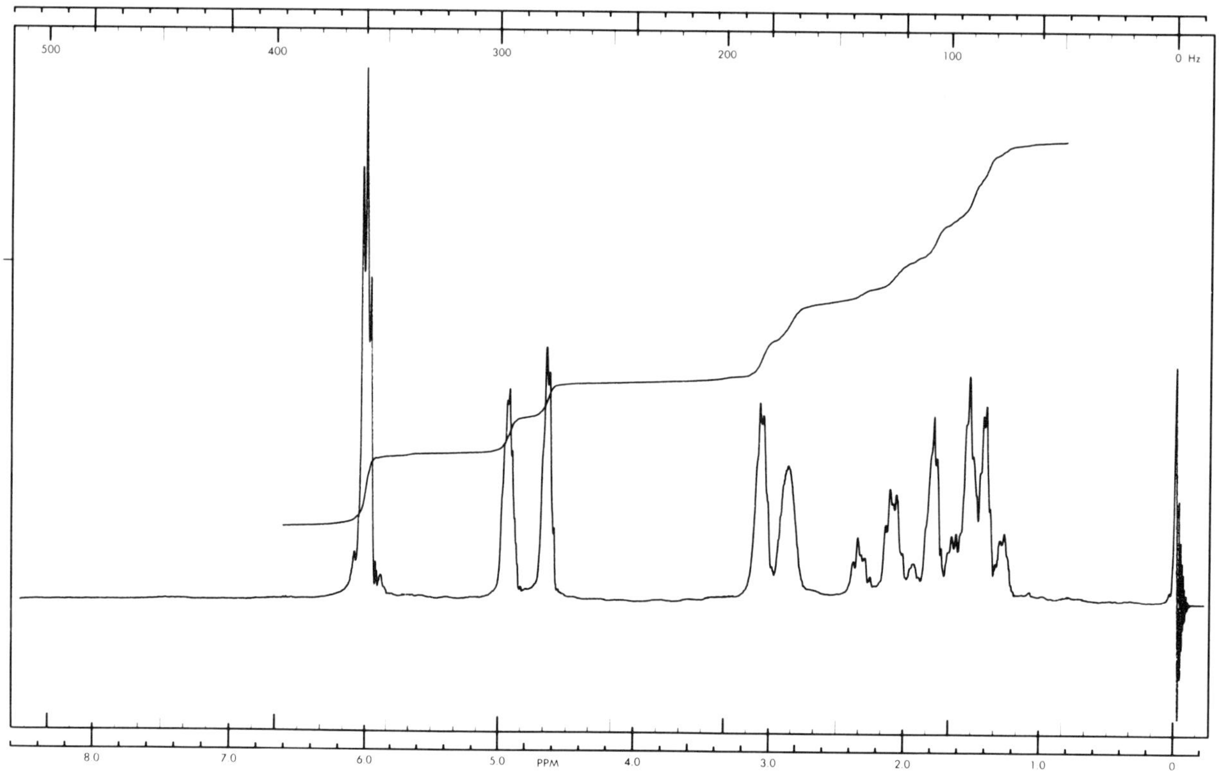

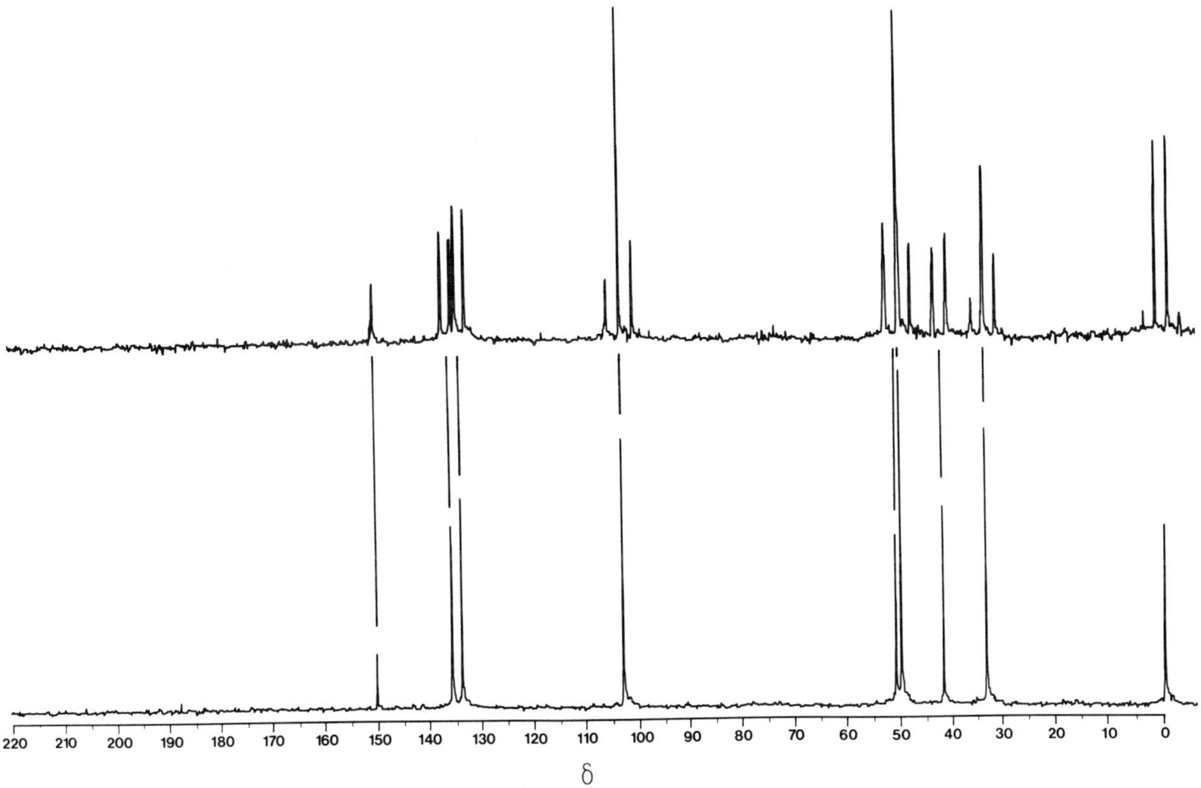
220
210
200
190
180
170
160
150
140
130
120
110
100
90
80
70
60
50
40
30
20
10
0
δ

Problem 2-34

2-34
$C_8H_{12}O_4$

MS: m172, 127, 60, 55
IR: 2900 (broad), 1700 cm^{-1}
UV: none
PMR: d_6-Me_2CO + TMS; upper scan offset 105 Hz
CMR: d_6-Me_2CO + TMS

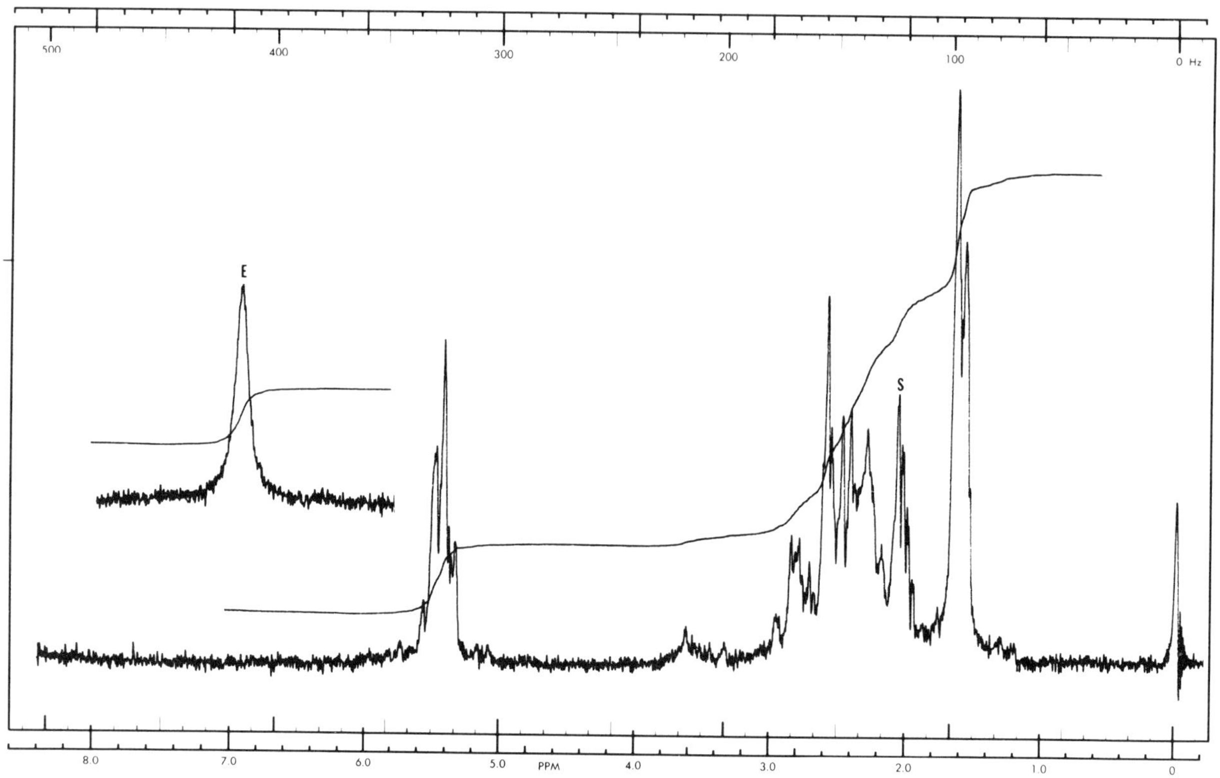

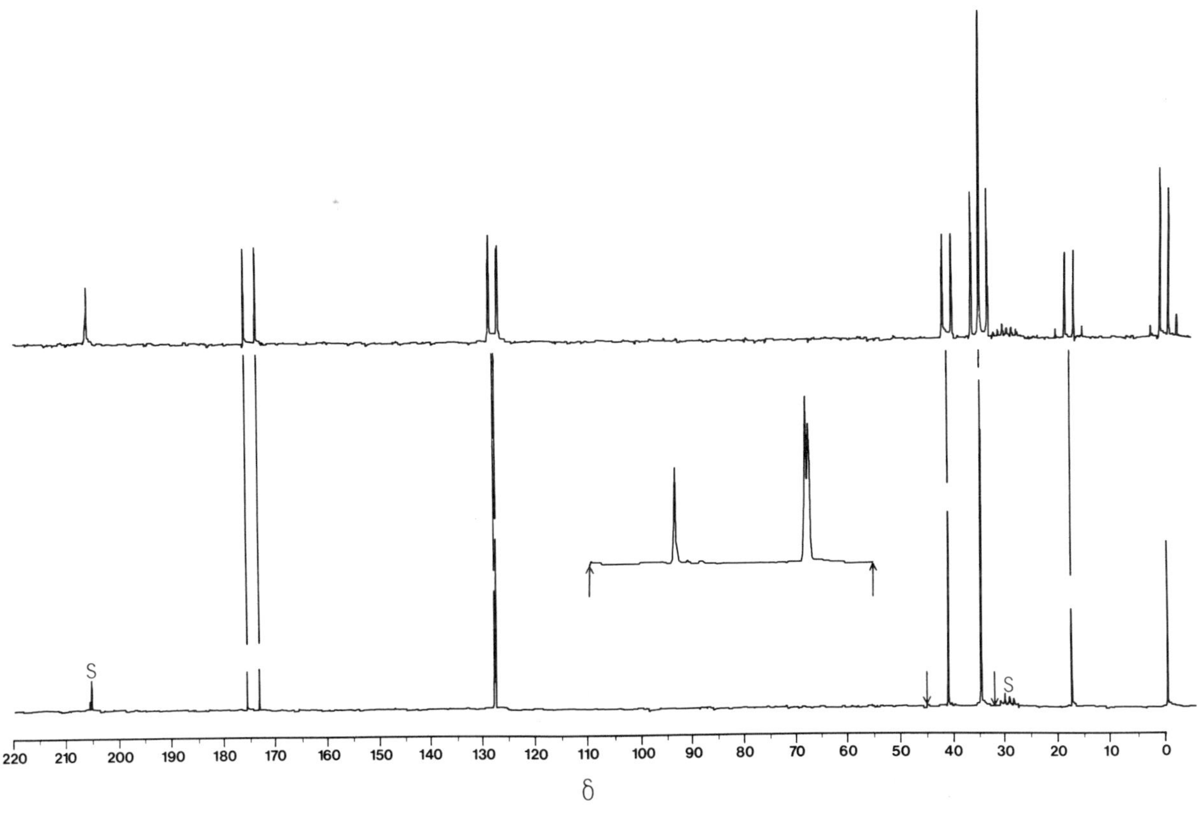
S
S
220
210
200
190
180
170
160
150
140
130
120
110
100
90
80
70
60
50
40
30
20
10
0
δ

Problem 2-35

2-35

$C_6H_{14}O_3$

MS: m134(w), 116, 103, 61, 31

IR: 3380 cm^{-1}

UV: none

PMR: D_2O + DSS

CMR: D_2O + dioxane

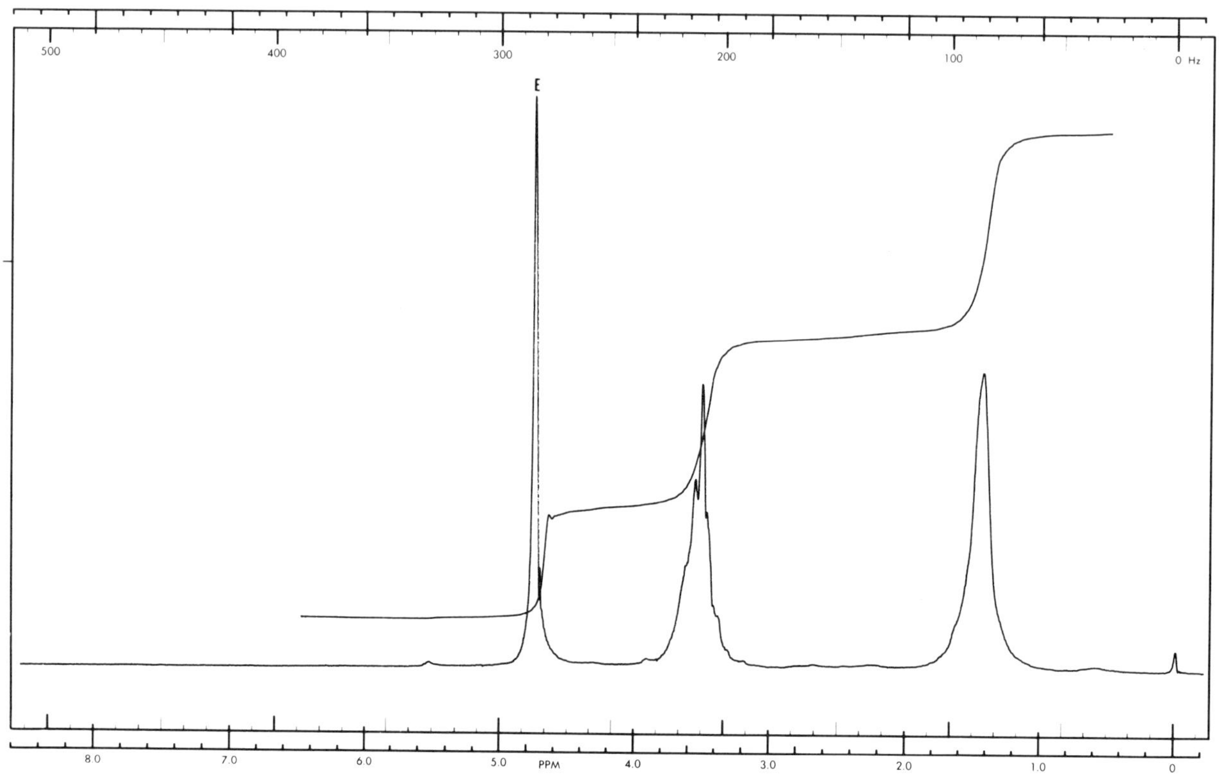

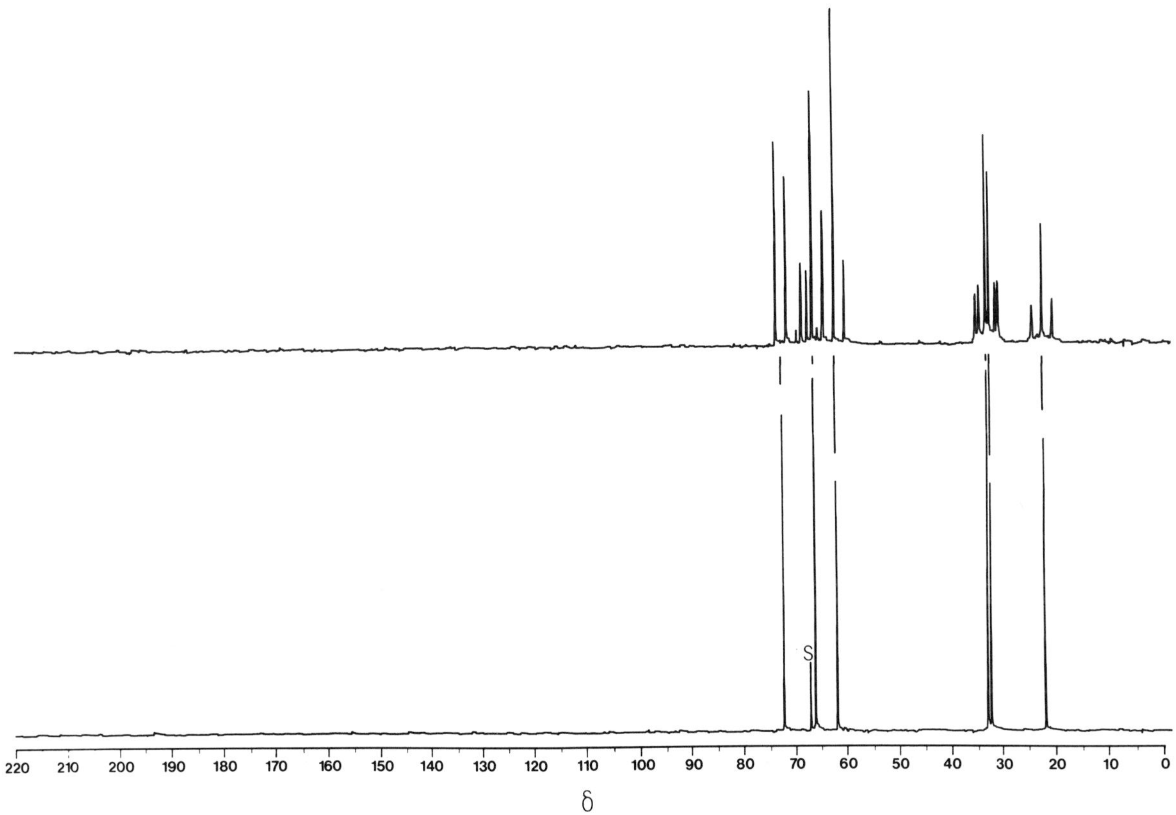
S
220
210
200
190
180
170
160
150
140
130
120
110
100
90
80
70
60
50
40
30
20
10
0
δ

Problem 2-36

2-36
$C_8H_{15}N$

MS: m125, 124
IR: 3250 cm^{-1}
UV: none

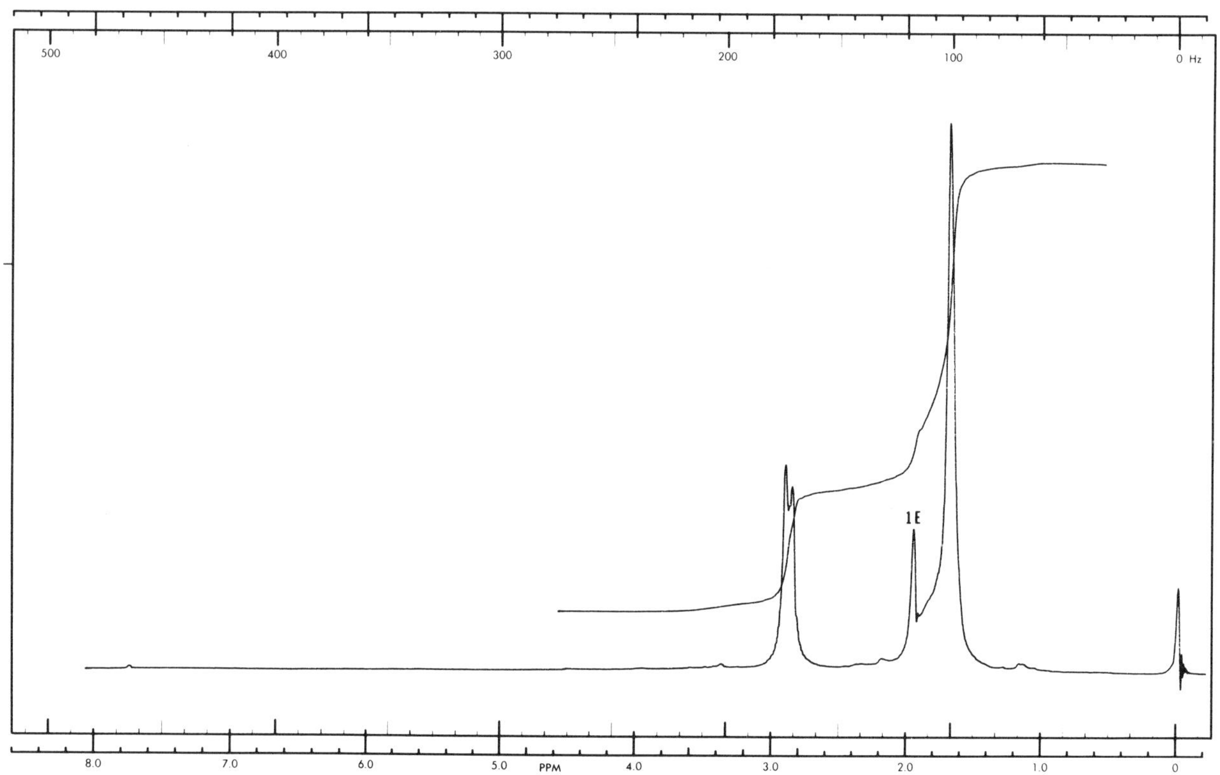

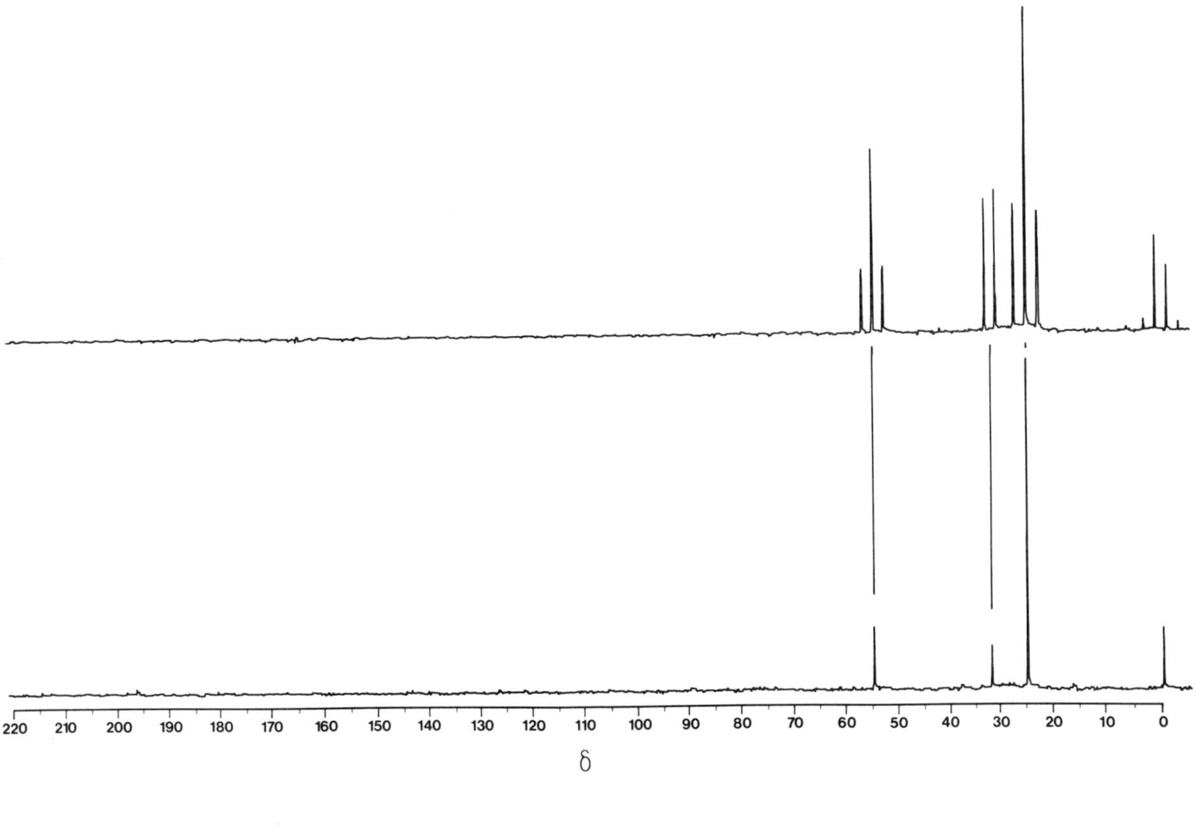
220
210
200
190
180
170
160
150
140
130
120
110
100
90
80
70
60
50
40
30
20
10
0
δ

Problem 2-37

2-37

$C_{10}H_{14}O$

MS: m150(s), 122

IR: 1720 cm^{-1}

UV: 285 (ε 14)

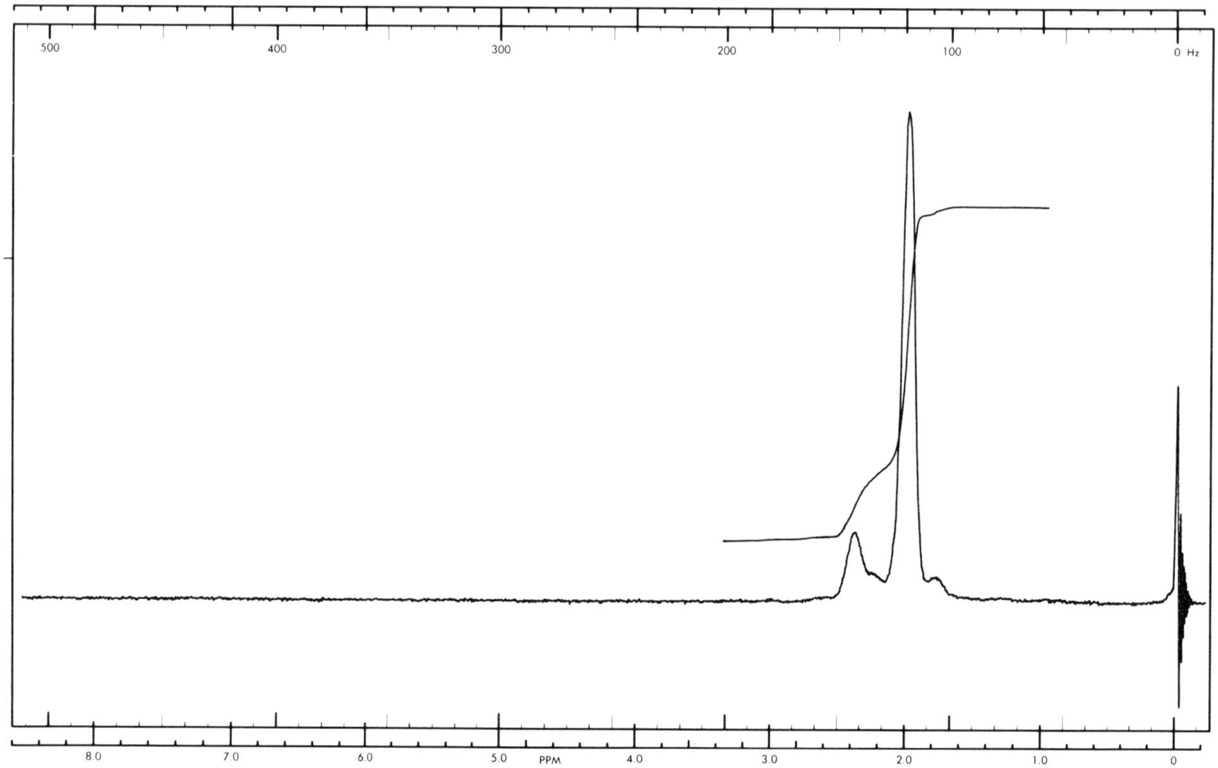

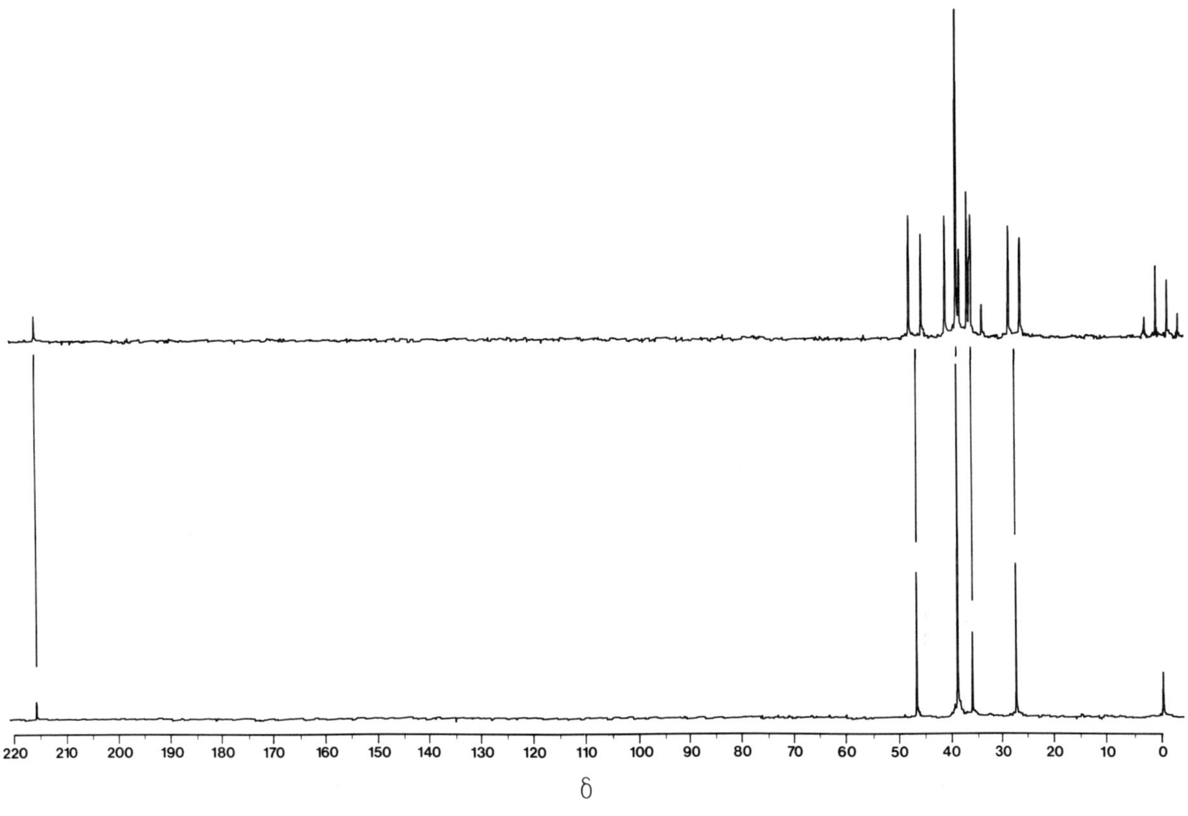
220
210
200
190
180
170
160
150
140
130
120
110
100
90
80
70
60
50
40
30
20
10
0
δ

Problem 2-38

2-38

$C_{10}H_{14}O$

MS: m150(w), 83, 67

IR: 1635 cm^{-1}

UV: none

CMR: 10 sec delay

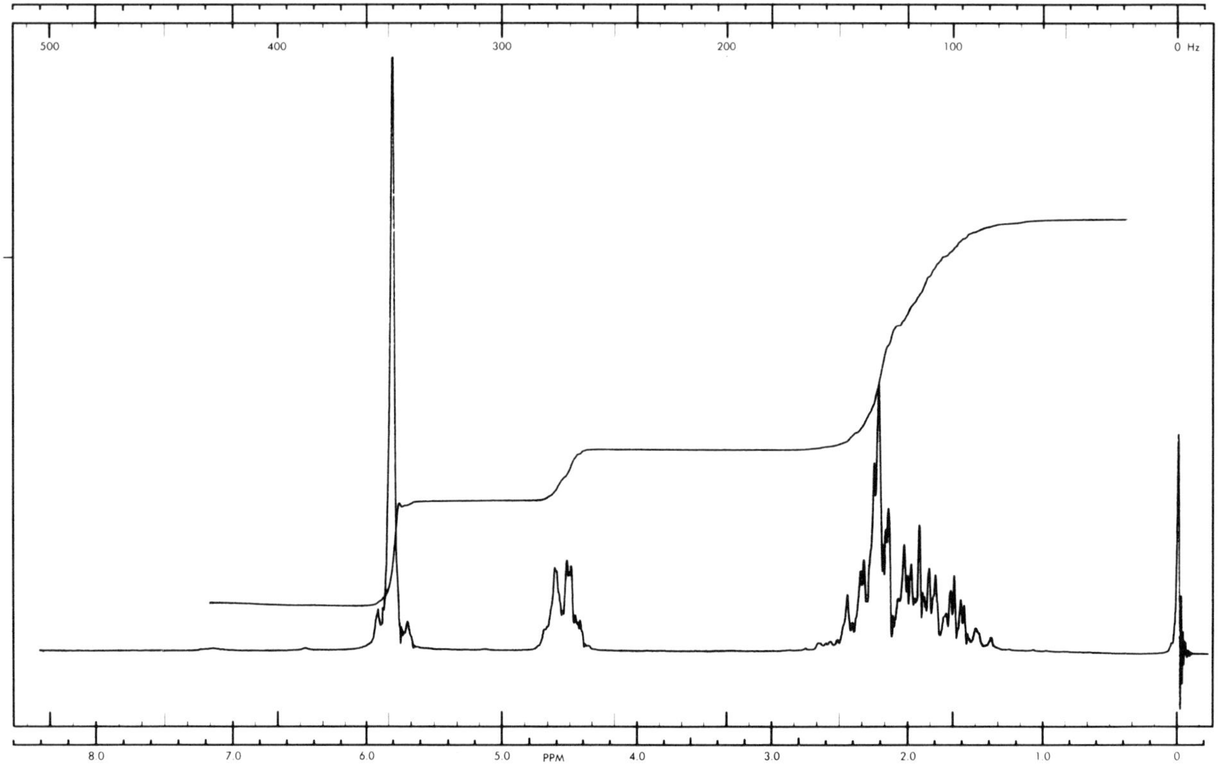

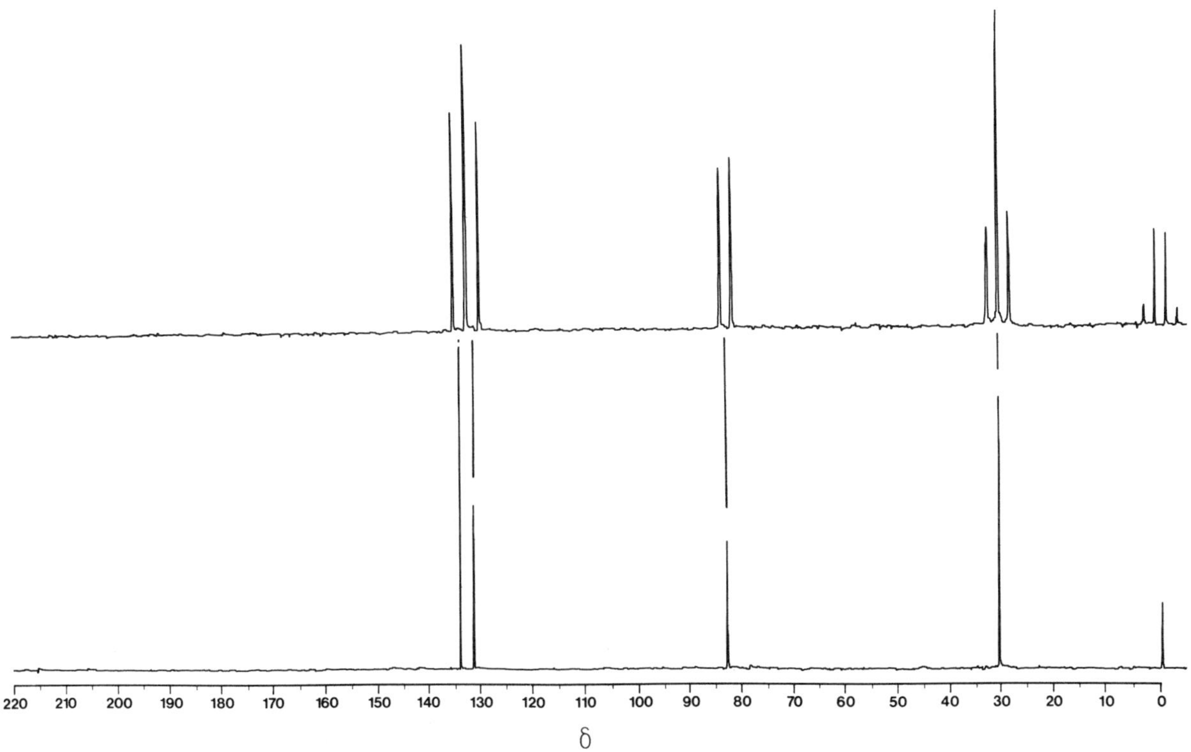
220
210
200
190
180
170
160
150
140
130
120
110
100
90
80
70
60
50
40
30
20
10
0
δ

Problem 2-39

2-39

$C_5H_2F_6O_4$

MS: m240(w), 221, 195

IR: 3100 (broad), 1775 (broad) cm^{-1}

UV: none

PMR: D_2O + DSS

CMR: D_2O + dioxane; 25 sec delay

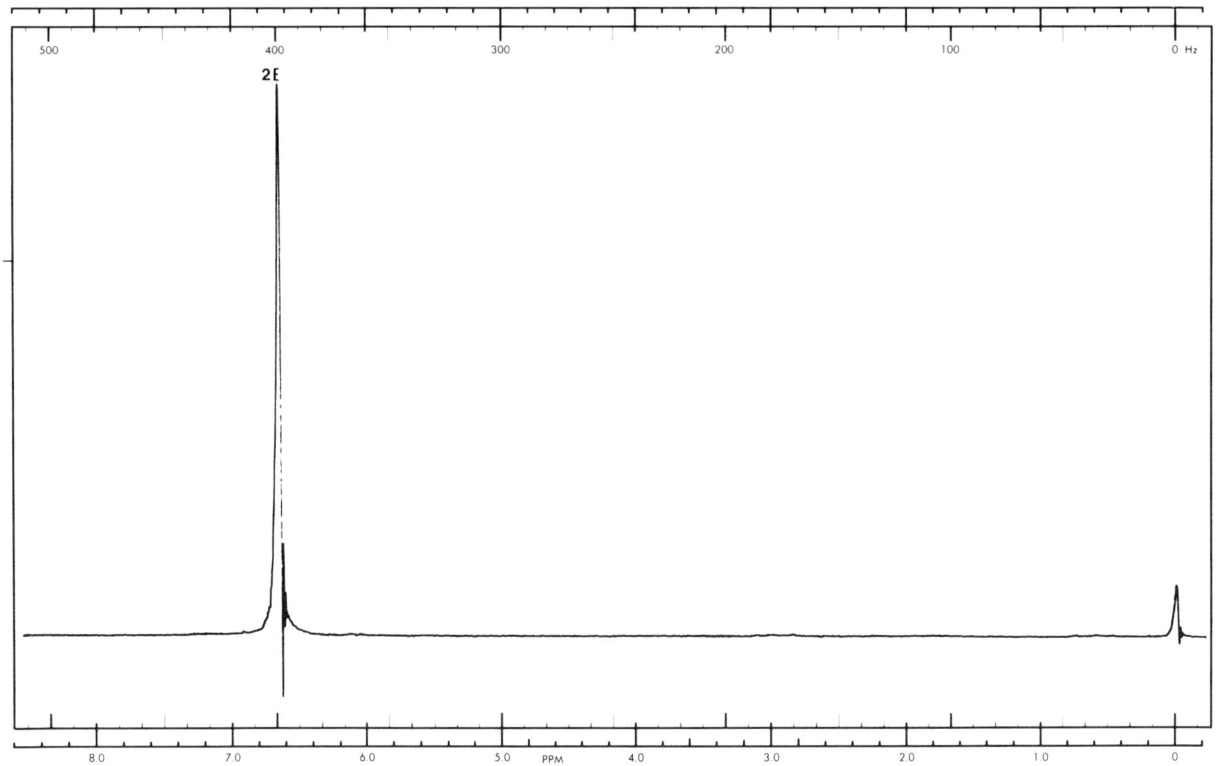

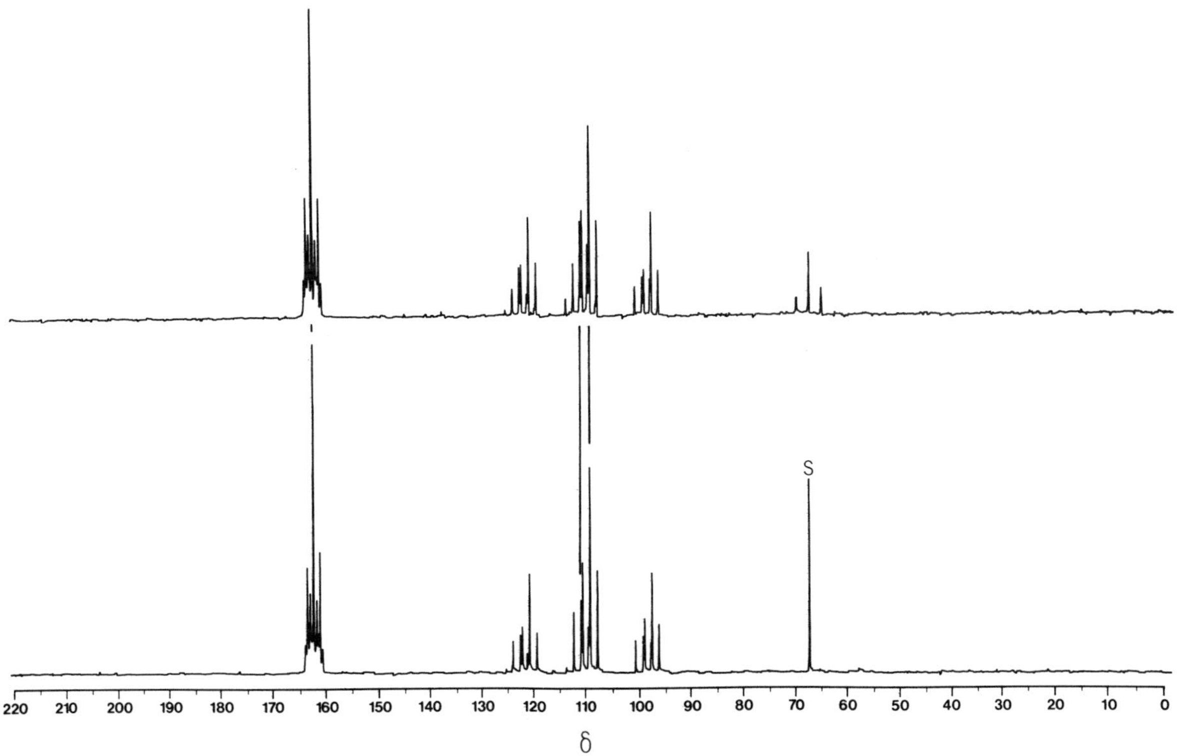
S
220
210
200
190
180
170
160
150
140
130
120
110
100
90
80
70
60
50
40
30
20
10
0
δ

Problem 2-40

2-40 MS: m231, 230, 126, 105, 77

$C_{13}H_{10}ClNO$ IR: 3400, 3300, 1625 cm^{-1}

UV: Strong absorption

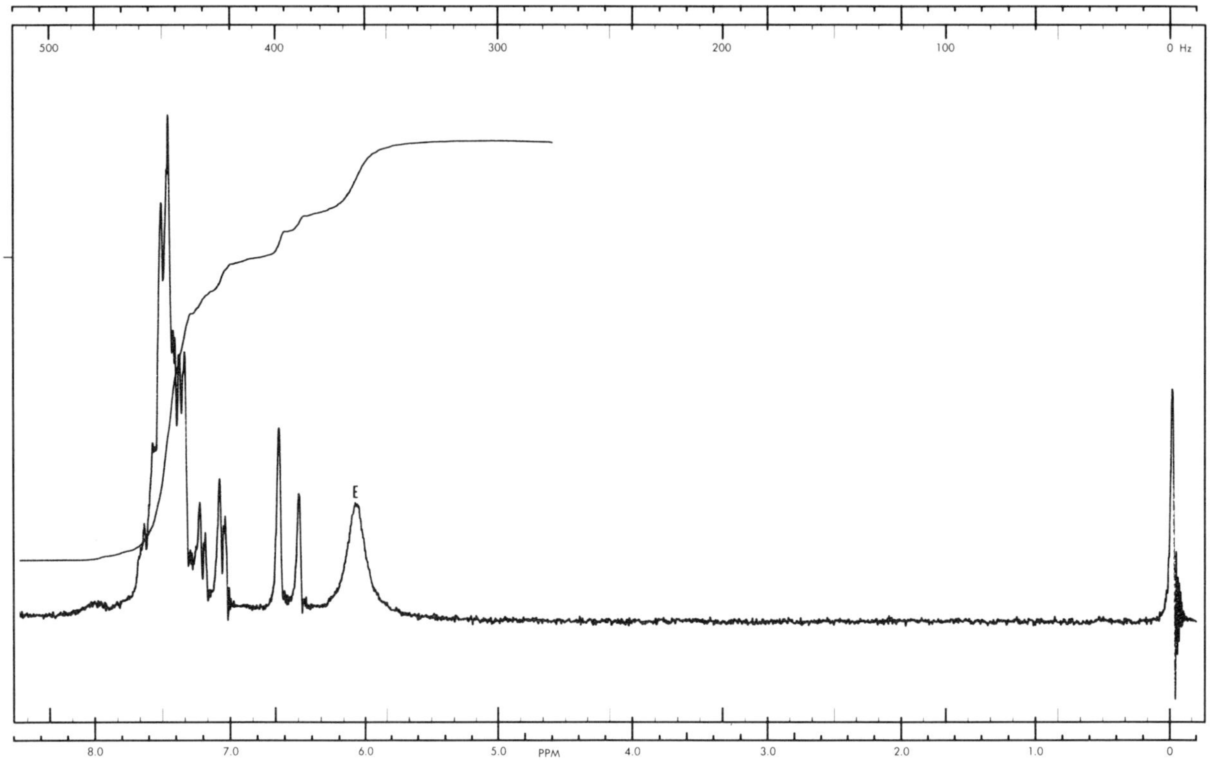

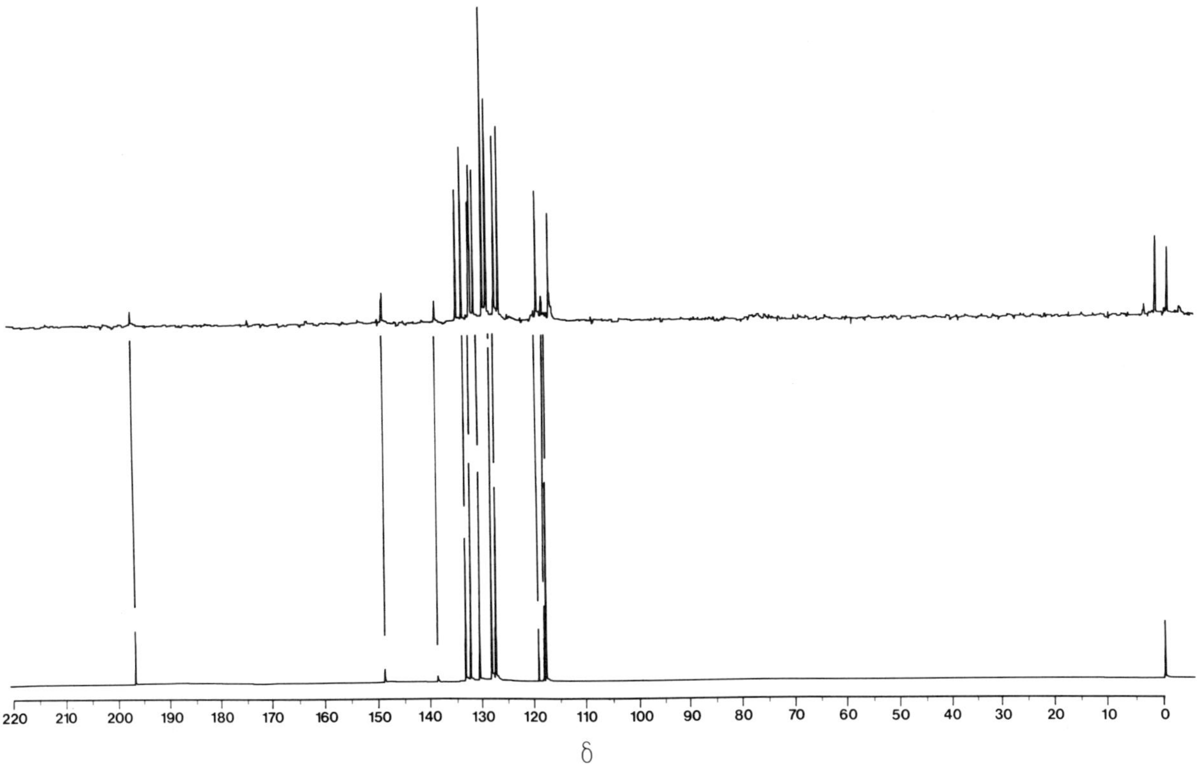
220
210
200
190
180
170
160
150
140
130
120
110
100
90
80
70
60
50
40
30
20
10
0
δ

Problem 2-41

2-41 $C_{10}H_{18}$

MS: m138(w), 123, 69, 55, 39

IR: 1650 cm^{-1}

UV: none

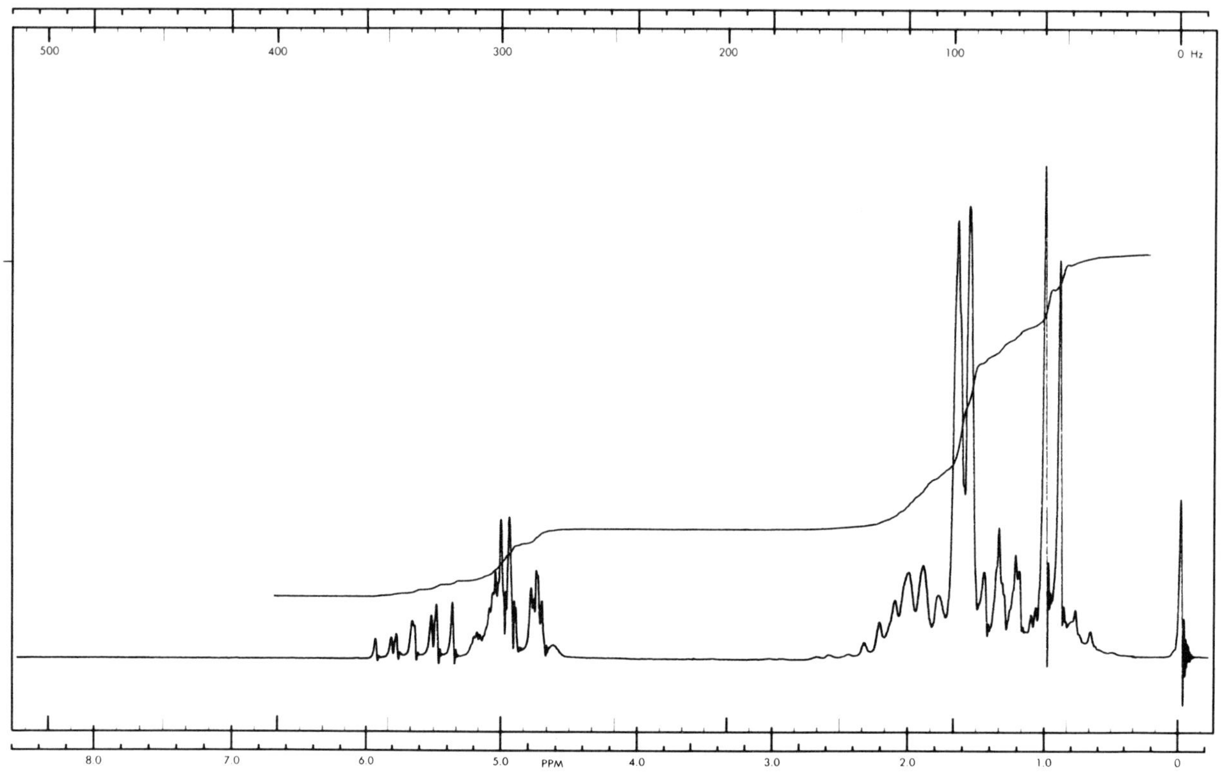

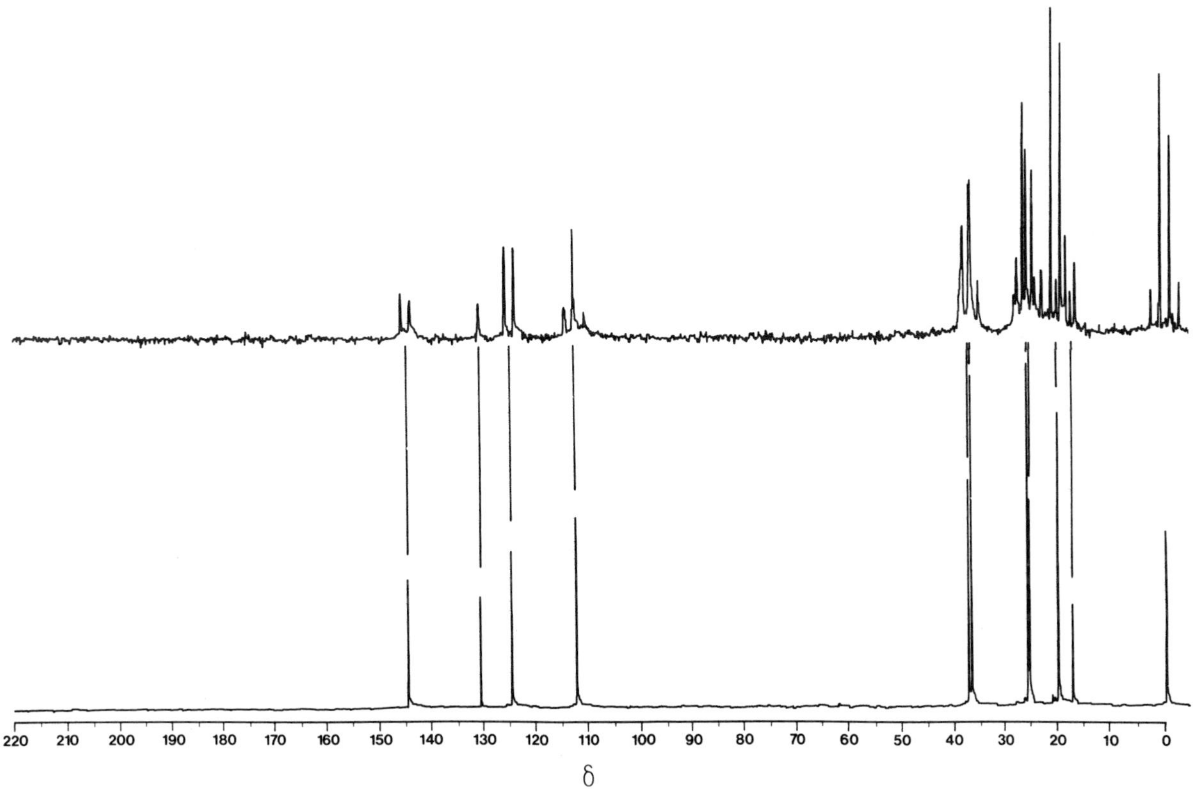
220 210 200 190 180 170 160 150 140 130 120 110 100 90 80 70 60 50 40 30 20 10 0
δ

Problem 2-42

2-42

$C_{10}H_{18}$

MS: m138(w), 123, 83, 55, 41, 39

IR: 1660 cm^{-1}

UV: none

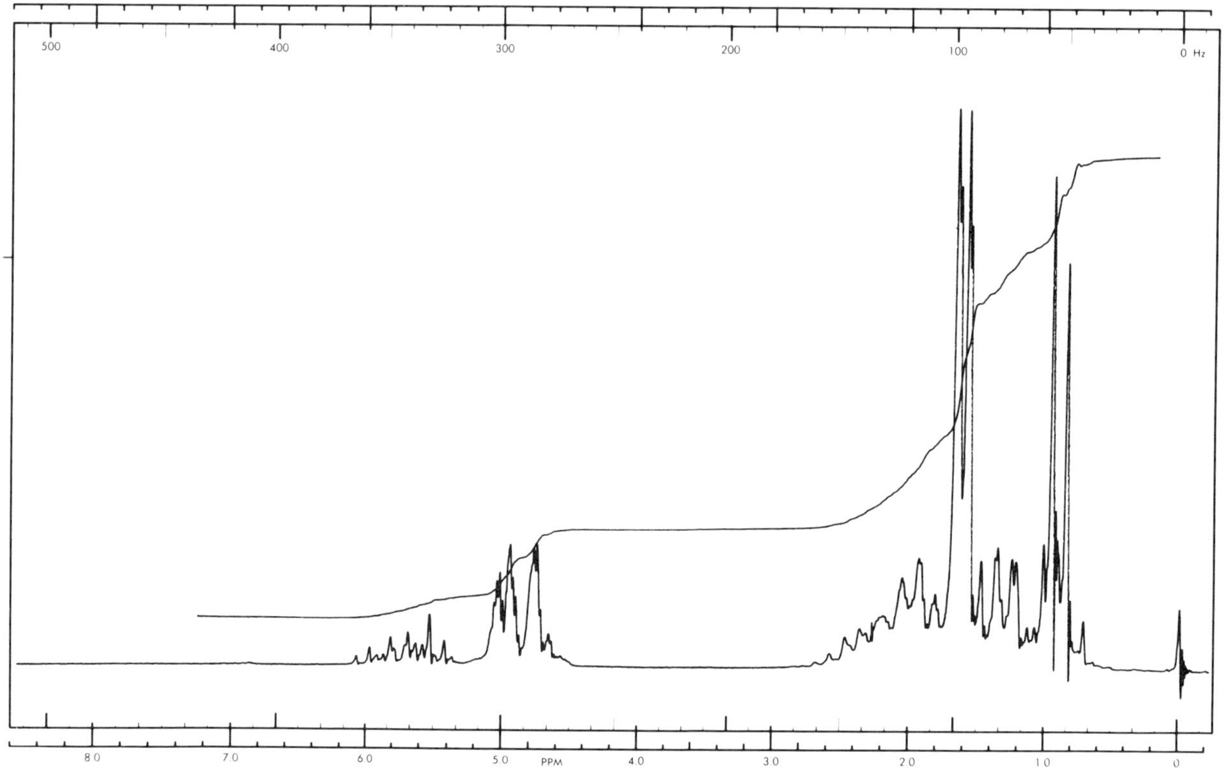

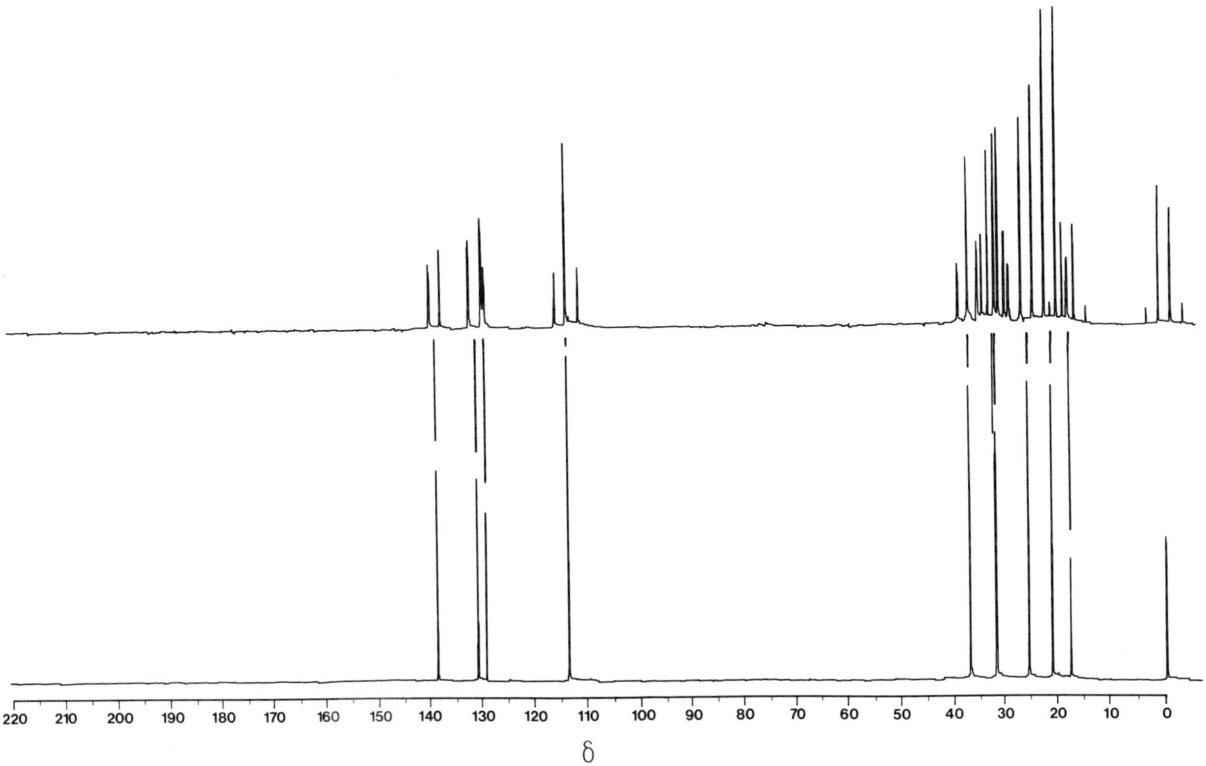
220
210
200
190
180
170
160
150
140
130
120
110
100
90
80
70
60
50
40
30
20
10
0
δ

Problem 2-43

2-43 MS: m152, 137, 124, 110, 42

$C_{10}H_{16}O$ IR: 1750 cm^{-1}

UV: 285 (ε 20)

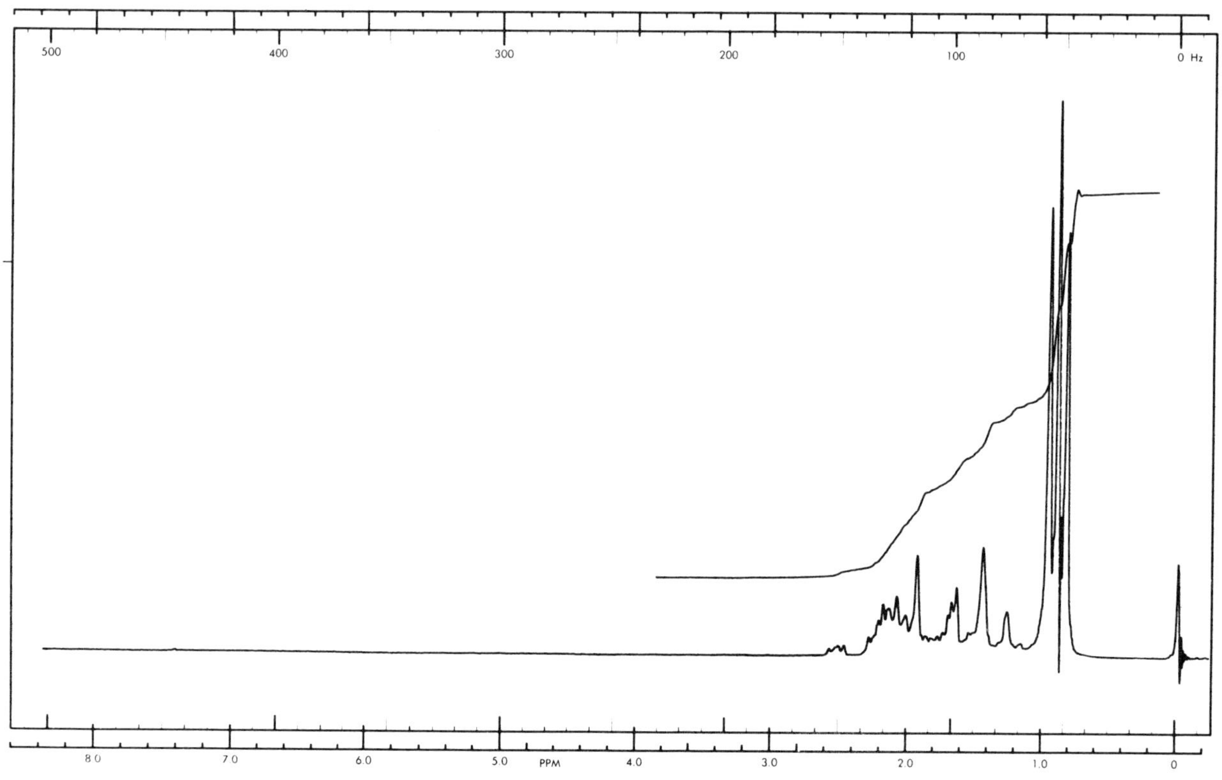

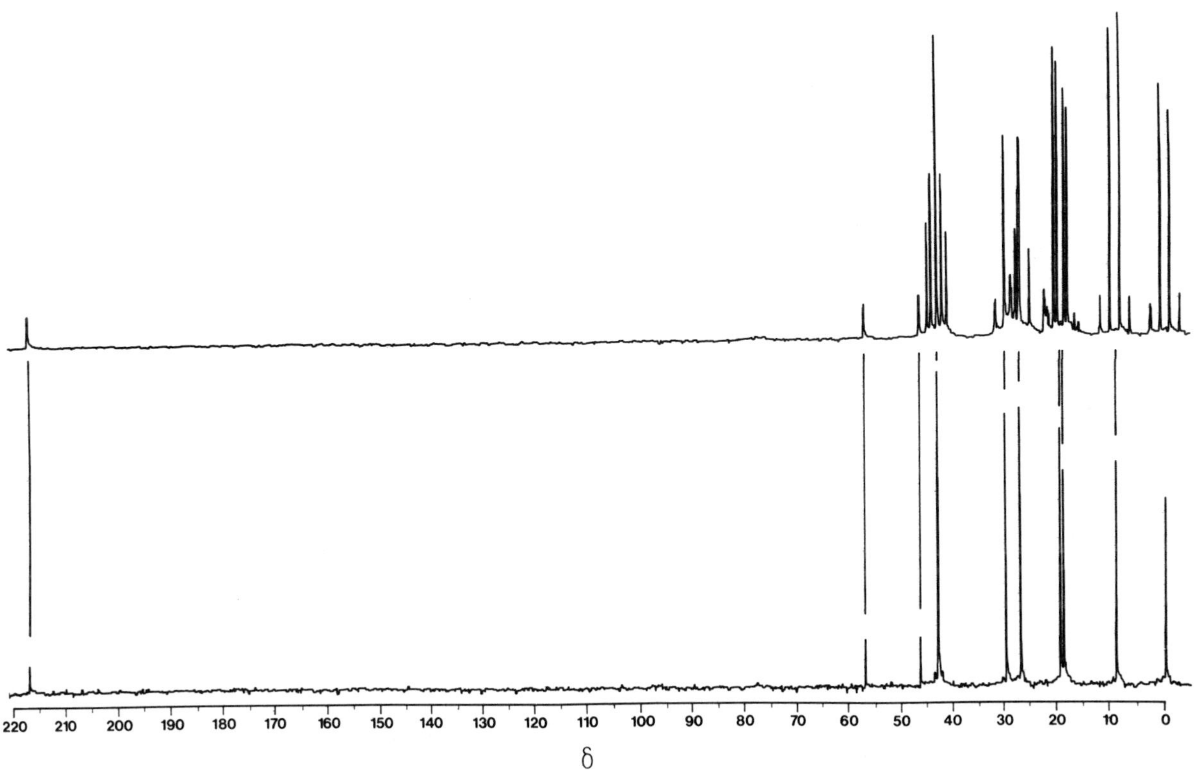
220
210
200
190
180
170
160
150
140
130
120
110
100
90
80
70
60
50
40
30
20
10
0
δ

Problem 2-44

2-44

$C_{17}H_{18}O_3$

MS: m270, 163, 121, 91

IR: 1740 cm^{-1}

UV: Strong absorption

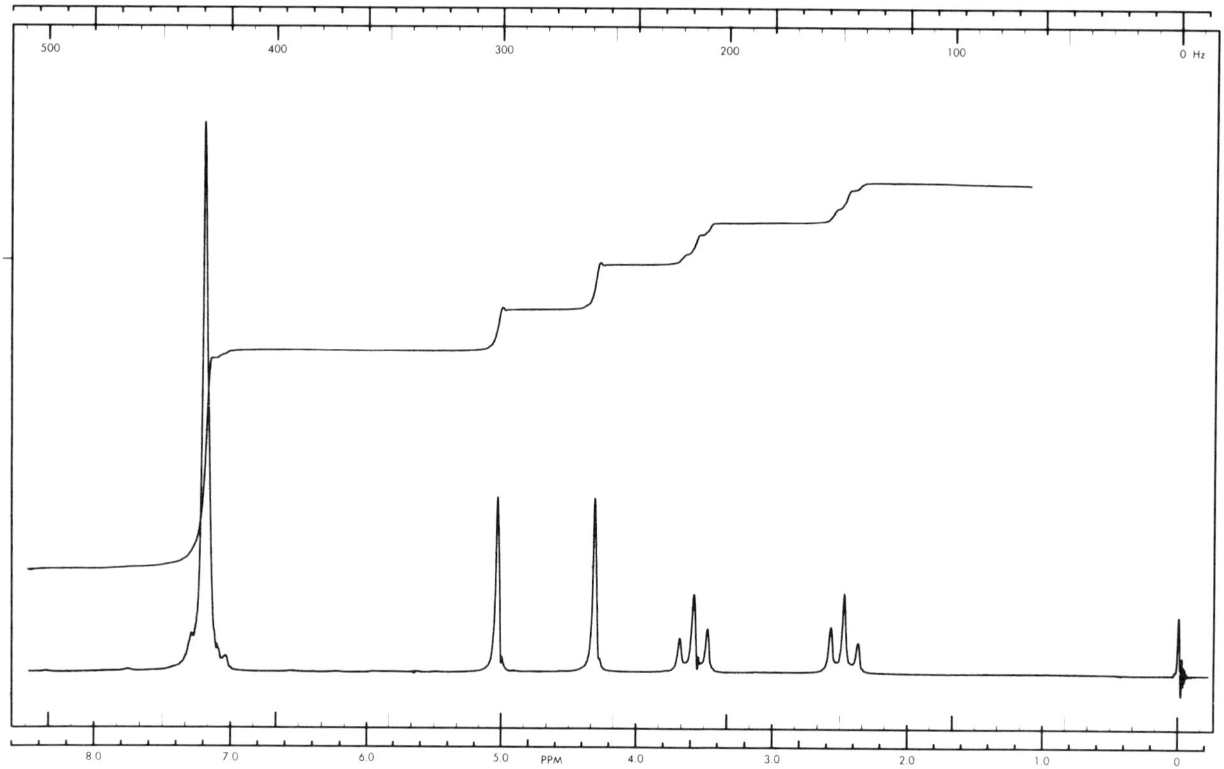

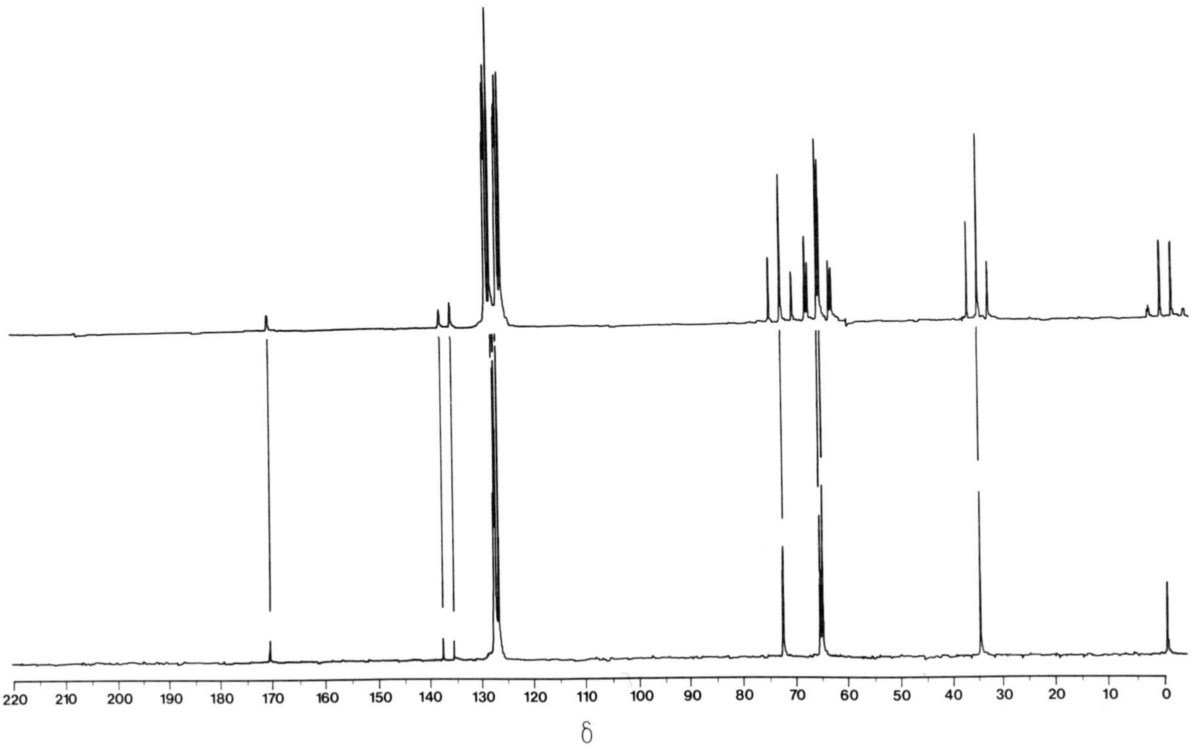
220
210
200
190
180
170
160
150
140
130
120
110
100
90
80
70
60
50
40
30
20
10
0
δ

Problem 2-45

2-45	MS:	Free acid m219, 202, 131, 31
$(C_9H_{16}NO_5)_2Ca$	IR:	3200, 1650, 1600 cm^{-1}
	UV:	none
	PMR:	D_2O + DSS
	CMR:	D_2O + dioxane

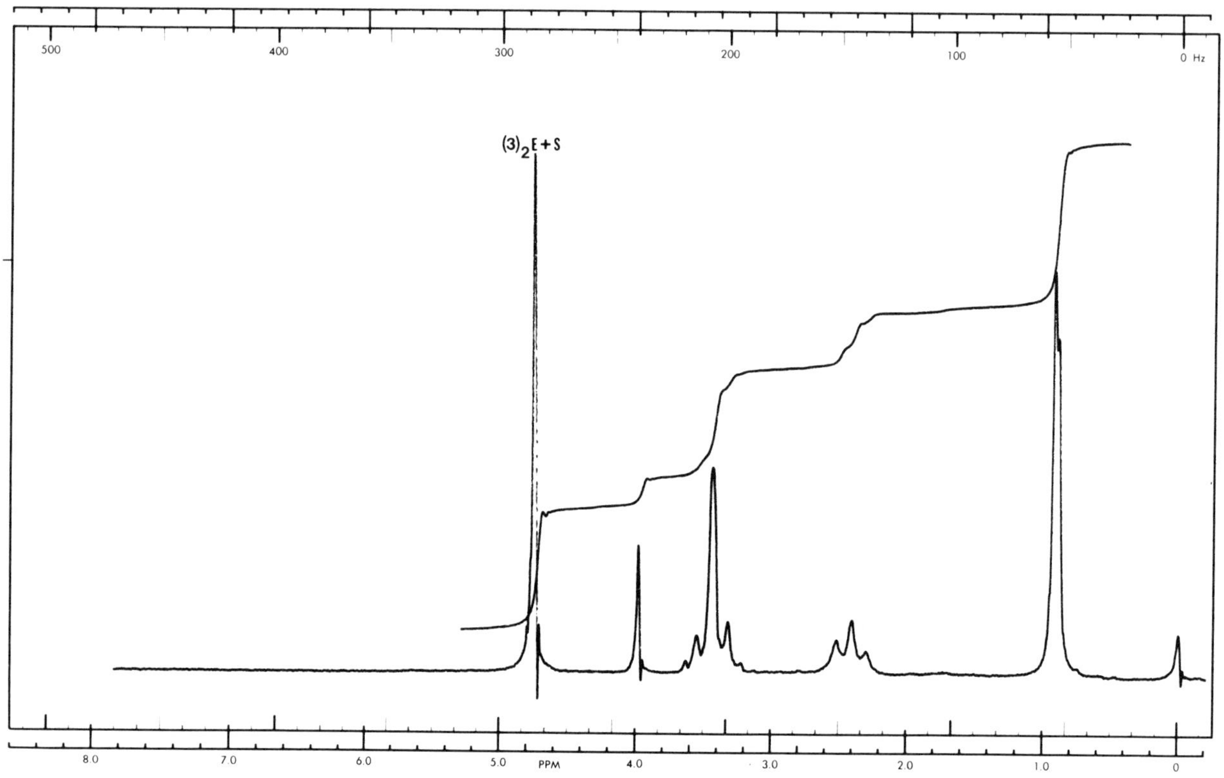

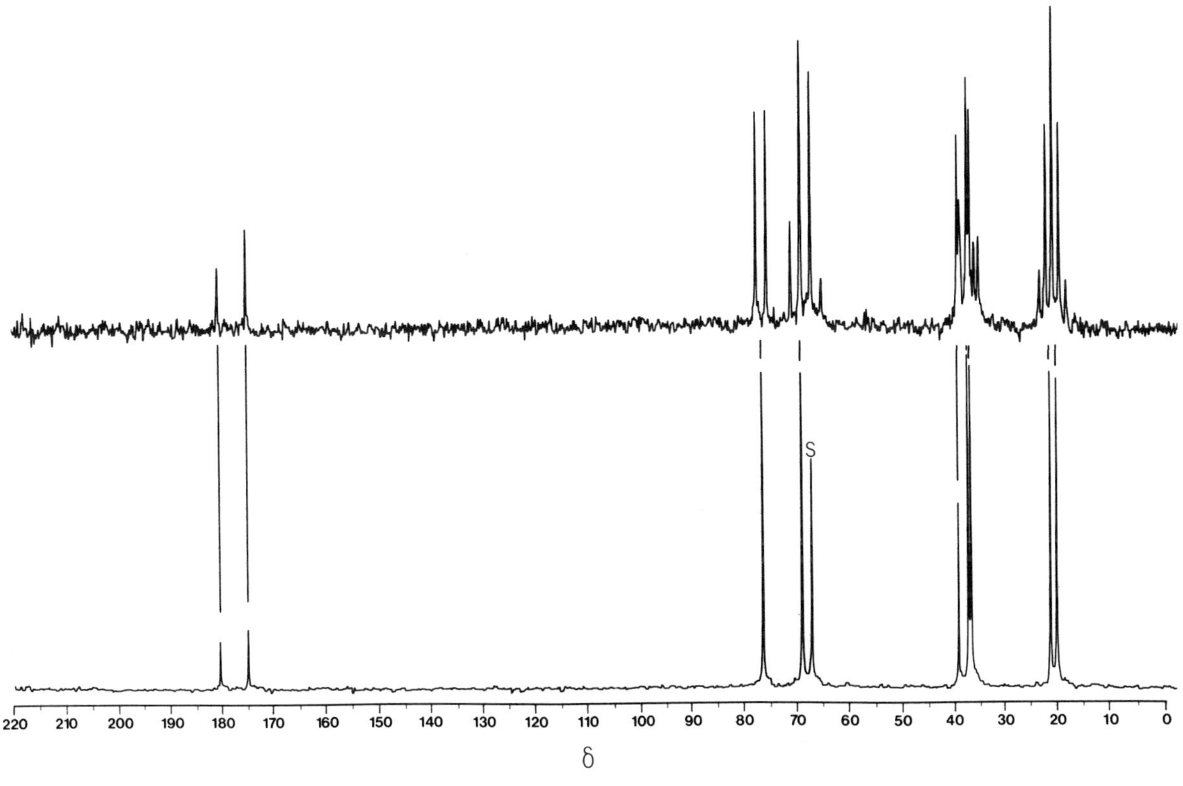
S
220
210
200
190
180
170
160
150
140
130
120
110
100
90
80
70
60
50
40
30
20
10
0
δ

Problem 2-46

2-46

$C_9H_{12}N_2O_6$

MS: m244(w), 226, 133, 111

IR: 3200, 1700 cm^{-1}

UV: 205 (ε 9800), 261 (ε 10,000)

PMR: D_2O + DSS

CMR: D_2O + dioxane

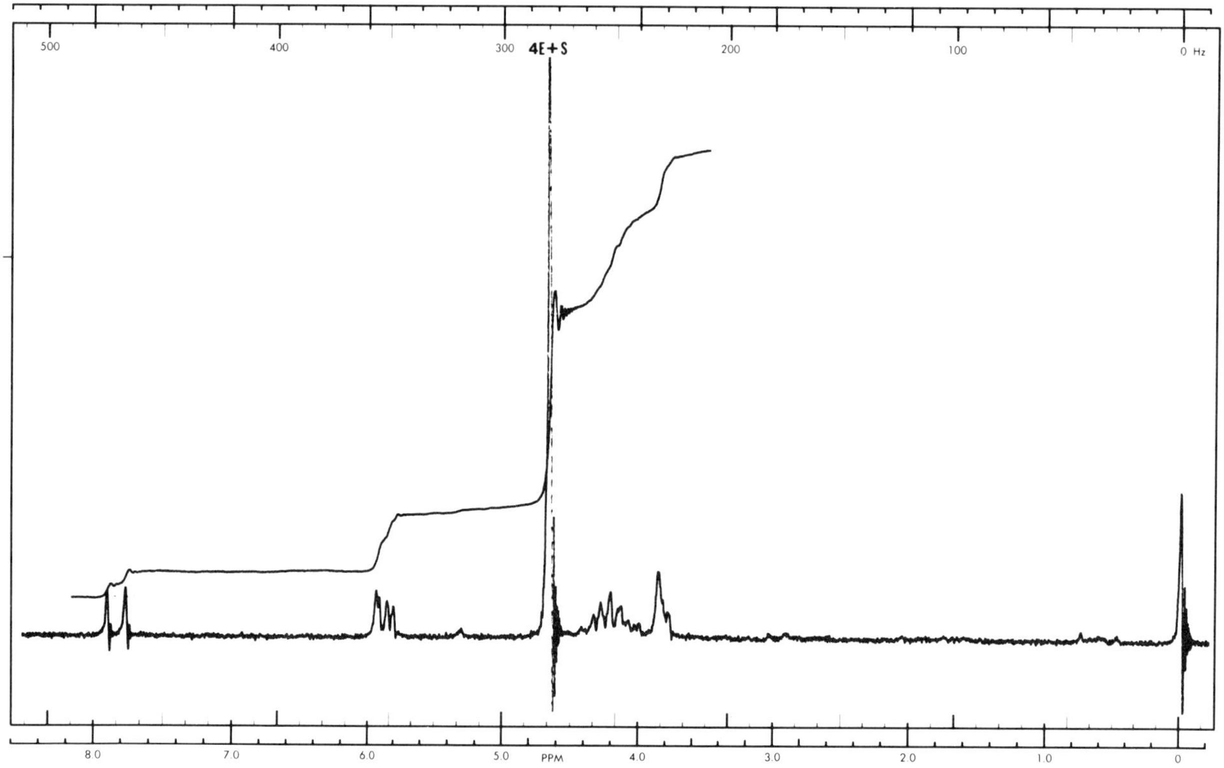

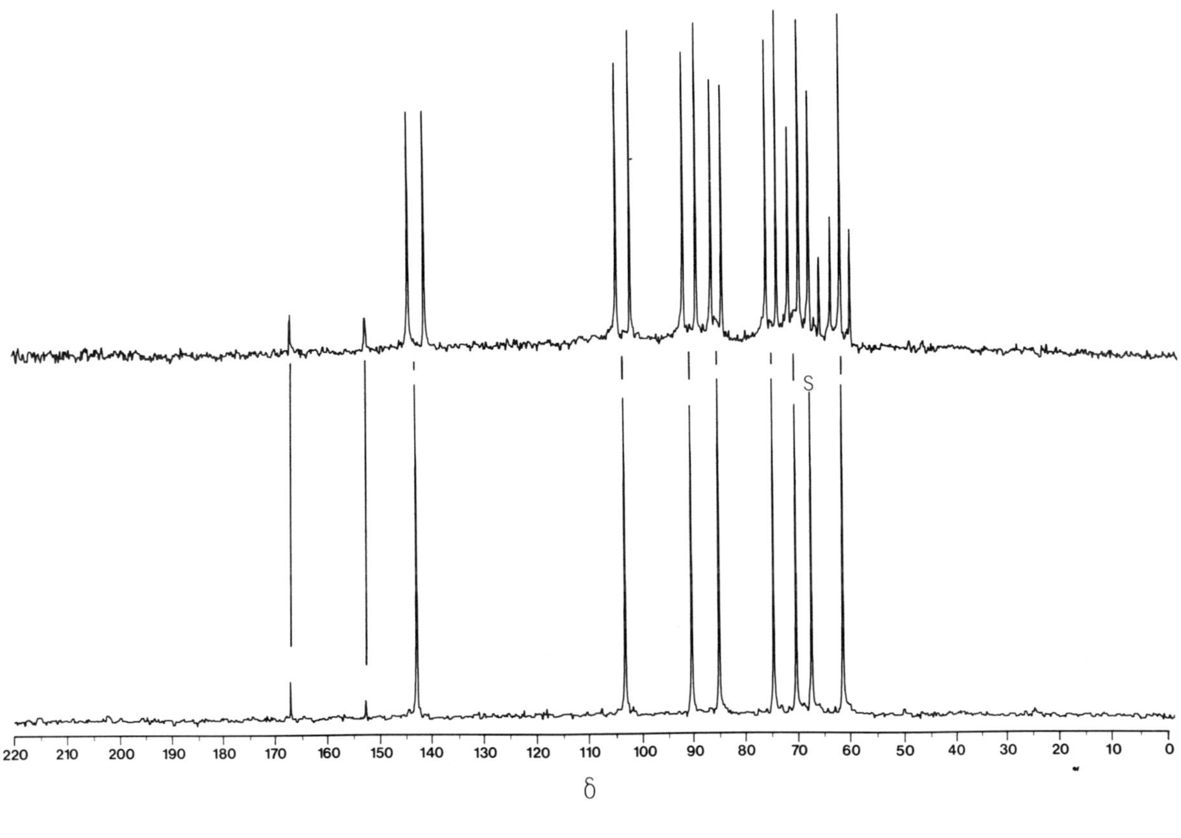
S
220
210
200
190
180
170
160
150
140
130
120
110
100
90
80
70
60
50
40
30
20
10
0
δ

Problem 2-47

2-47
$C_9H_{13}N_3O_5$

MS: m243(w), 225, 133, 110
IR: 3200, 1700 cm^{-1}
UV: 271 (ε 9100) at pH 8, 280 (ε 13,400) at pH 2
PMR: D_2O + DSS
CMR: D_2O + dioxane

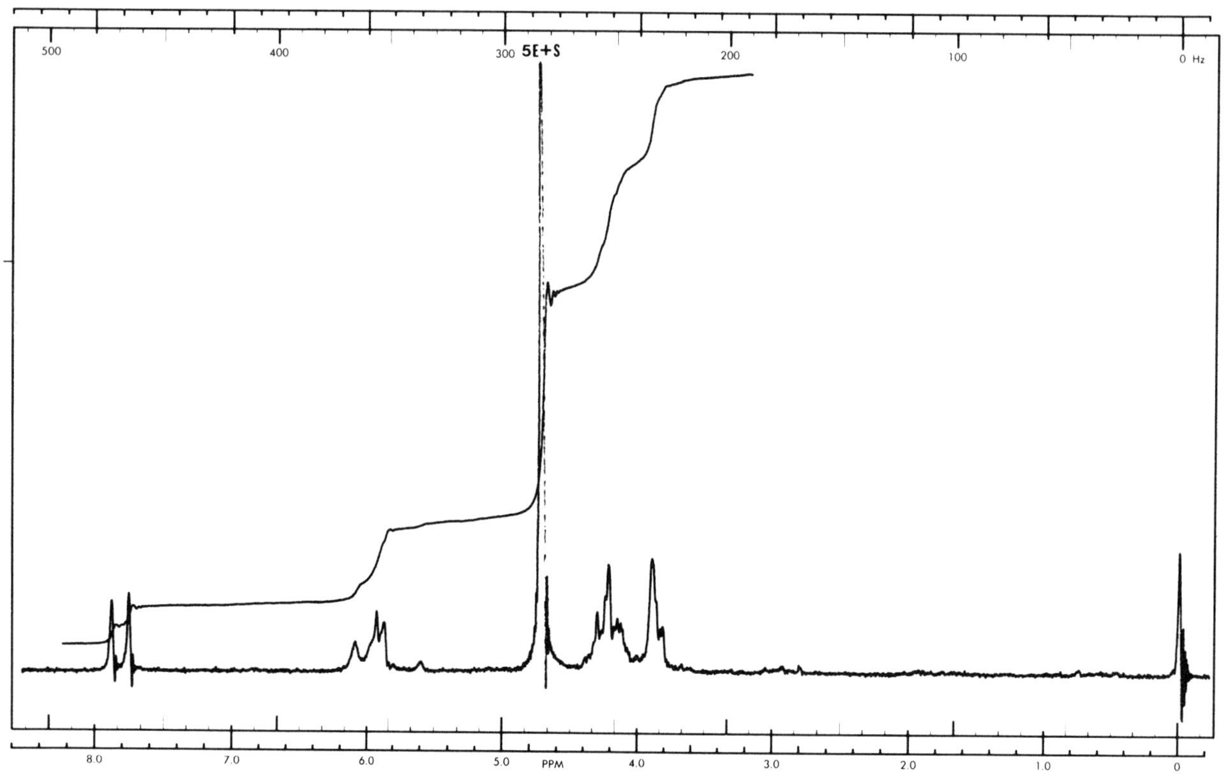

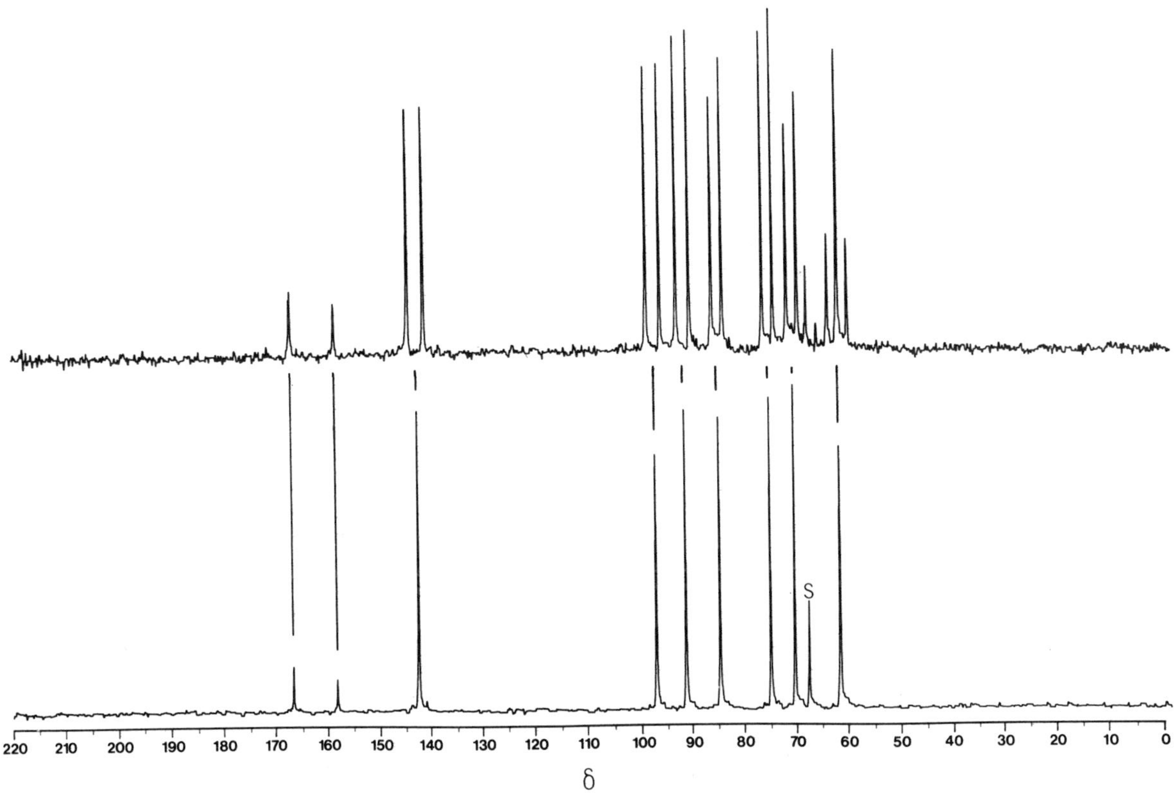
S
220
210
200
190
180
170
160
150
140
130
120
110
100
90
80
70
60
50
40
30
20
10
0
δ

Problem 2-48

2-48

$C_{14}H_{10}O_2$

MS: m210(s), 195, 182, 167

IR: 1700 cm^{-1}

UV: Strong absorption

CMR: 25 sec delay

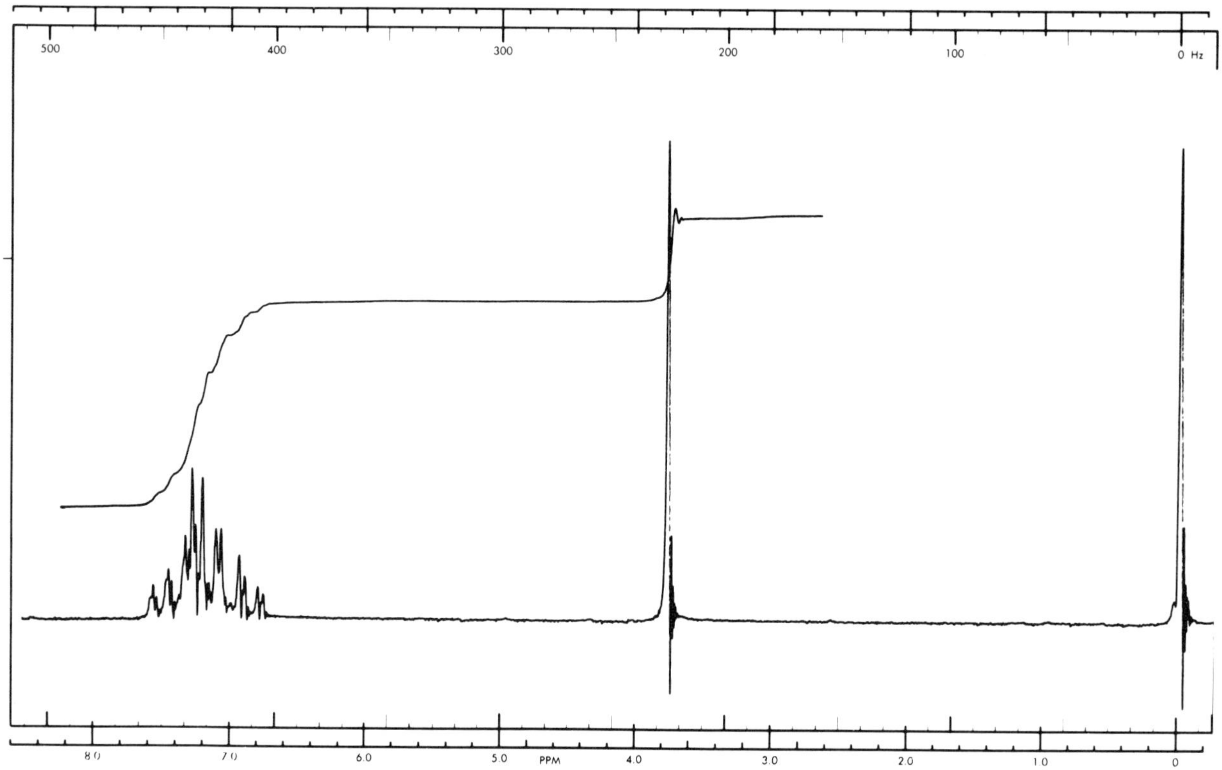

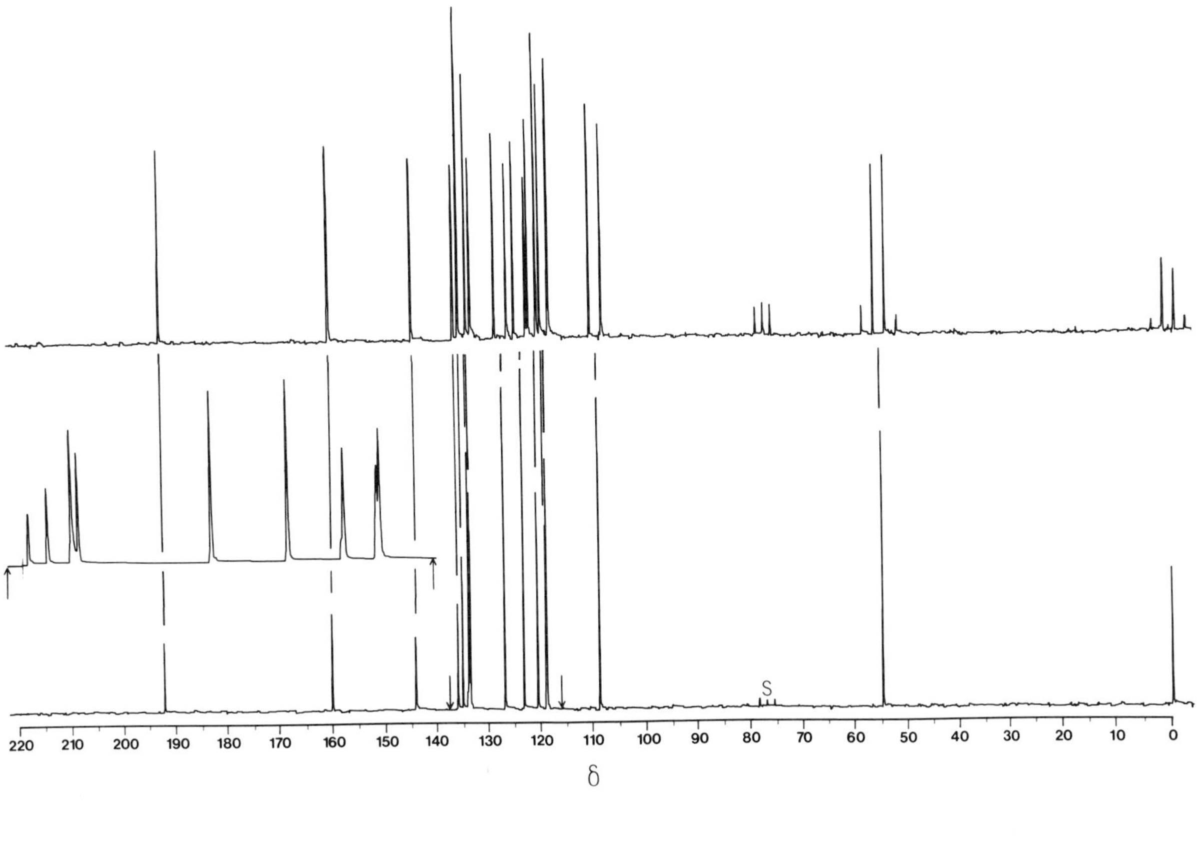
S
220
210
200
190
180
170
160
150
140
130
120
110
100
90
80
70
60
50
40
30
20
10
0
δ

Problem 2-49

2-49 $C_{11}H_{10}Cl_6$

MS: m352(w), 323, 321, 319, 317, 276, 274, 272, 270

IR: -

UV: none

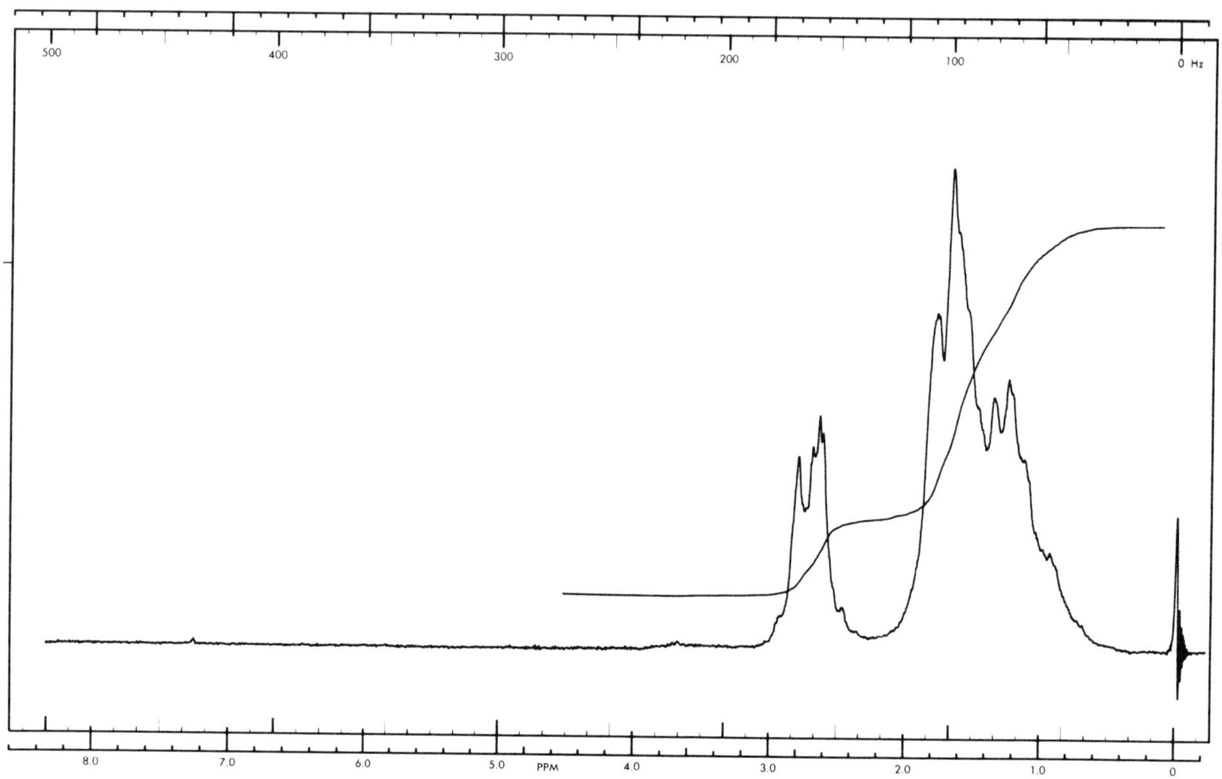

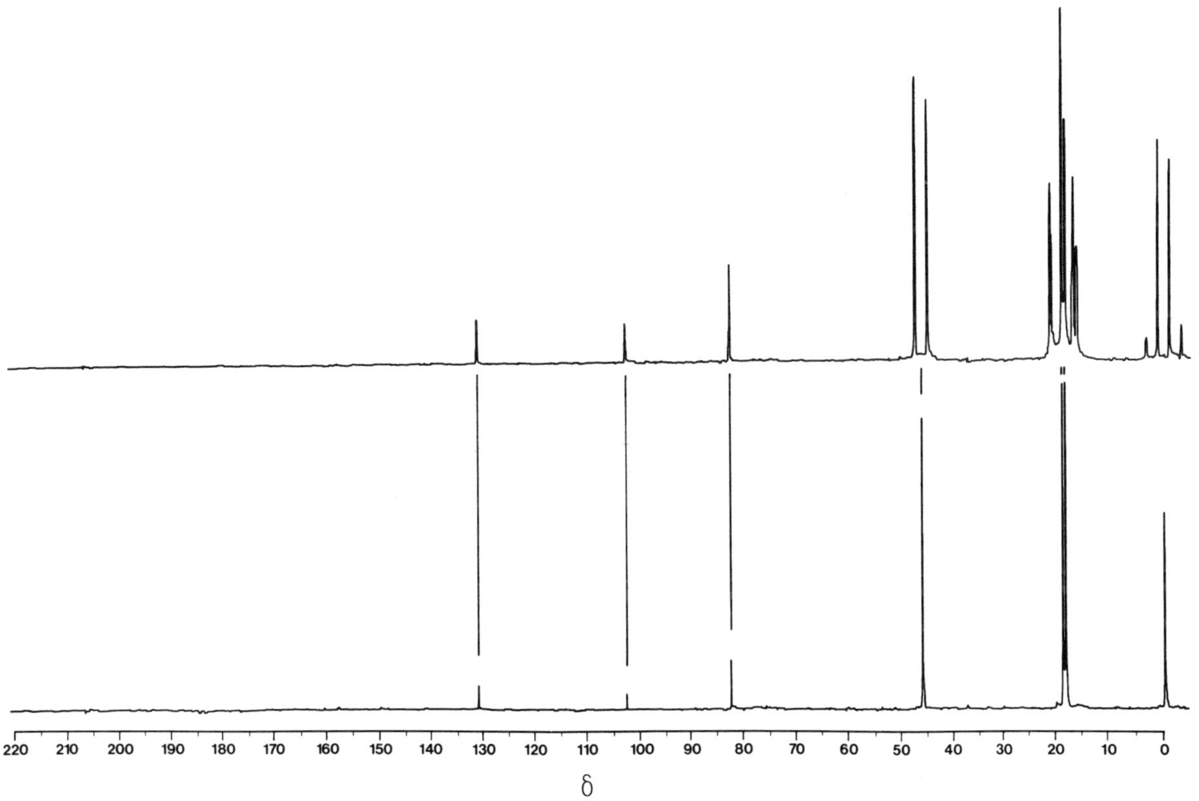
220 210 200 190 180 170 160 150 140 130 120 110 100 90 80 70 60 50 40 30 20 10 0
δ

Problem 2-50

2-50

$C_{10}H_{16}$

MS: m136, 121, 68

IR: 1655 cm^{-1}

UV: none

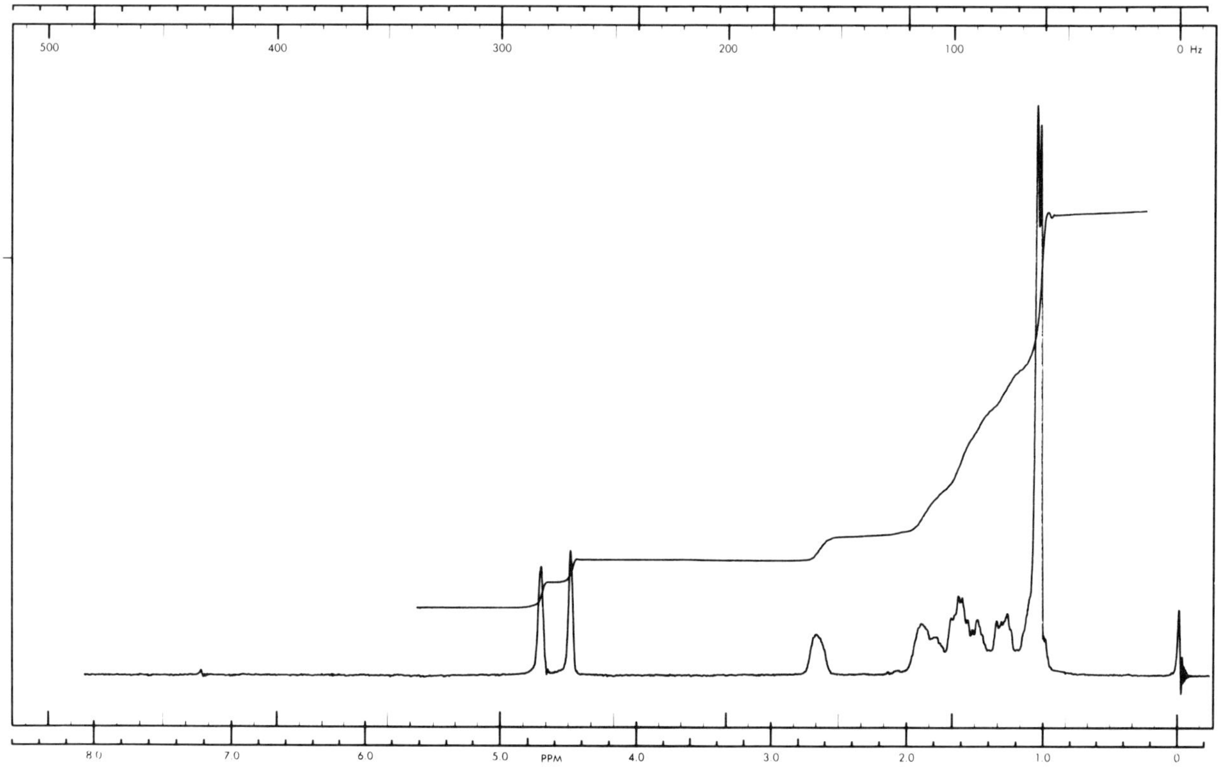

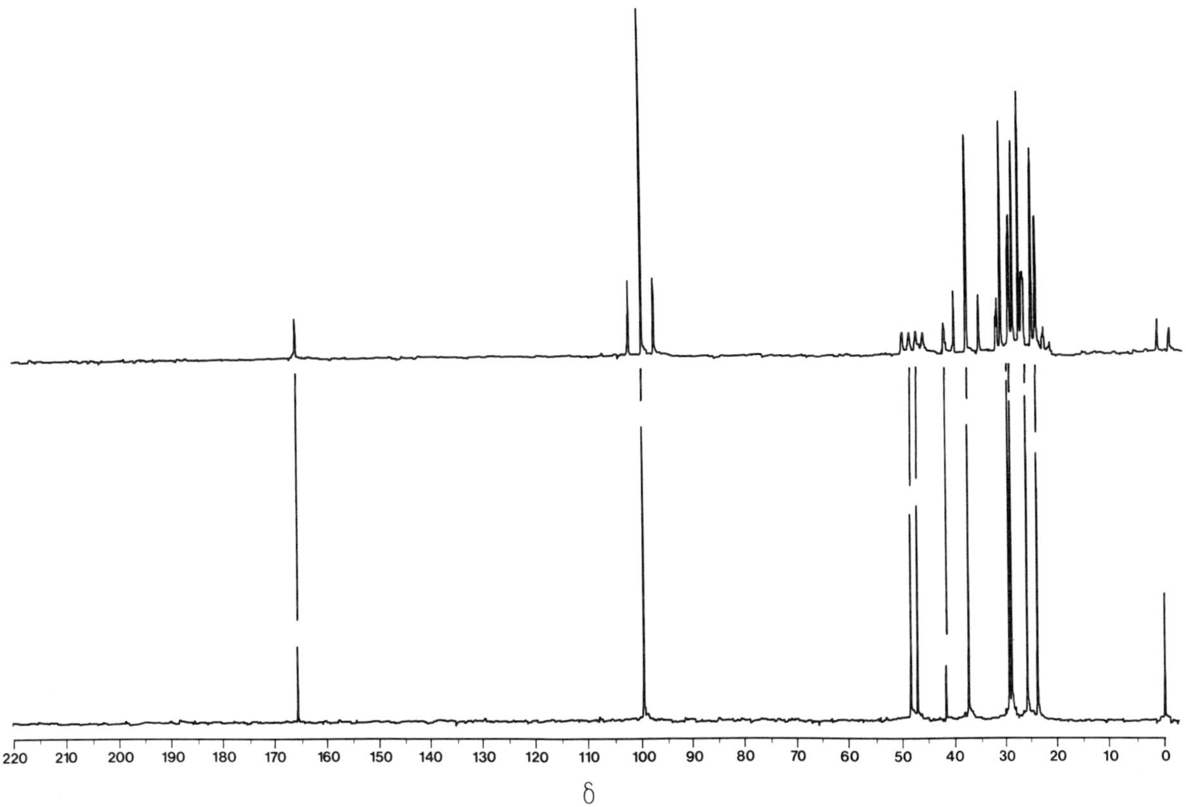
220 210 200 190 180 170 160 150 140 130 120 110 100 90 80 70 60 50 40 30 20 10 0
δ

Problem 2-51

2-51

$C_{10}H_{16}$

MS: m136, 121

IR: 1645 cm^{-1}

UV: none

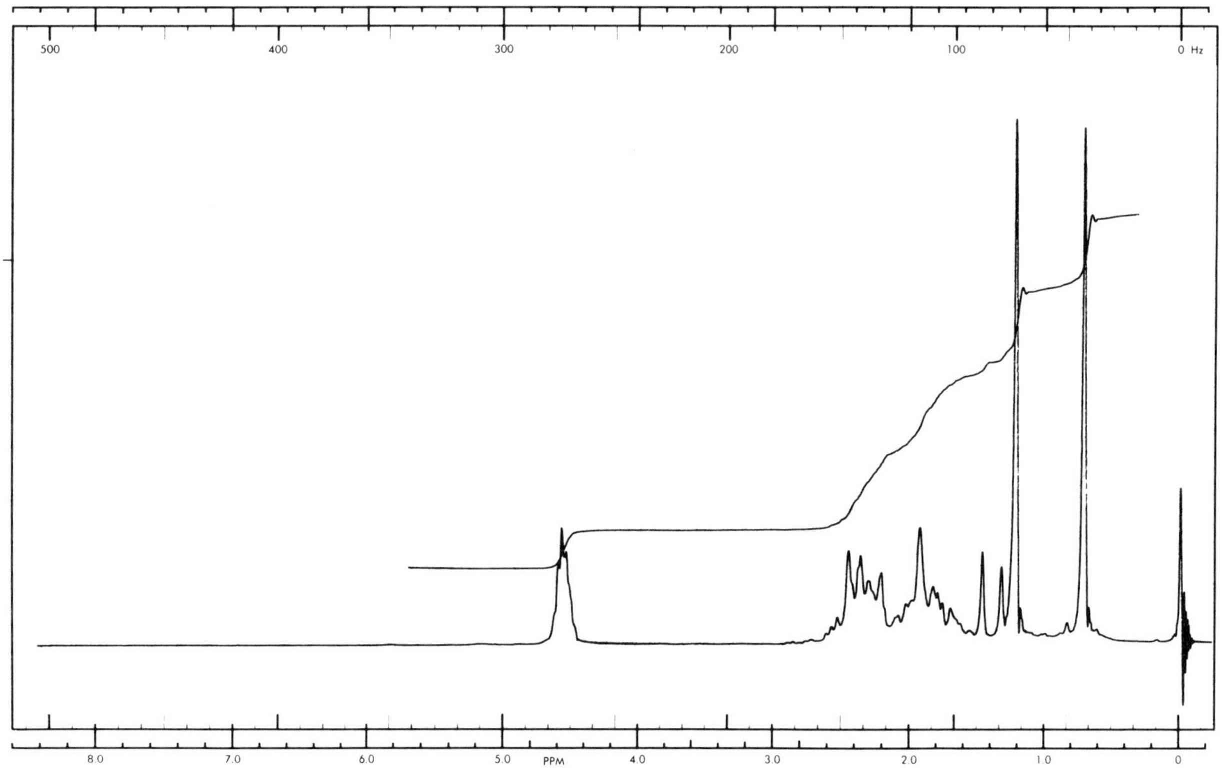

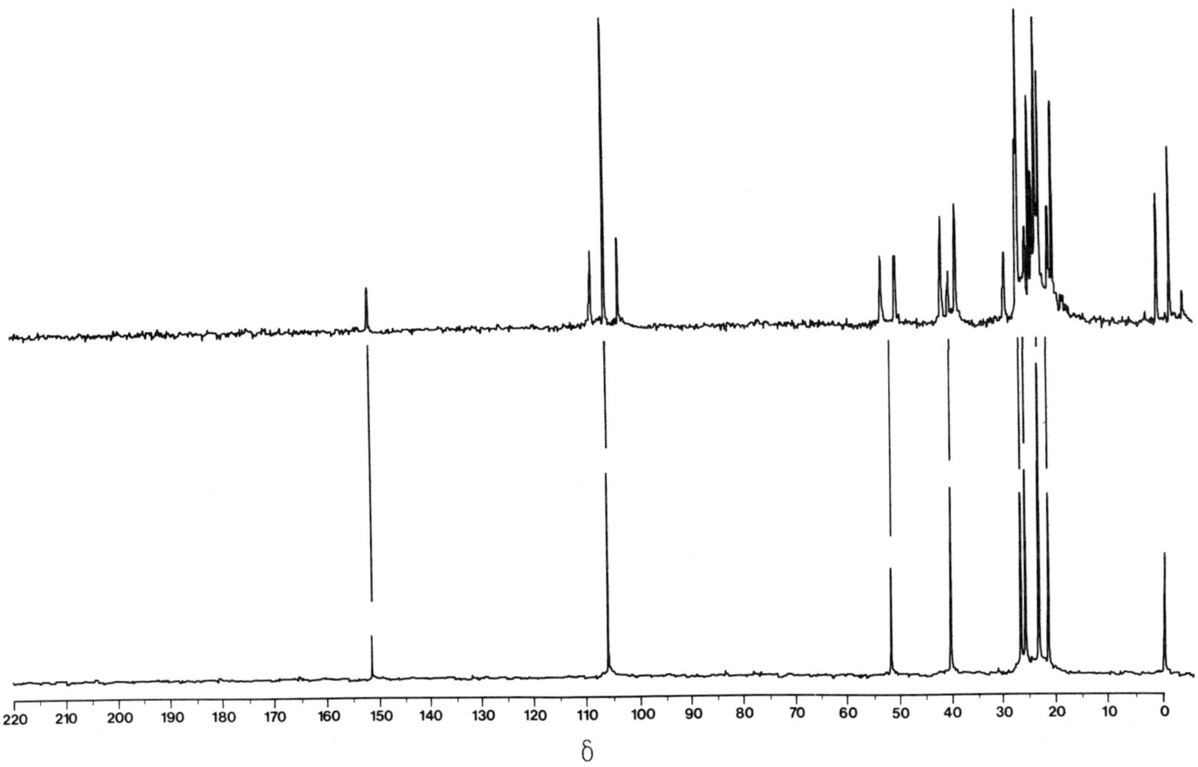

220 210 200 190 180 170 160 150 140 130 120 110 100 90 80 70 60 50 40 30 20 10 0
δ

Problem 2-52

2-52

$C_{11}H_{16}O$

MS: m164(w), 149, 91, 65, 57

IR: -

UV: none

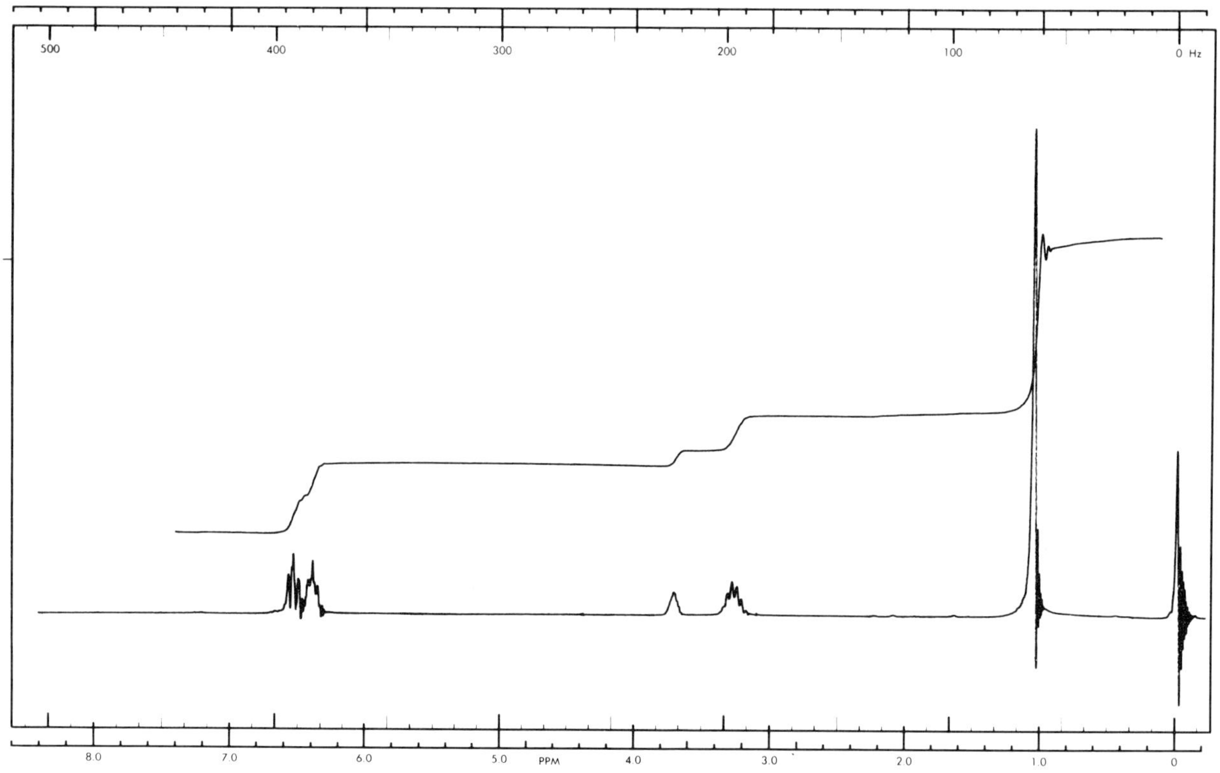

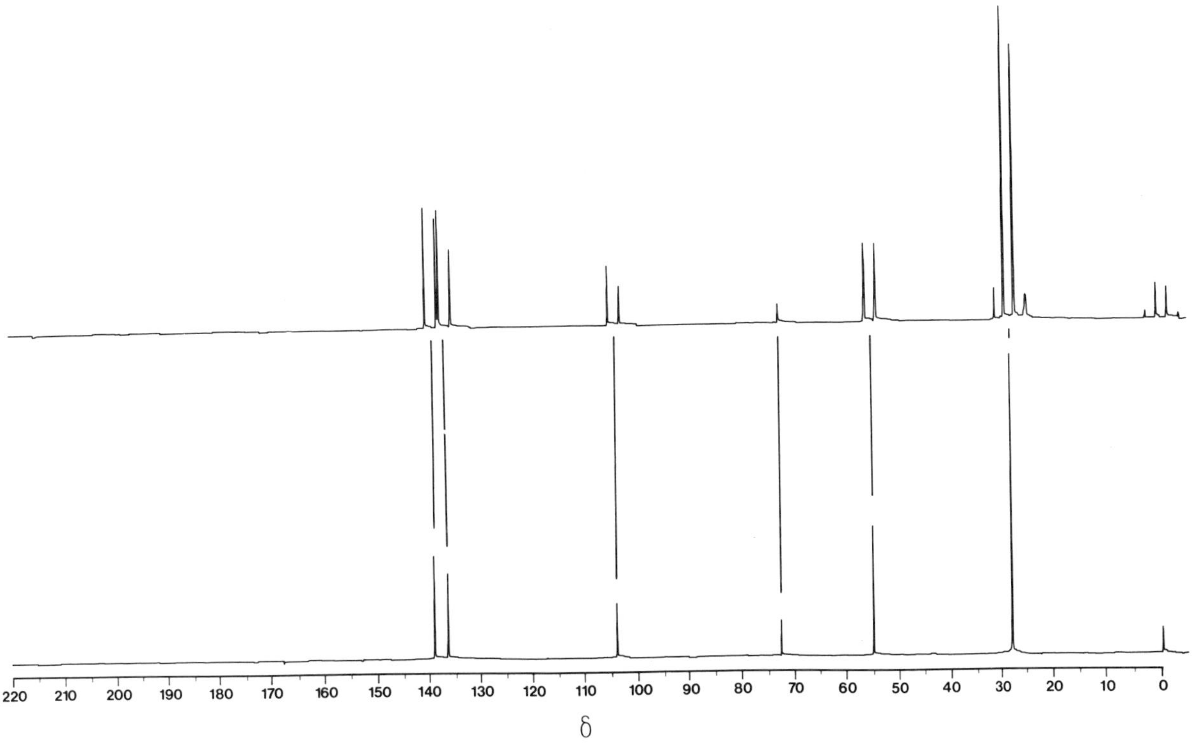
220
210
200
190
180
170
160
150
140
130
120
110
100
90
80
70
60
50
40
30
20
10
0
δ

Problem 2-53

2-53

$C_{16}H_{22}O_4$

MS: m278(w), 193, 85

IR: -

UV: Strong absorption

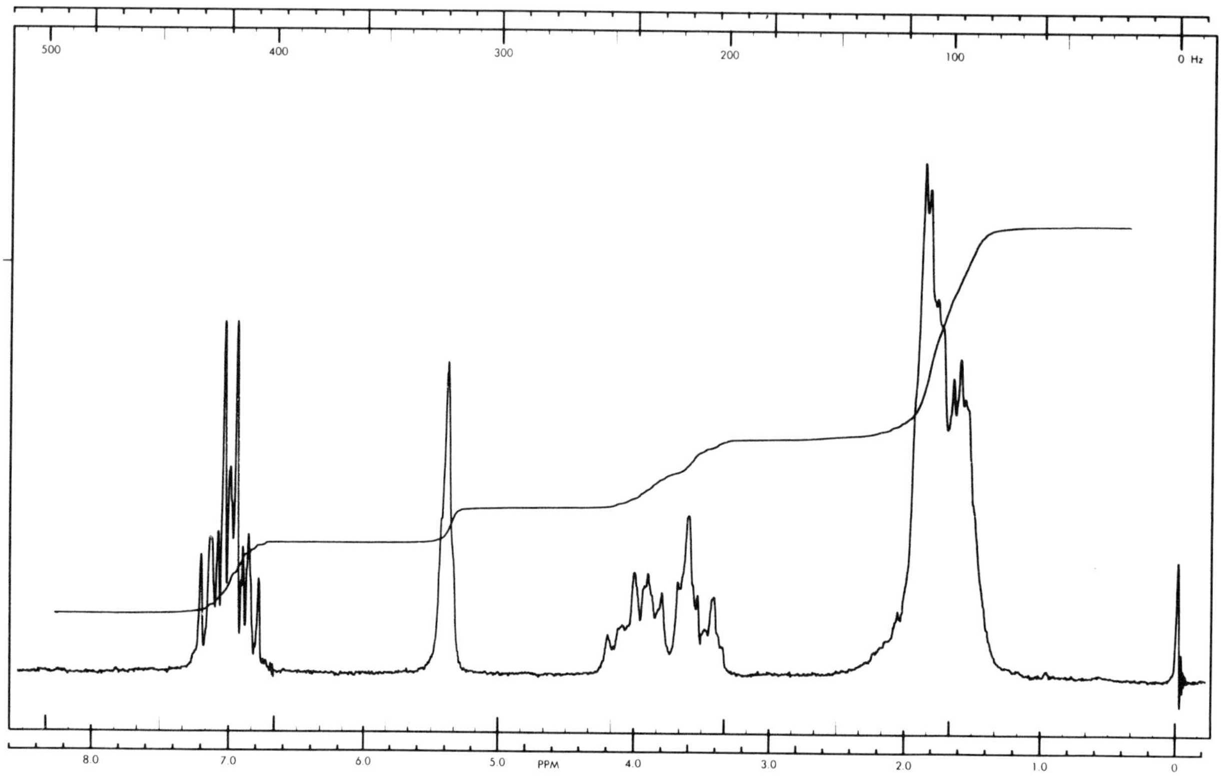

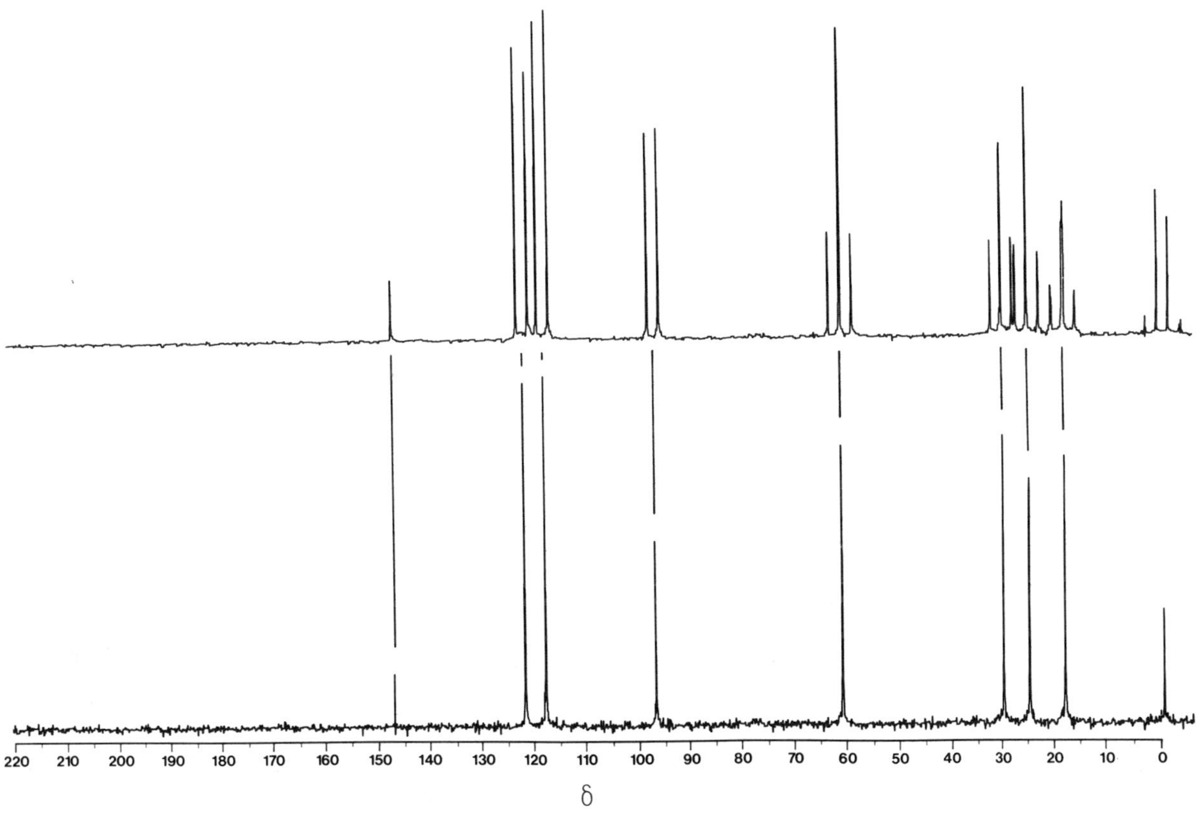
220
210
200
190
180
170
160
150
140
130
120
110
100
90
80
70
60
50
40
30
20
10
0
δ

Problem 2-54

2-54

C_6H_9NOS

MS: m143, 125, 112, 31

IR: 3300 cm^{-1}

UV: Strong absorption

PMR and CMR: d_6-Me_2CO + TMS

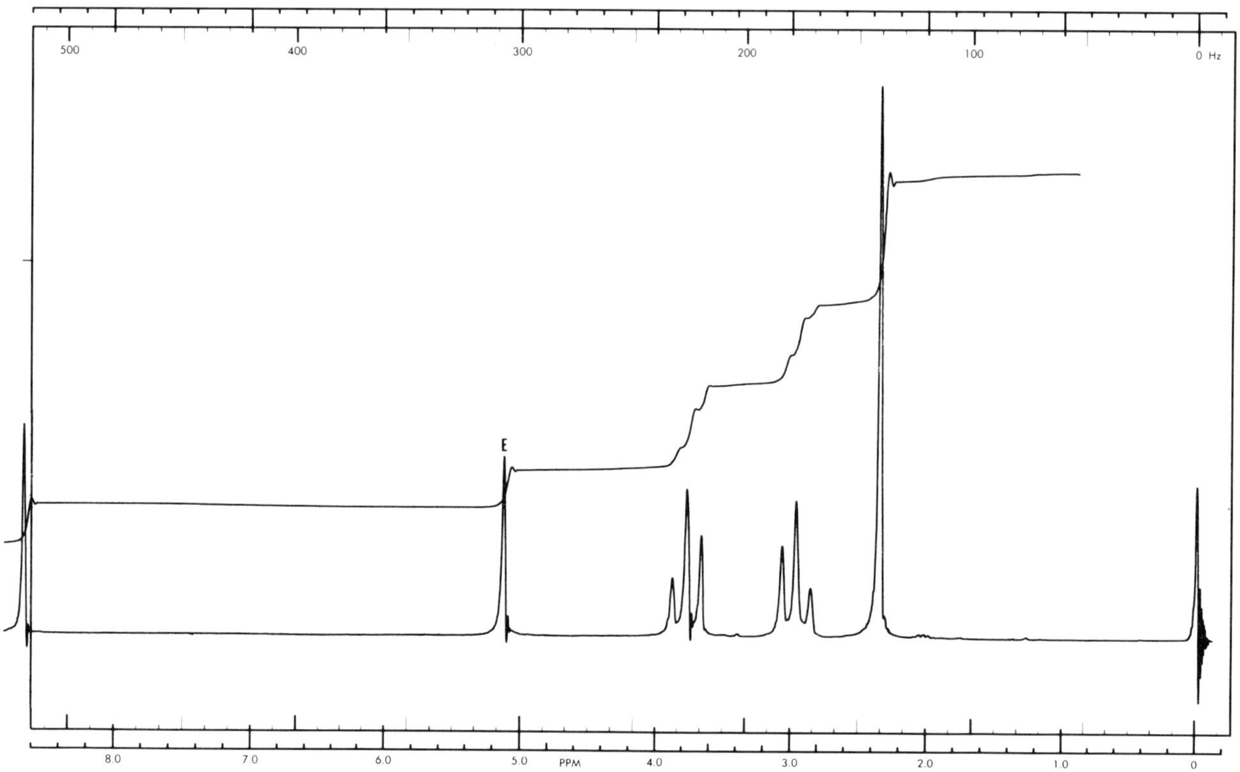

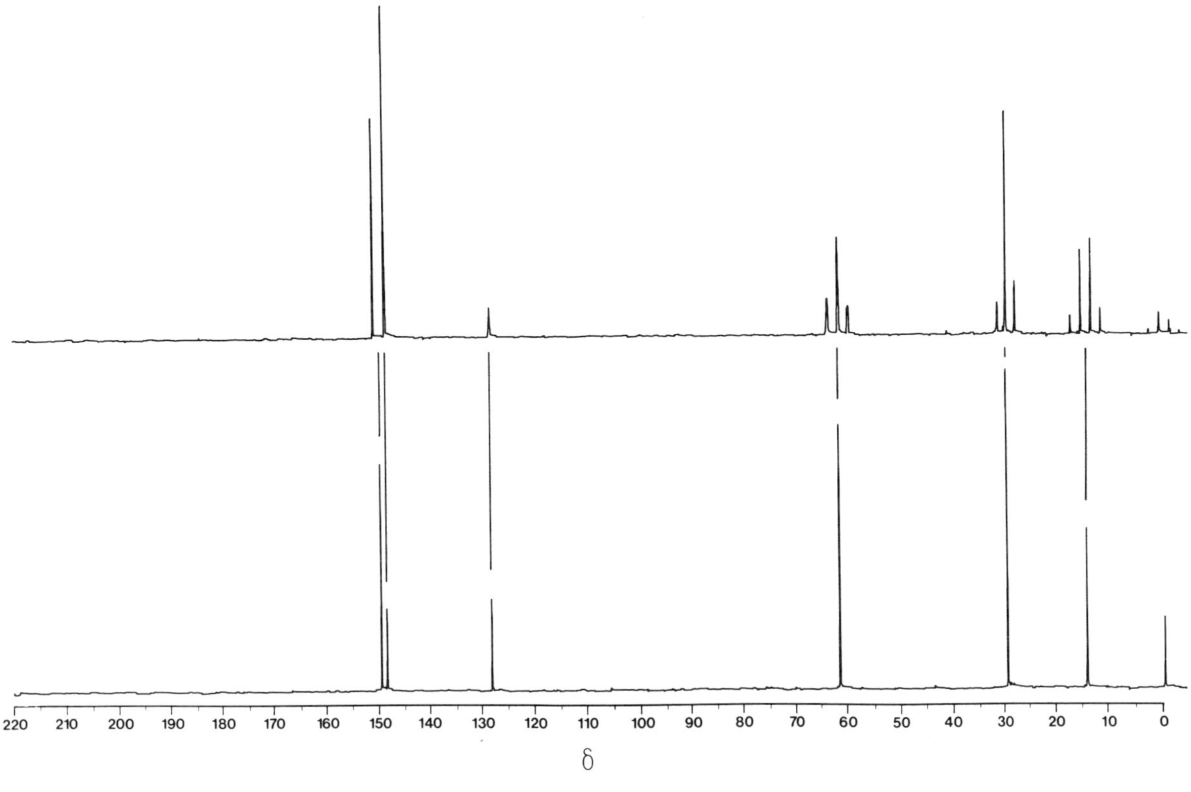
220
210
200
190
180
170
160
150
140
130
120
110
100
90
80
70
60
50
40
30
20
10
0
δ

Problem 2-55

2-55	MS:	m238, 221, 207, 157, 143, 31
$C_8H_{18}N_2O_4S$	IR:	2900 cm^{-1} (broad)
	UV:	none
	PMR:	D_2O + DSS
	CMR:	D_2O + dioxane

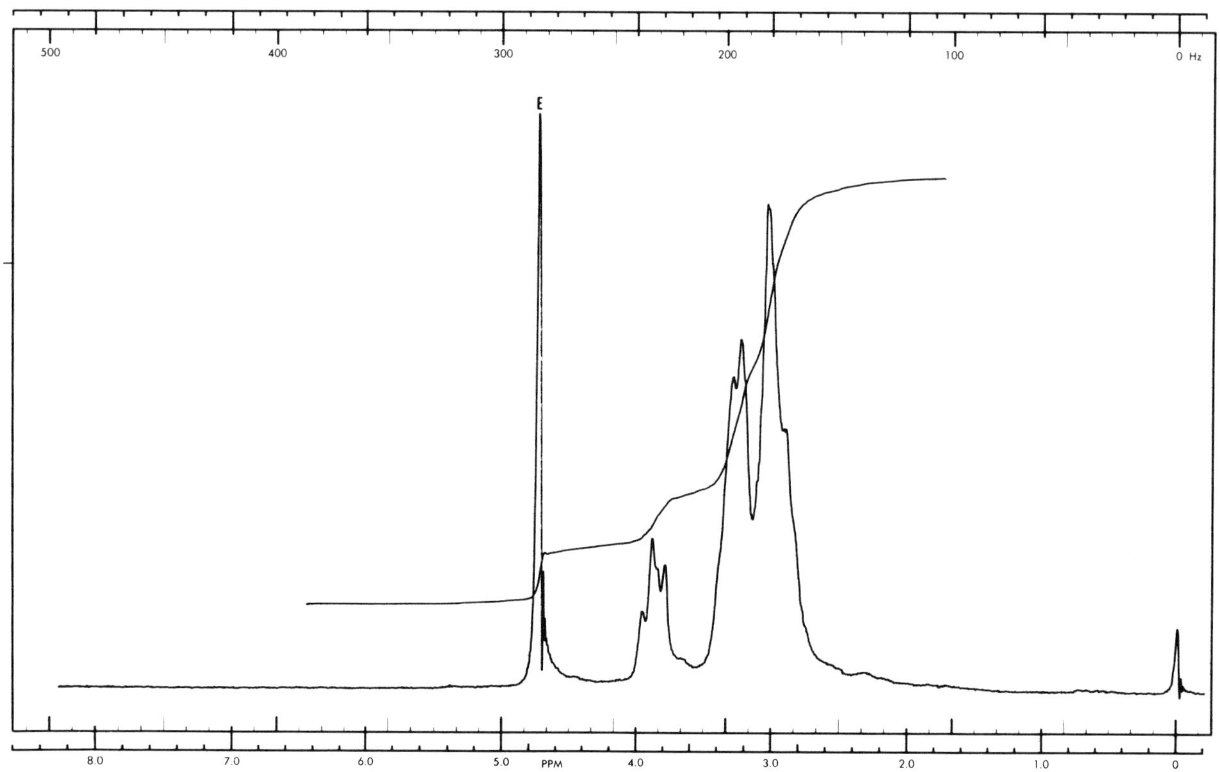

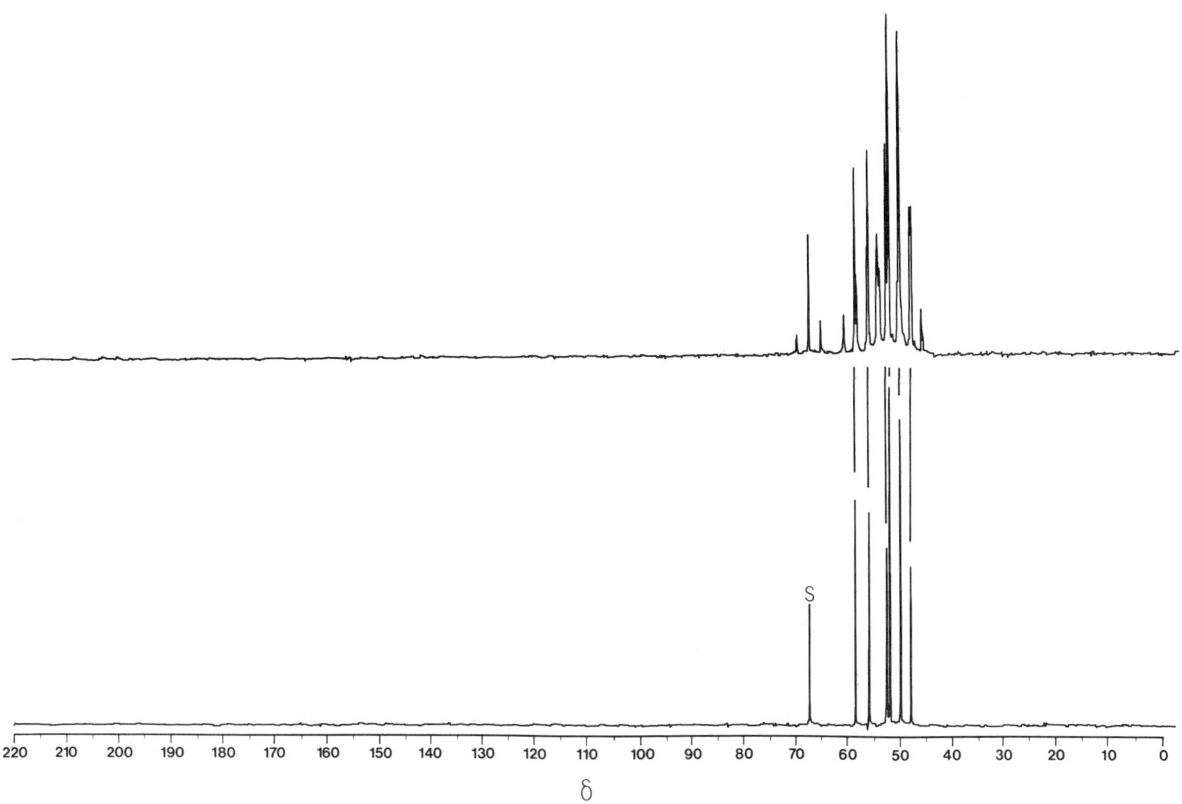
S
220 210 200 190 180 170 160 150 140 130 120 110 100 90 80 70 60 50 40 30 20 10 0
δ

Problem 2-56

2-56

$C_{10}H_{14}N_5O_{13}P_3Na_2$

MS: Not volatile

IR: 3600-2700 (broad), 1690, 1620 cm^{-1}

UV: 260 (ε 15,100)

PMR: D_2O + DSS; upper scan offset 150 Hz

CMR: D_2O + dioxane

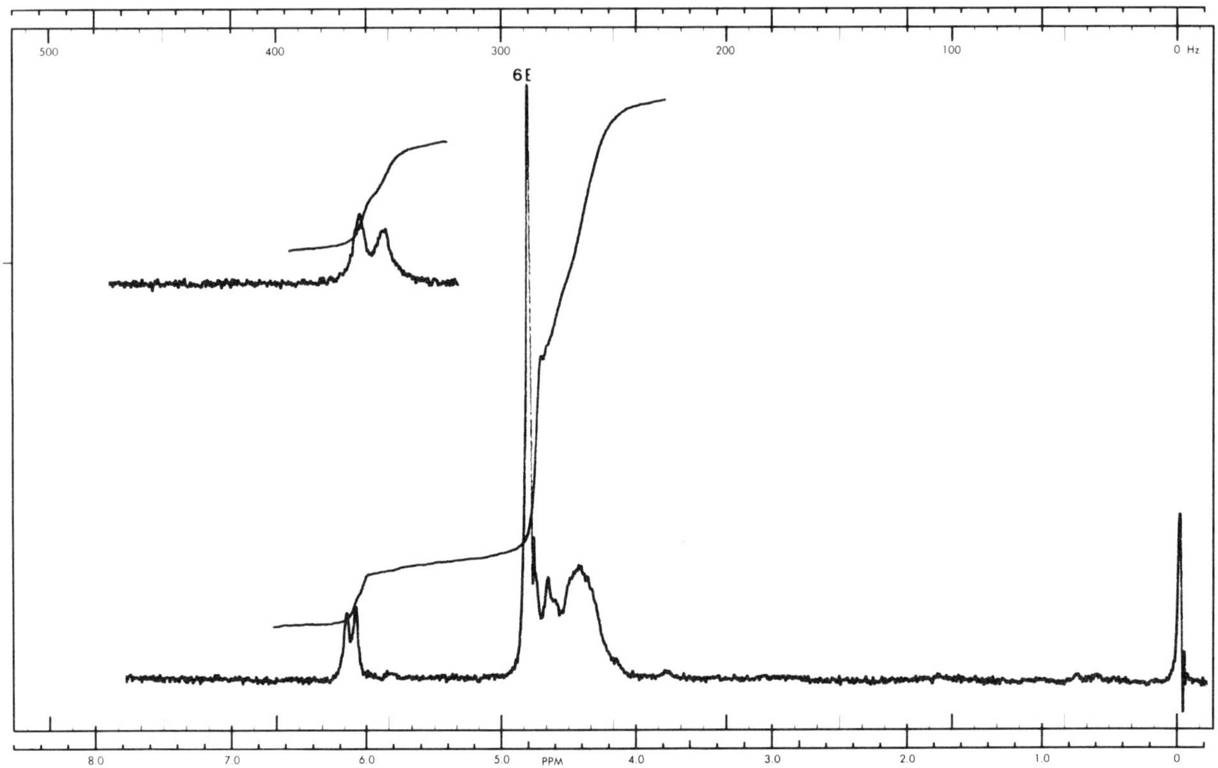

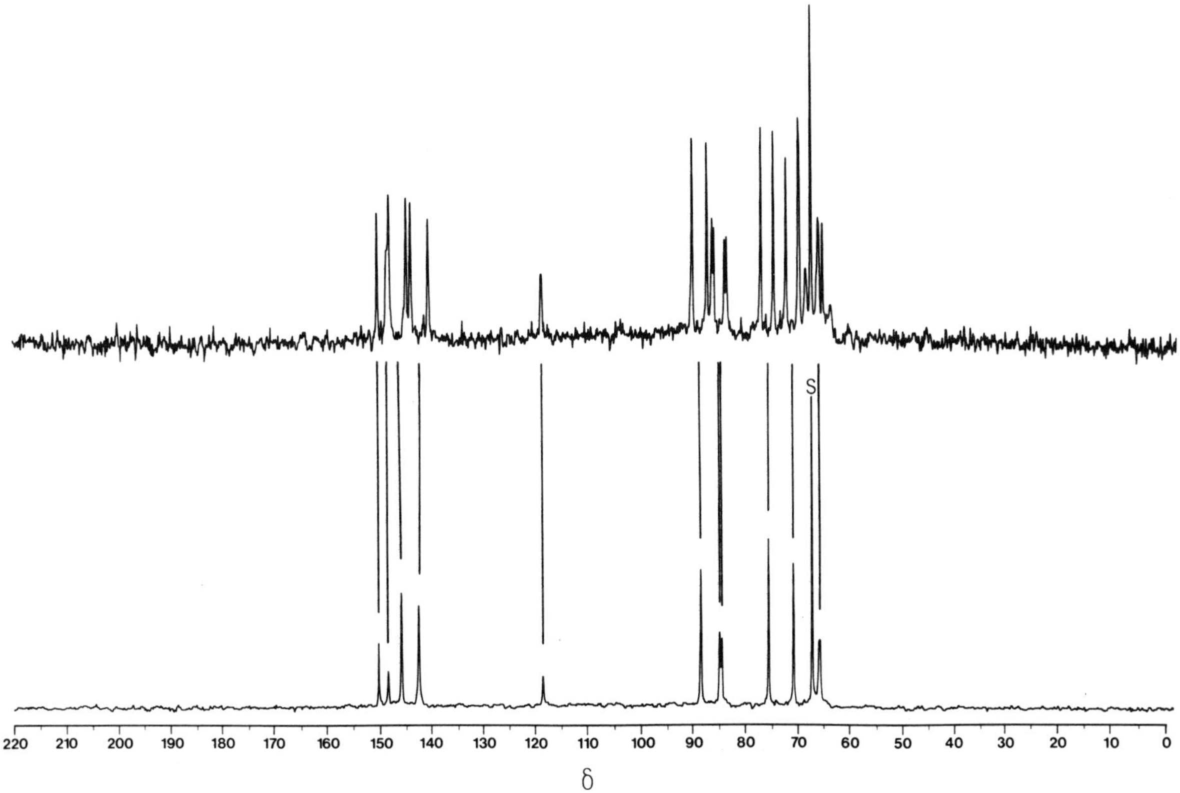
S
220
210
200
190
180
170
160
150
140
130
120
110
100
90
80
70
60
50
40
30
20
10
0
δ

Problem 2-57

2-57

$C_8H_{10}N_4O_2$

MS: m194, 179, 167, 166, 137

IR: 1710, 1670 cm^{-1}

UV: Strong absorption

CMR: 10 sec delay

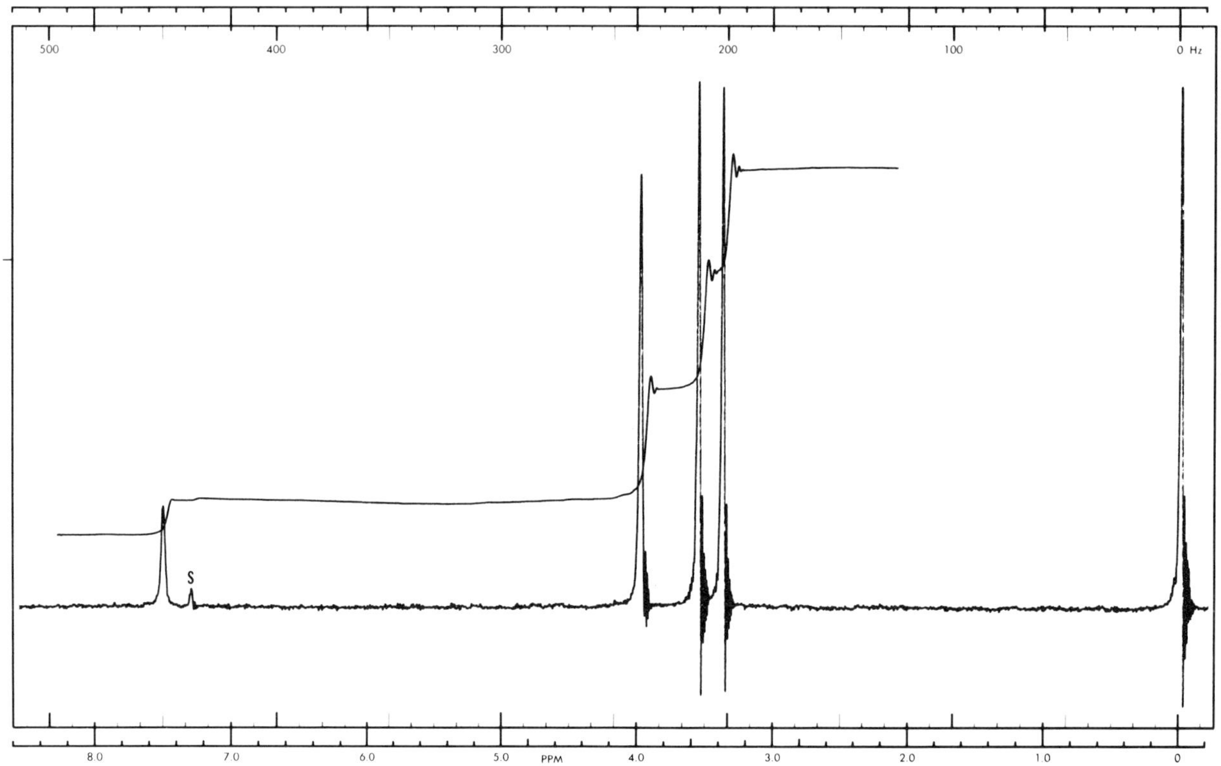

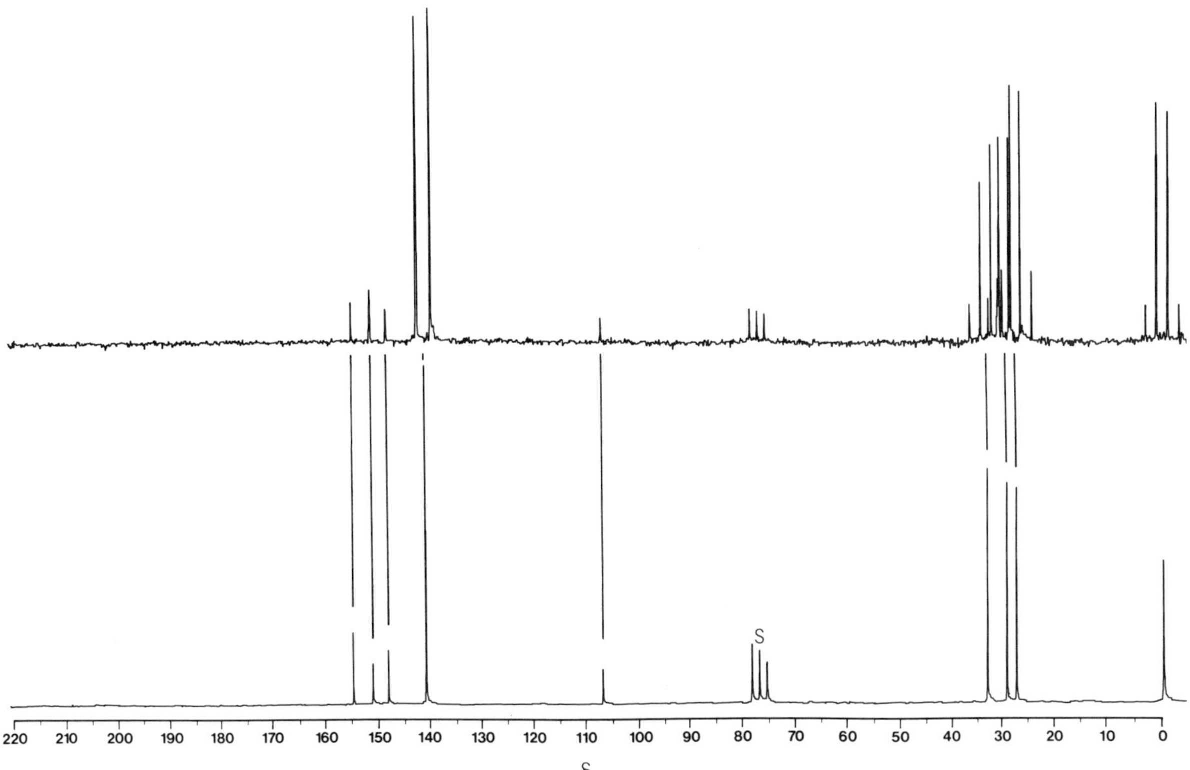
S
220
210
200
190
180
170
160
150
140
130
120
110
100
90
80
70
60
50
40
30
20
10
0
δ

Problem 2-58

2-58

$C_{21}H_{29}N_2O_5^+Cl^-$

MS: Free base m388, 195, 178, 167, 58

IR: 2600 (broad), 1640, 1600 cm^{-1}

UV: 213 (ε 47,100), 260 (ε 12,400)

PMR: D_2O + DSS

CMR: D_2O + dioxane

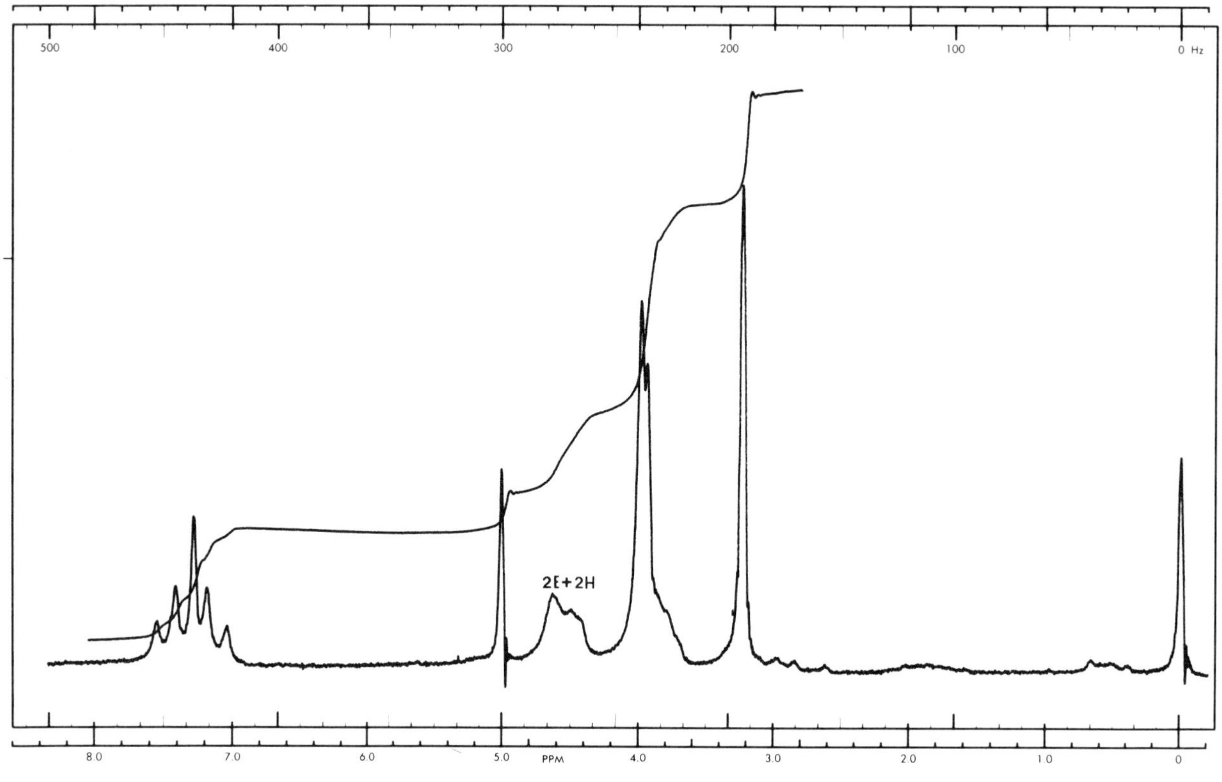

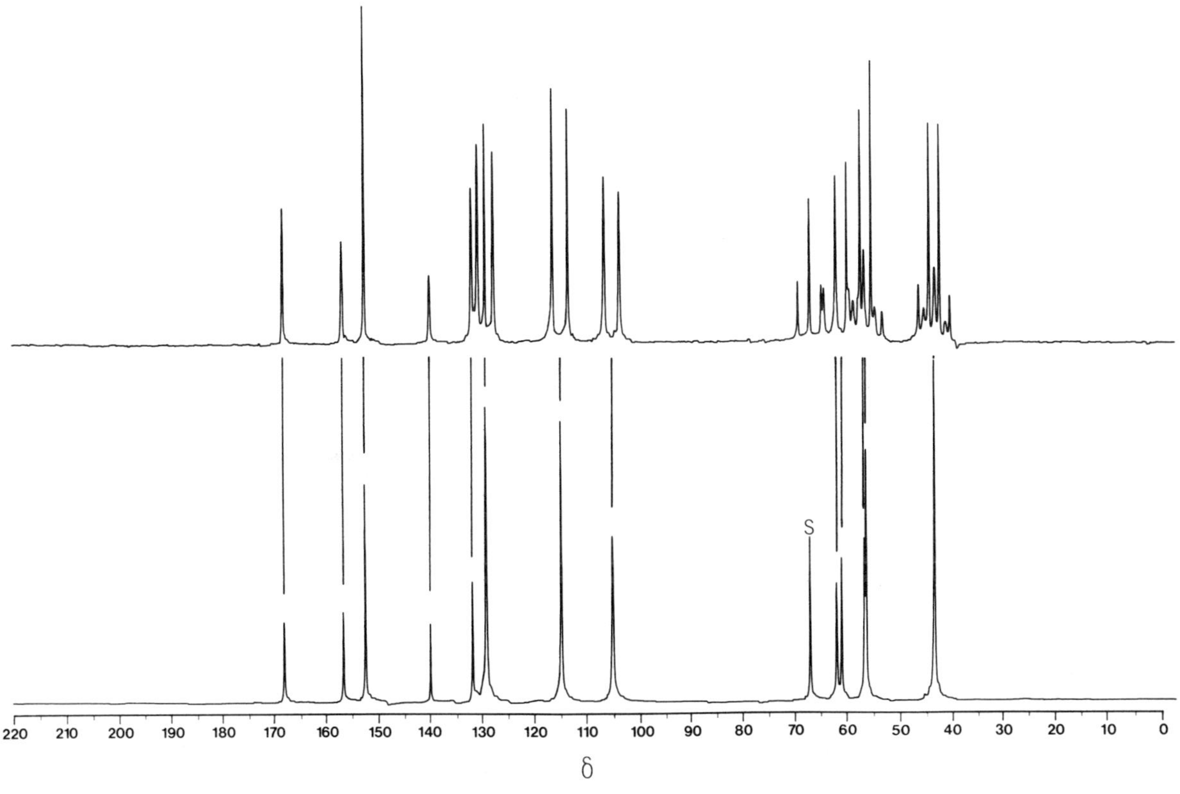
S
220
210
200
190
180
170
160
150
140
130
120
110
100
90
80
70
60
50
40
30
20
10
0
δ

Problem 2-59

2-59

$C_{15}H_{25}N_2O_2^+Cl^-$

MS: Free base m264, 176, 58, 57

IR: 3400, 2650, 1700, 1600 cm^{-1}

UV: 226 (ε 7090), 309 (ε 28,200)

PMR: D_2O + DSS

CMR: D_2O + dioxane

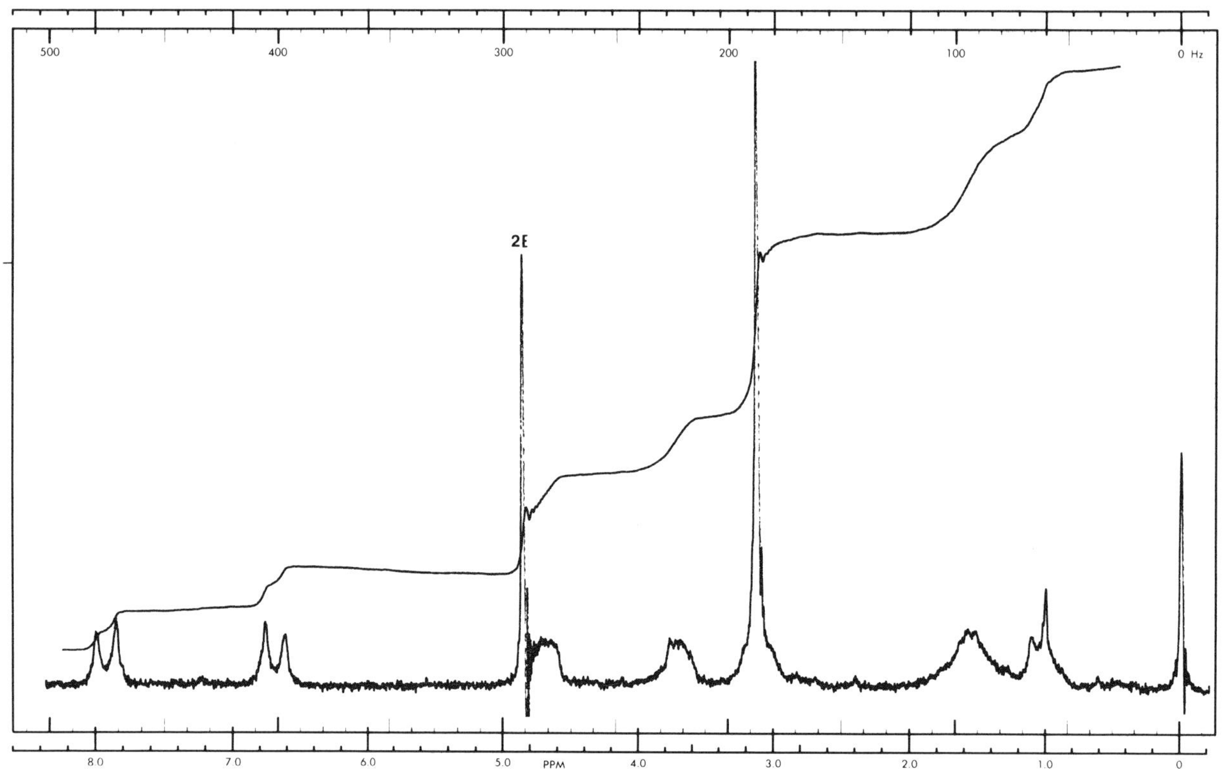

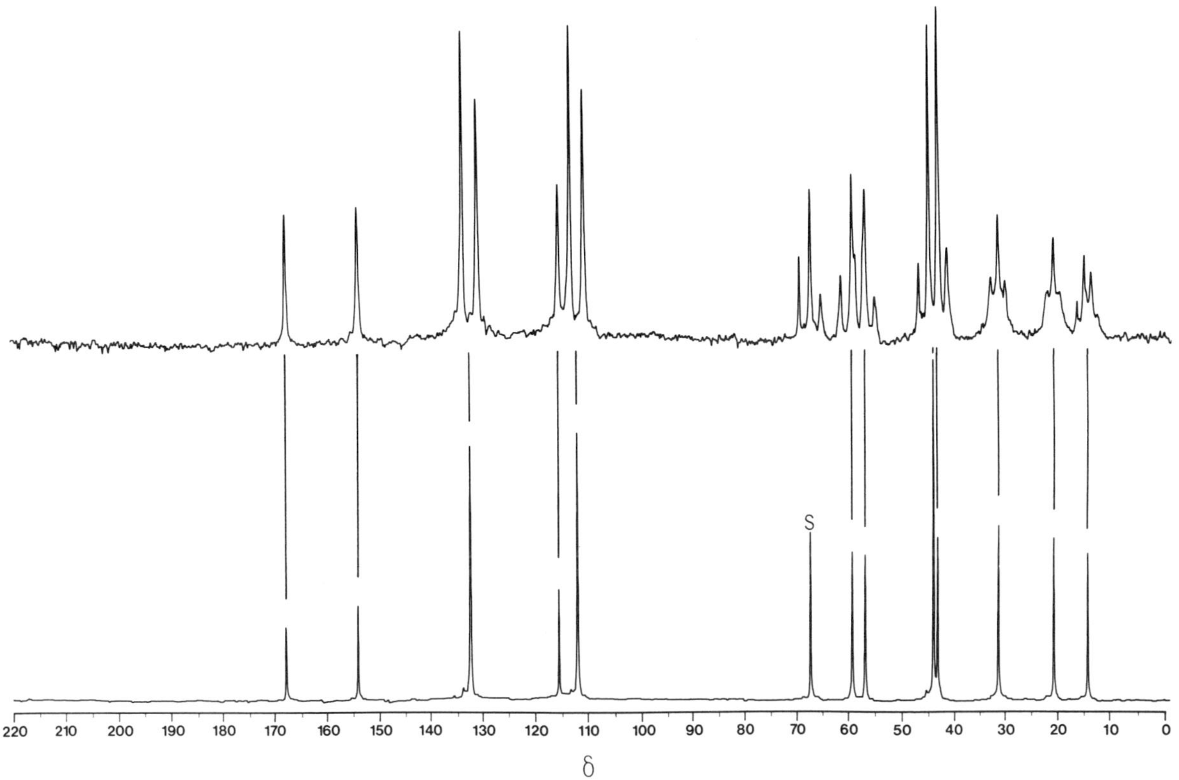
S
220 210 200 190 180 170 160 150 140 130 120 110 100 90 80 70 60 50 40 30 20 10 0
δ

Problem 2-60

2-60

$C_{19}H_{24}O_2$

MS: m284, 256, 255

IR: 1743 cm^{-1}

UV: Strong absorption

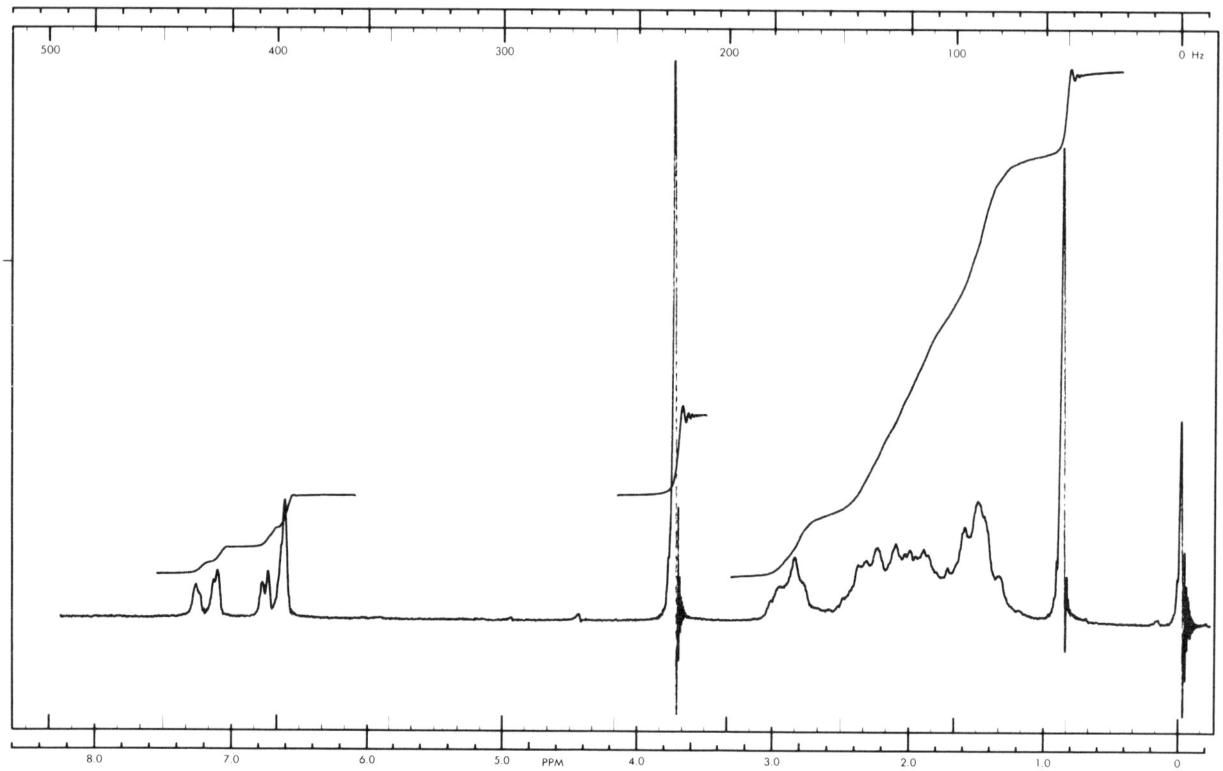

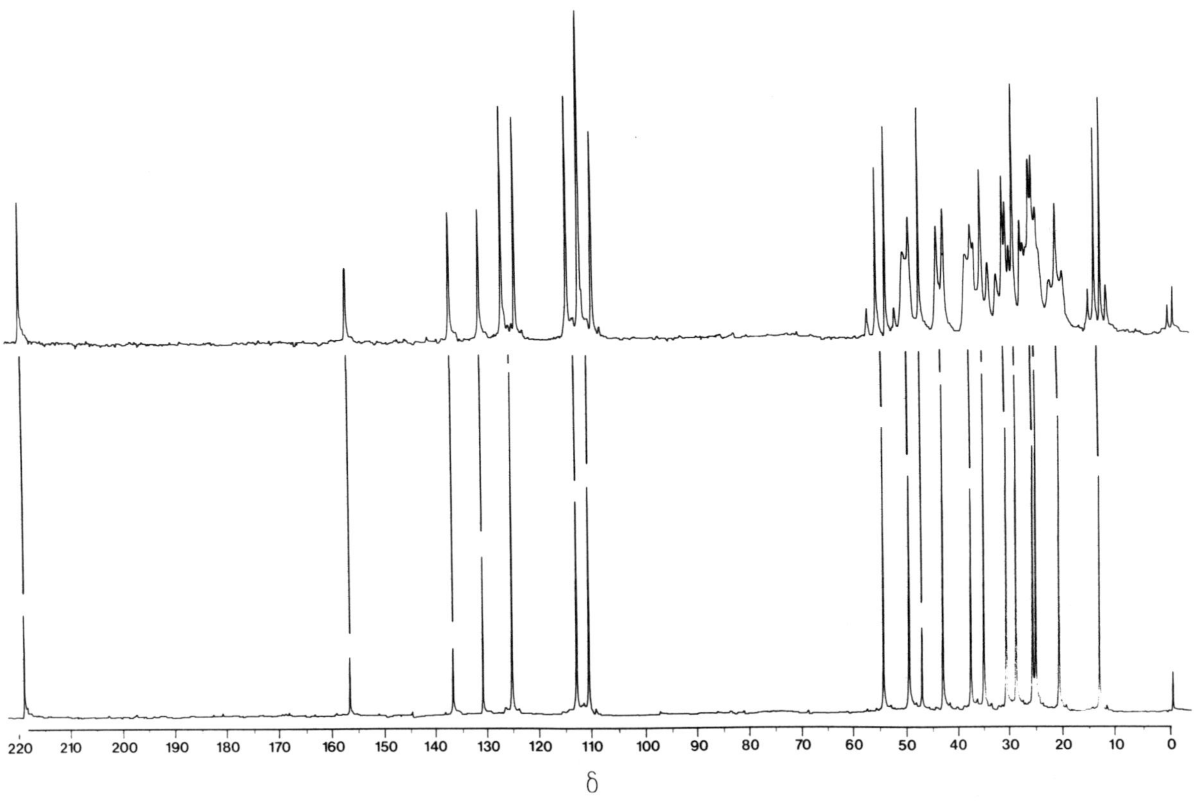
220 210 200 190 180 170 160 150 140 130 120 110 100 90 80 70 60 50 40 30 20 10 0
δ

Problem 2-61

2-61 MS: m332, 299, 297, 168, 166

$C_{18}H_{18}Cl_2N_2$ IR: -

UV: Strong absorption

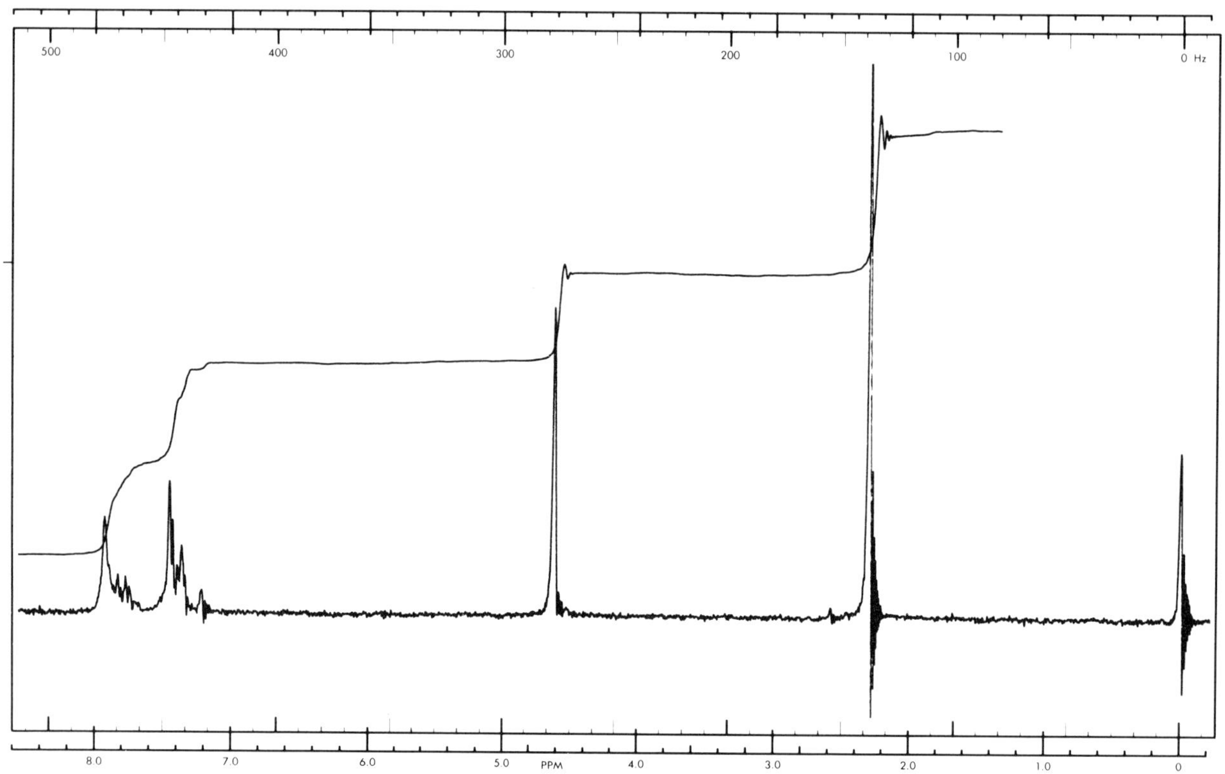

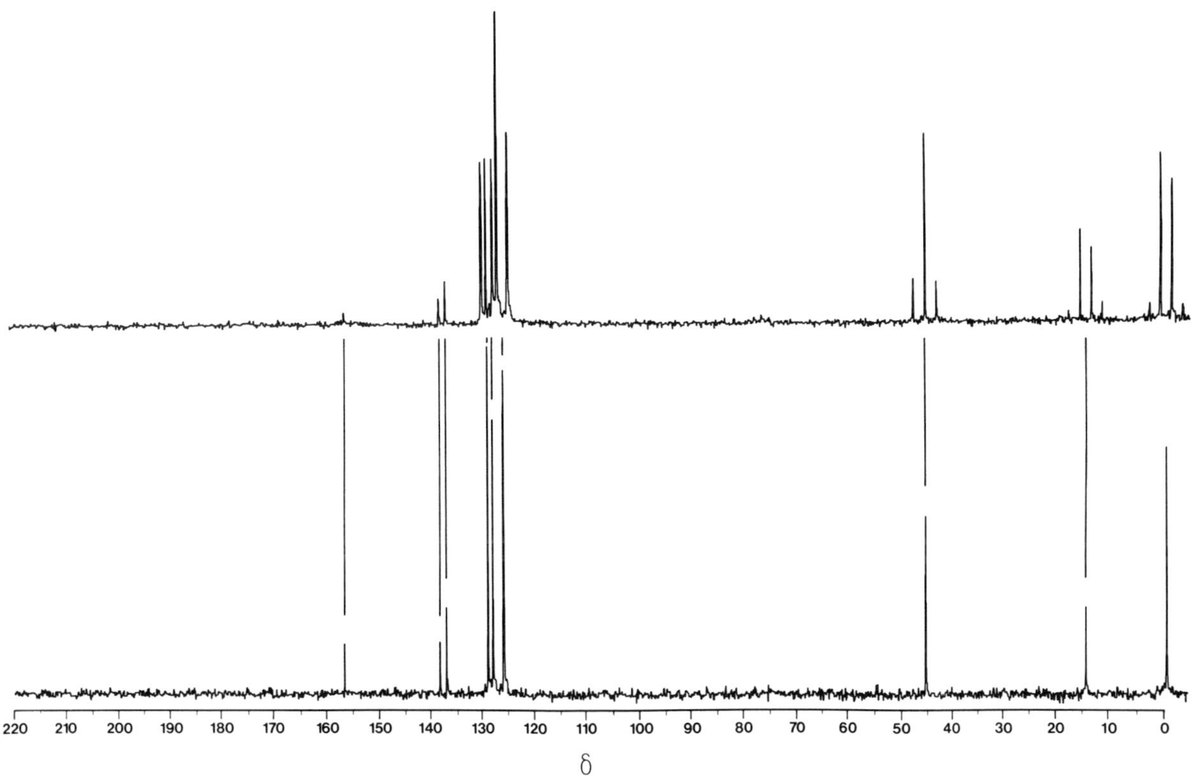
220
210
200
190
180
170
160
150
140
130
120
110
100
90
80
70
60
50
40
30
20
10
0
δ

Problem 2-62

2-62

$C_{20}H_{20}N_2$

MS: m288(s), 273, 211, 185, 103, 77

IR: 1625 cm^{-1}

UV: 248 (ε 12,000)

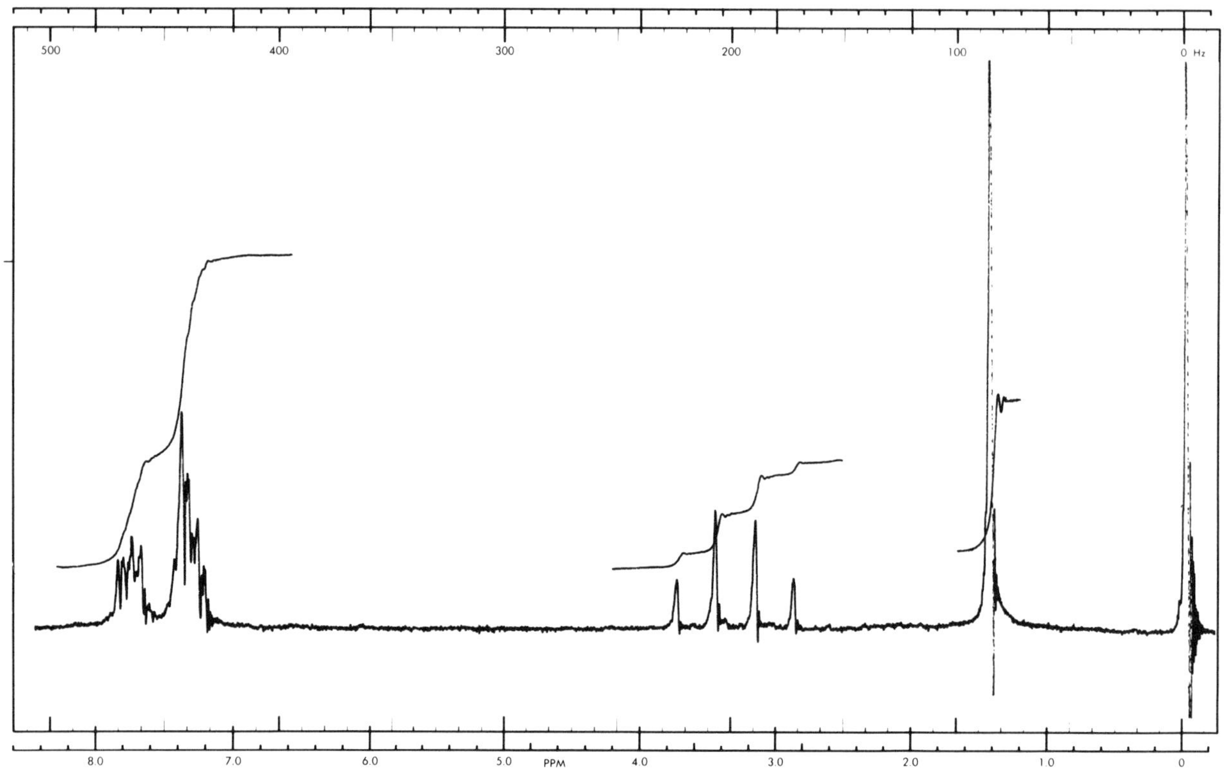

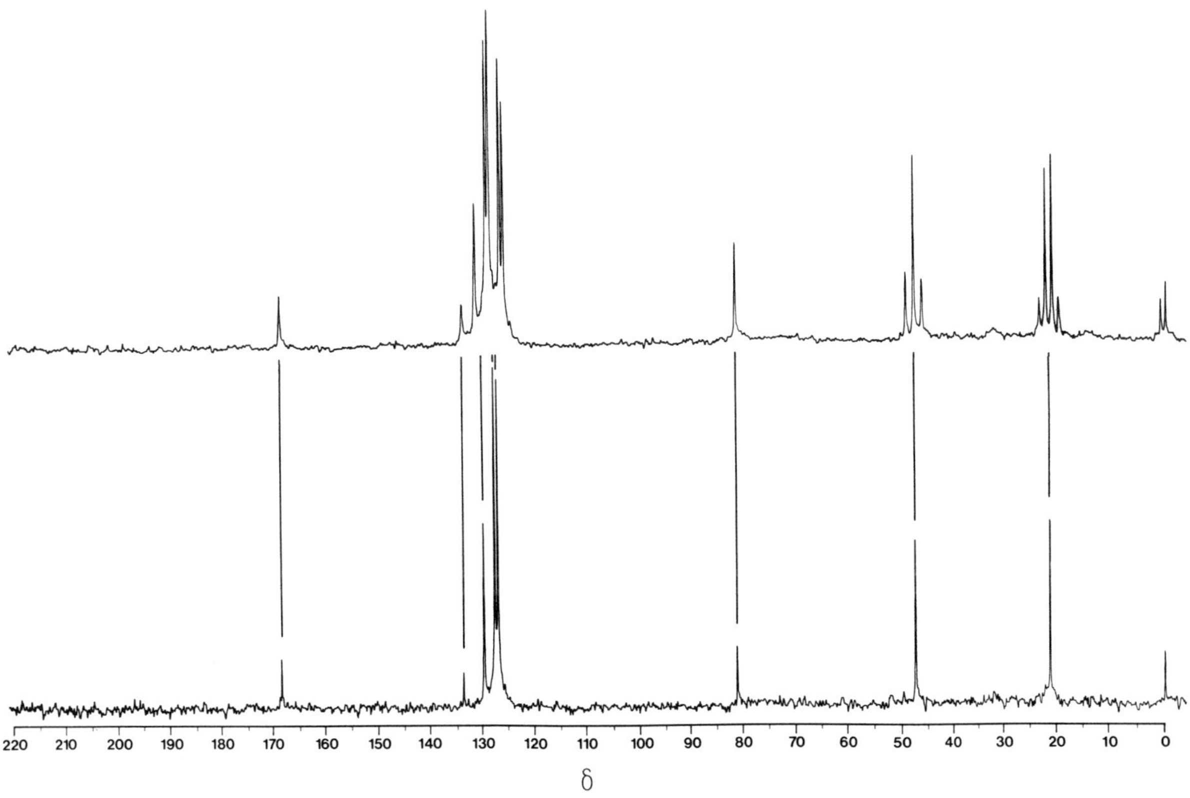
220
210
200
190
180
170
160
150
140
130
120
110
100
90
80
70
60
50
40
30
20
10
0
δ

Problem 2-63

2-63

$C_{10}H_{16}O_4S$

MS: m232, 190, 151, 42

IR: 3000-1600 (very broad), 1740 cm^{-1}

UV: 284 (ε 20)

PMR: D_2O + DSS

CMR: D_2O + dioxane; 5 sec delay

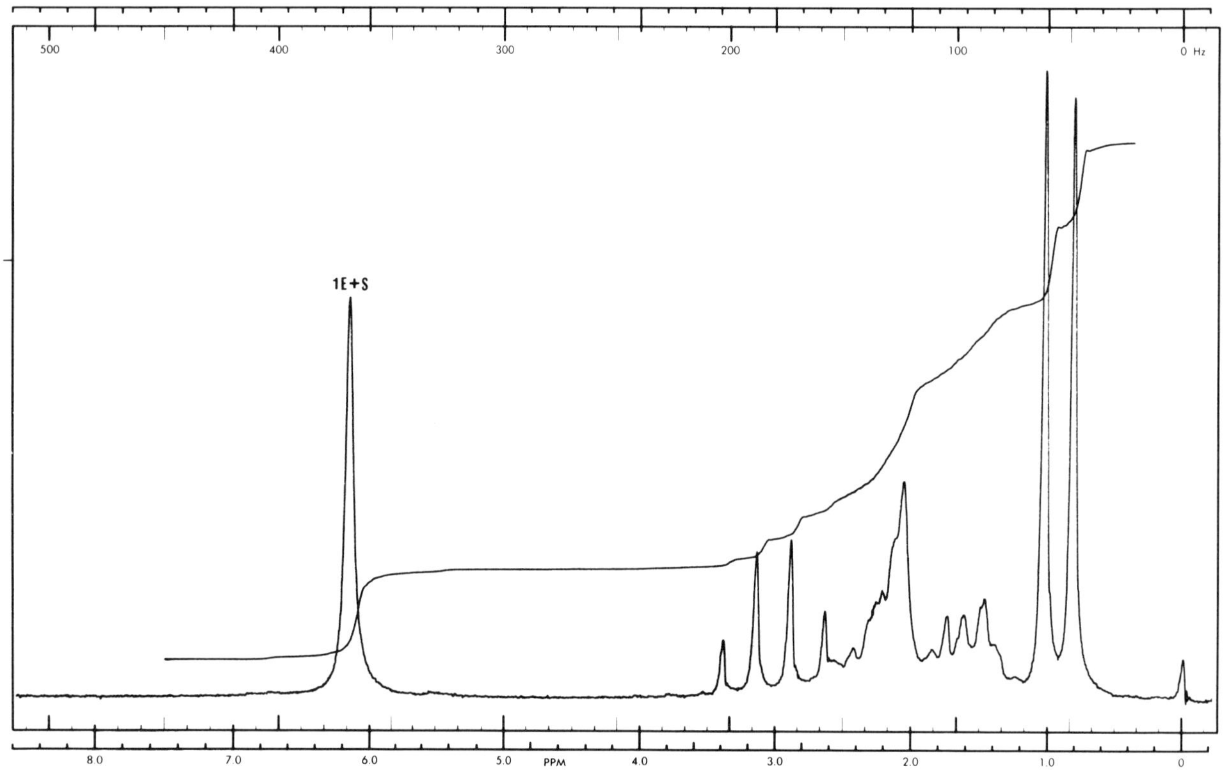

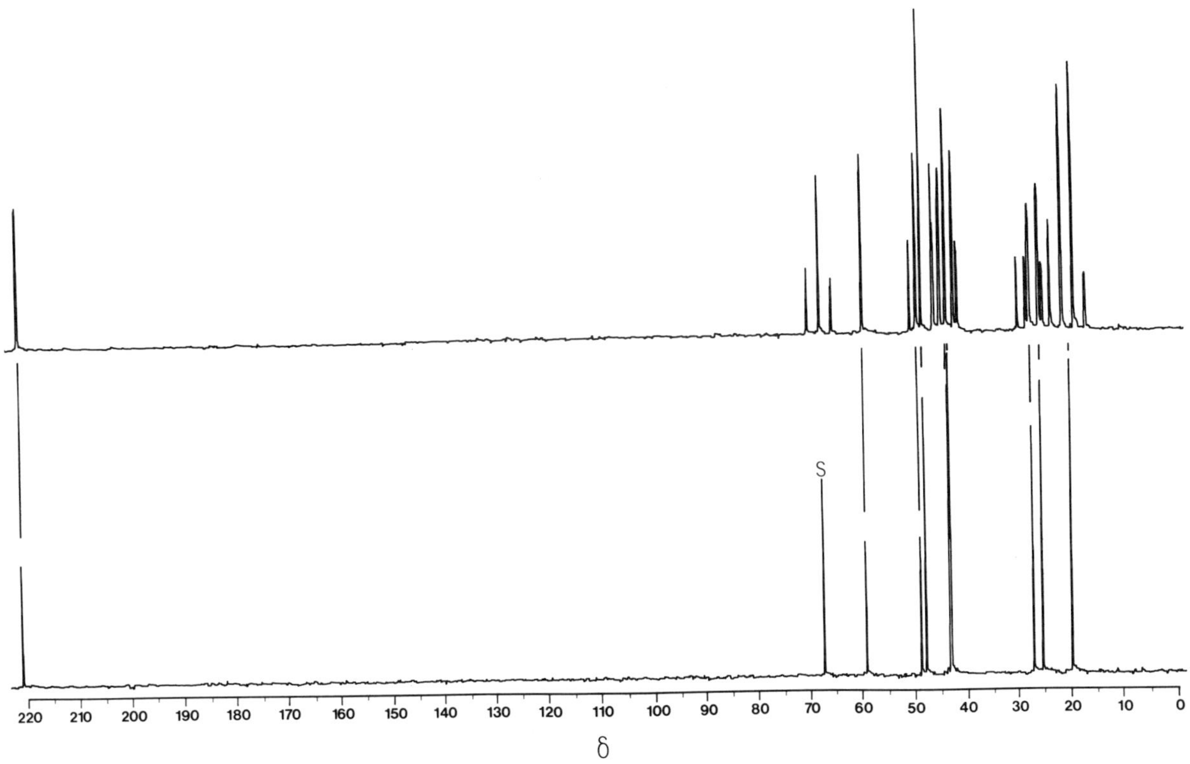
S
220
210
200
190
180
170
160
150
140
130
120
110
100
90
80
70
60
50
40
30
20
10
0
δ

Problem 2-64

2-64

$C_{12}H_{18}N_4OS^{2+}2Cl^-$

MS: Not volatile

IR: 3400, 3000 (broad), 1650, 1600 cm^{-1}

UV: 243 (ε 11,300), 260 (ε 10,500)

PMR: D_2O + DSS; upper scan offset 100 Hz

CMR: D_2O + dioxane

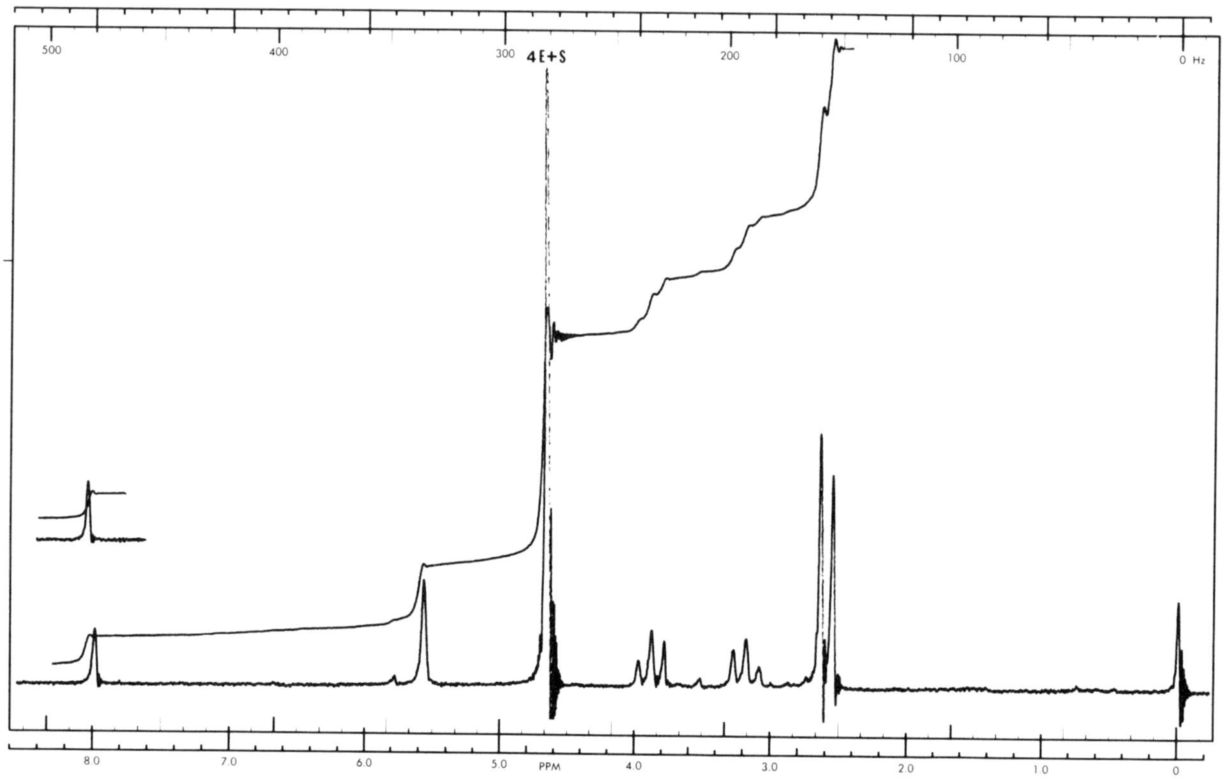

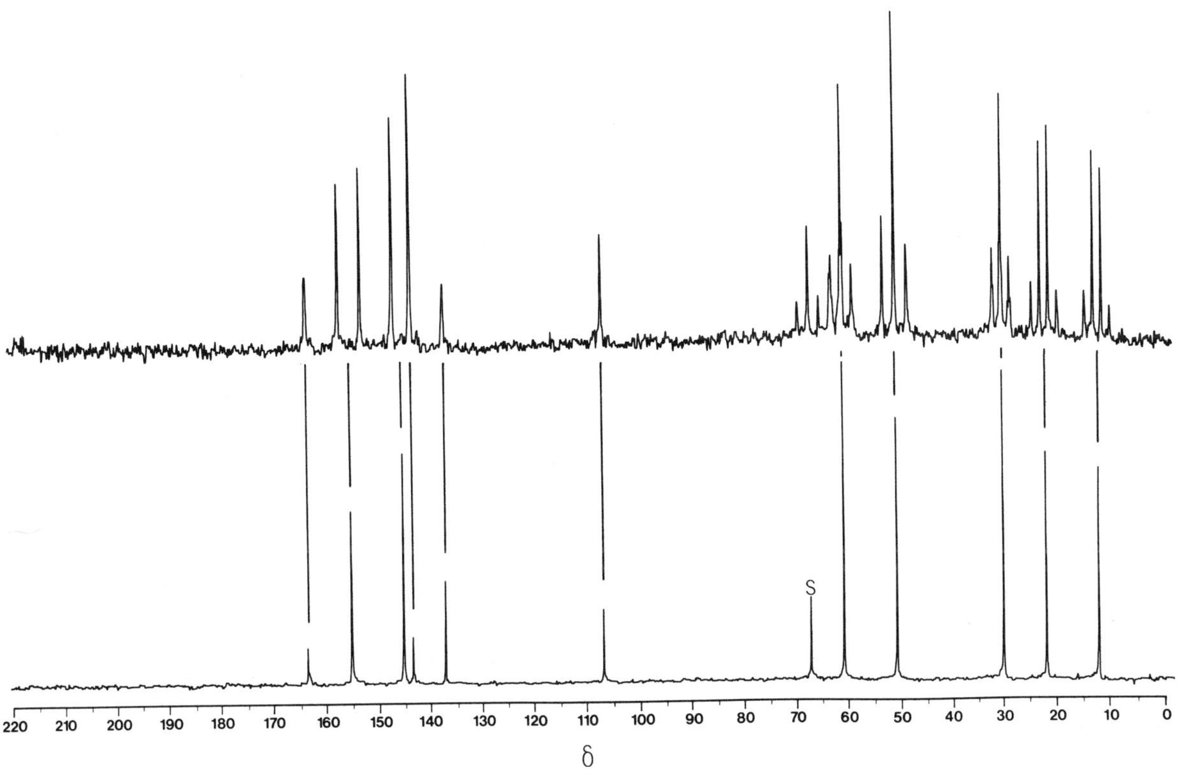
S
220 210 200 190 180 170 160 150 140 130 120 110 100 90 80 70 60 50 40 30 20 10 0
δ

Problem 2-65

2-65

$C_{13}H_{16}O_4$

MS: m236(w), 191, 120, 116

IR: 2900 (broad), 1700 cm^{-1}

UV: none

PMR and CMR: d_6-DMSO + TMS

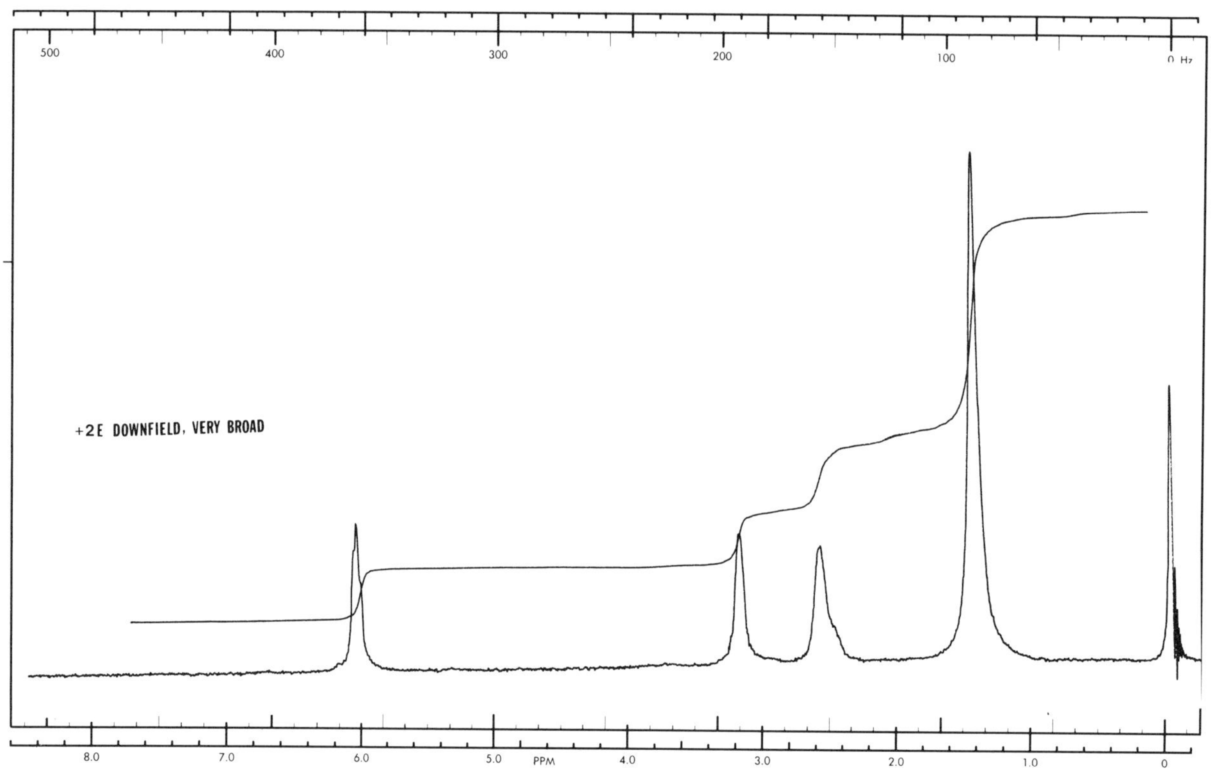

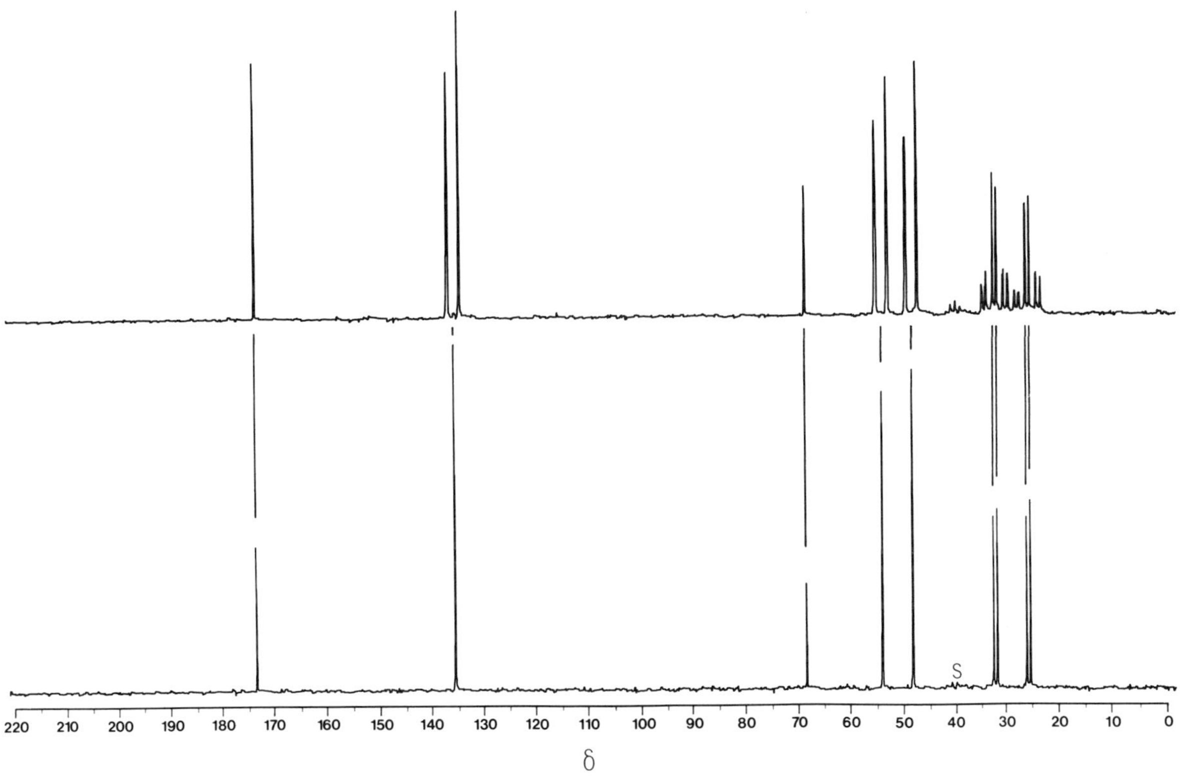
S
220
210
200
190
180
170
160
150
140
130
120
110
100
90
80
70
60
50
40
30
20
10
0
δ

References to Answers

The first reference given is to a volume of *Chemical Abstracts, Collective Formula Index,* if the compound was found there. The *Chemical Abstracts* registry number given in brackets at the end of the reference may help to confirm that the correct answer has been located in these *Indices.* **A** refers to Aldrich Chemical Co., Inc., catalog; the molecular formula index of this catalog will generally provide the quickest way to find the compound. For those compounds included, the formula and *Chemical Abstracts* registry number indices of the *Merck Index* will also be fast.

Problem No.	Reference
1-1	**7,** 511F, column 2, 44th cmpd.
1-2	**8,** 1553F, column 2, 20th cmpd. [504-15-4]; **A** 0-820-8.
1-3	**8,** 8th cmpd. [2938-98-9].
1-4	**8,** column 2, 3rd cmpd. [280-57-9]; **A** D2,780-2.
1-5	**8,** 2048F, column 2, 2nd cmpd. [1124-11-4]; **A** 18,393-8.
1-6	**8,** 5th cmpd. [15290-29-6].
1-7	**8,** 1113F, 13th cmpd. [120-80-9]; **A** 13,501-1.
1-8	**8,** 9th from last cmpd. [294-93-9]; **A** 19,490-5.
1-9	**8,** 2nd cmpd. [16177-48-3].
1-10	**8,** 4th cmpd. [486-25-9]; **A** F150-6.
1-11	**8,** 1223F, column 1, next to last cmpd. [279-49-2]; **A** 0-870-4.
1-12	**8,** 1223F, column 1, last cmpd. [286-20-4]; **A** C10,250-4.
1-13	**8,** 9th cmpd. [69-65-8]; **A** M235-7.
1-14	**8,** 1st cmpd. [930-57-4].
1-15	**8,** 1786F, 3rd cmpd. [4733-50-0]; **A** 14,113-5.
1-16	**8,** 23rd cmpd. [112-49-2]; **A** T5,980-3.
1-17	**8,** 3rd cmpd. [132-65-0]; **A** D3,220-2.
1-18	**8,** 867F, 7th from last line [9003-31-0]; **A** 18,214-1.

1-19 **8,** column 1, last cmpd. [98-92-0]; **A** N755-9.
1-20 **8,** 3rd cmpd. [3459-94-7].
1-21 **8,** 2nd cmpd. [4358-63-8].
1-22 **8,** 1st cmpd. [102-76-1]; **A** T4,370-2.
1-23 **8,** 4th cmpd. [6373-50-8].
1-24 **8,** 5875F, column 3, 1st cmpd. [128-37-0]; **A** D4,740-4.
1-25 **8,** 5th from last cmpd. [89-01-0]; **A** P5,620-8.
1-26 **8,** 4784F, column 1, 2nd cmpd. [585-48-8].
1-27 **8,** 1422F, column 3, 1st cmpd. [19456-53-2].
1-28 **8,** 3192F, column 1, 10th cmpd. [28795-95-1] except *E* configuration.
1-29 **8,** 1600F, column 2, 20th cmpd. [497-38-1]; **A** N3,260-1.
1-30 **8,** 1599F, column 3, 18th cmpd. [17356-19-3]; **A** 13,086-9.
1-31 **8,** 6th from last cmpd. [2379-55-7]; **A** D18,497-7.
1-32 **8,** 4th cmpd. [3698-83-7].
1-33 **8,** 1640F, column 2, 1st cmpd. [583-60-8]; **A** M3,840-0.
1-34 **8,** 2178F, 1st cmpd., 2nd subheading [3236-48-4].
1-35 **8,** 18th cmpd. [576-83-0]; **A** B7,160-8.
1-36 **8,** 4th cmpd. [3030-47-5].
1-37 **7,** 1133F, column 2, 60th cmpd. (*E* configuration).
1-38 **8,** 6618F, column 2, 18th cmpd. from bottom [6263-83-8].
1-39 **8,** 3rd cmpd. [95-14-7]; **A** B1,140-0.
1-40 **8,** 7th cmpd. [632-56-4]; **A** 13,188-1.
1-41 From $CH_2{=}C(CH_2^-)_2$ and 2 moles of acetone.
1-42 **8,** 30th cmpd. [134-96-3]; **A** S760-2.
1-43 **8,** 19th from last cmpd. [89-83-8]; **A** 11,209-7.
1-44 **8,** 8th cmpd. [1503-49-7].
1-45 **8,** 2117F, 34th cmpd. [110-93-0]; **A** M4,880-5.
1-46 **8,** 3062F, column 1, 5th from last cmpd. [98-29-3]; **A** 12, 424-9.
1-47 **8,** 15th cmpd. [94-59-7]; **A** S20-8.
1-48 **8,** 3544F, column 1, 17th cmpd., 1st subheading [2052-14-4].
1-49 **8,** 1314F, column 3, 25th cmpd., 10th subheading [87-89-8]; **A** I-665-2.
1-50 **8,** 7792F, column 2, last cmpd. [744-45-6].
1-51 **8,** 2565F, column 2, 12th cmpd. from bottom [78-59-1]; **A** I-1,870-9.

1-52 **7,** 1097F, 27th cmpd.; **A** 14,937-3.
1-53 **8,** 2375F, 6th cmpd. [50-78-2]; **A** 13,292-6.
1-54 **8,** 5th cmpd. [52-90-4]; **A** 16,814-9.
1-55 **8,** 1724F, column 2, 15th cmpd. from bottom [60-31-1]; **A** 13,535-6.
1-56 **8,** [12078-28-3].
1-57 **8,** *Subject* Index, 29969S, column 2, 7th cmpd. from bottom [5332-35-4].
1-58 **8,** 4168F, column 2, 7th cmpd. from bottom [706-31-0]; **A** C9,748-2.
1-59 **8,** 3rd cmpd. [371-40-4]; **A** F380-0.
1-60 **8,** last cmpd. [12108-13-3].

2-1 **8,** last cmpd. [91-19-0]; **A** Q160-3.
2-2 **8,** 1st cmpd. [501-65-5]; **A** D20,480-3.
2-3 **8,** 3870F, column 1, 2nd cmpd. [92-69-3]; **A** 13,434-1.
2-4 **8,** 6557F, column 3, 4th cmpd. from bottom [538-58-9].
2-5 **8,** 1570F. 4th cmpd. (HCl salt) [2687-43-6].
2-6 **8,** 6229F, column 2, 12th cmpd. from bottom [2026-27-9].
2-7 **8,** 4529F, column 3, 29th cmpd. [15936-45-5].
2-8 **7,** 15th cmpd.
2-9 **8,** 924F, column 2, last cmpd. [147-85-3]; **A** 13,154-7.
2-10 **8,** 9369F, column 2, 7th cmpd. from bottom, 13th subheading [117-84-0].
2-11 **8,** 4th cmpd. [883-93-2].
2-12 **8,** 3929, column 2, 2nd from last cmpd. [534-85-0]; **A** P2,835-2.
2-13 **8,** 2800F, column 2, 1st cmpd. [83-56-7]; **A** D11,560-6.
2-14 **8,** 2800F, column 3, 1st cmpd. [582-17-2]; **A** D11,640-8.
2-15 **8,** 3140F, column 2, 4th cmpd. [5453-94-1].
2-16 **8,** 7th cmpd. [18014-00-1].
2-17 **8,** 2306F, column 1, last cmpd., 7th subheading [584-84-9]; **A** T3,985-3.
2-18 **8,** 2067F, column 3, 33rd cmpd. [6493-77-2]; **A** 10,531-7.
2-19 **8,** 5341F, column 1, 19th cmpd. [10355-50-7].
2-20 **8,** 3186F, column 1, 2nd from last cmpd. [470-82-6].
2-21 **8,** 3187F, column 3, 12th cmpd. [106-24-1]; **A** 16,333-3.
2-22 **8,** 3187F, column 1, 1st cmpd., 5th subheading [89-79-2]; **A** I-2,800-3.
2-23 **8,** 3188F, column 1, 2nd from last cmpd. [27779-29-9].

2-24 **8,** 3698F, column 1, 6th from last cmpd. [112-38-9]; **A** 12,467-2.
2-25 **8,** 5th cmpd. [51-45-6].
2-26 **8,** 3941F, column 3, 4th cmpd. [57-30-7].
2-27 **8,** 2532F, 10th cmpd. [55-31-2].
2-28 **8,** 2nd cmpd. [50-81-7]; **A** A9,290-2.
2-29 **8,** 2nd cmpd. [51-21-8]; **A** 85,847-1.
2-30 **8,** 6743F, column 2, 10th cmpd. [147-24-0].
2-31 **8,** 4th cmpd. [74-79-3]; **A** A9,240-6.
2-32 **8,** 3240F, column 3, 5th cmpd. [1490-04-6]; **A** M278-0.
2-33 **8,** 1970F, column 2, 12th cmpd. [694-91-7] **A** 12,984-4.
2-34 **7,** 874F, column 2, 5th cmpd.
2-35 **8,** 1358F, column 2, 1st cmpd. [106-69-4]; **A** T6,620-6.
2-36 **8,** 6th cmpd. [283-24-9]; **A** A9,380-1.
2-37 **8,** 1st cmpd. [700-58-3]; **A** 14,604-8.
2-38 Minor product from cyclopentadiene + H_3O^+.
2-39 **8,** 1st cmpd. [376-73-8].
2-40 **8,** 17th cmpd. [719-59-5]; **A** A4,556-4.
2-41 **8,** 3168F, column 3, 5th from last cmpd. [2436-90-0].
2-42 **8,** 3168F, column 3, 4th from last cmpd. [6874-43-7].
2-43 **8,** 3129F, column 2, 3rd cmpd. from bottom [76-22-2]; **A** 14,807-5.
2-44 **7,** 3536F, column 2, 4th cmpd.
2-45 **8,** 2639F, column 1, 4th from last cmpd. [15773-29-2].
2-46 **8,** 2nd from last cmpd. [58-96-8]; **A** U288-1.
2-47 **8,** 3rd cmpd. [65-46-3]; **A** C12,210-6.
2-48 **8,** 4940F, column 3, 9th cmpd. [3133-07-1].
2-49 From cyclohexene and perchlorocyclopentadiene at 100° C.
2-50 **8,** 3103F, column 2, last cmpd. [79-92-5]; **A** C30-1.
2-51 **8,** 3105F, column 3, last cmpd. [127-91-3]; **A** 11,208-9.
2-52 **8,** 3606F, column 3, 24th cmpd. [877-06-5].
2-53 From pyrocatechol and 2 moles of dihydropyran; sharply melting, but stereochemistry unknown.
2-54 **8,** 21st cmpd. [137-00-8]; **A** 19,067-5.
2-55 **8,** 3rd cmpd. [7365-45-9]; **A** 16,371-6.

2-56 **8,** 3127F, column 3, 19th cmpd. [987-65-5]; **A** 2,620-9.
2-57 **8,** 9th cmpd. [58-08-2]; **A** C5-3.
2-58 **8,** 8424F, column 2, 10th cmpd. from bottom [554-92-7].
2-59 **8,** 5869F, column 2, 10th cmpd. from bottom [136-47-0].
2-60 **8,** 7626F, column 1, 18th cmpd. from bottom [1624-62-0].
2-61 **8,** 2nd cmpd. [22576-44-9].
2-62 From $(CH_2{=}CCH_2^-)_2$ and 2 moles of benzonitrile; contains no 4-membered rings.
2-63 **8,** 5th cmpd. [76-26-6]; **A** C210-7.
2-64 **8,** 4163F, column 2, 18th cmpd. [67-03-8]; **A** 10,917-7.
2-65 **7,** 2336F, column 2, 46th cmpd.

ALIPHATICS

$Q{-}\underset{q}{C}H_3$ $Q{-}\underset{t}{C}H_2{-}n{-}Bu$ $Q{-}\underset{d}{C}H{-}n{-}Bu_2$ $Q{-}\underset{s}{C}{-}n{-}Bu_3$

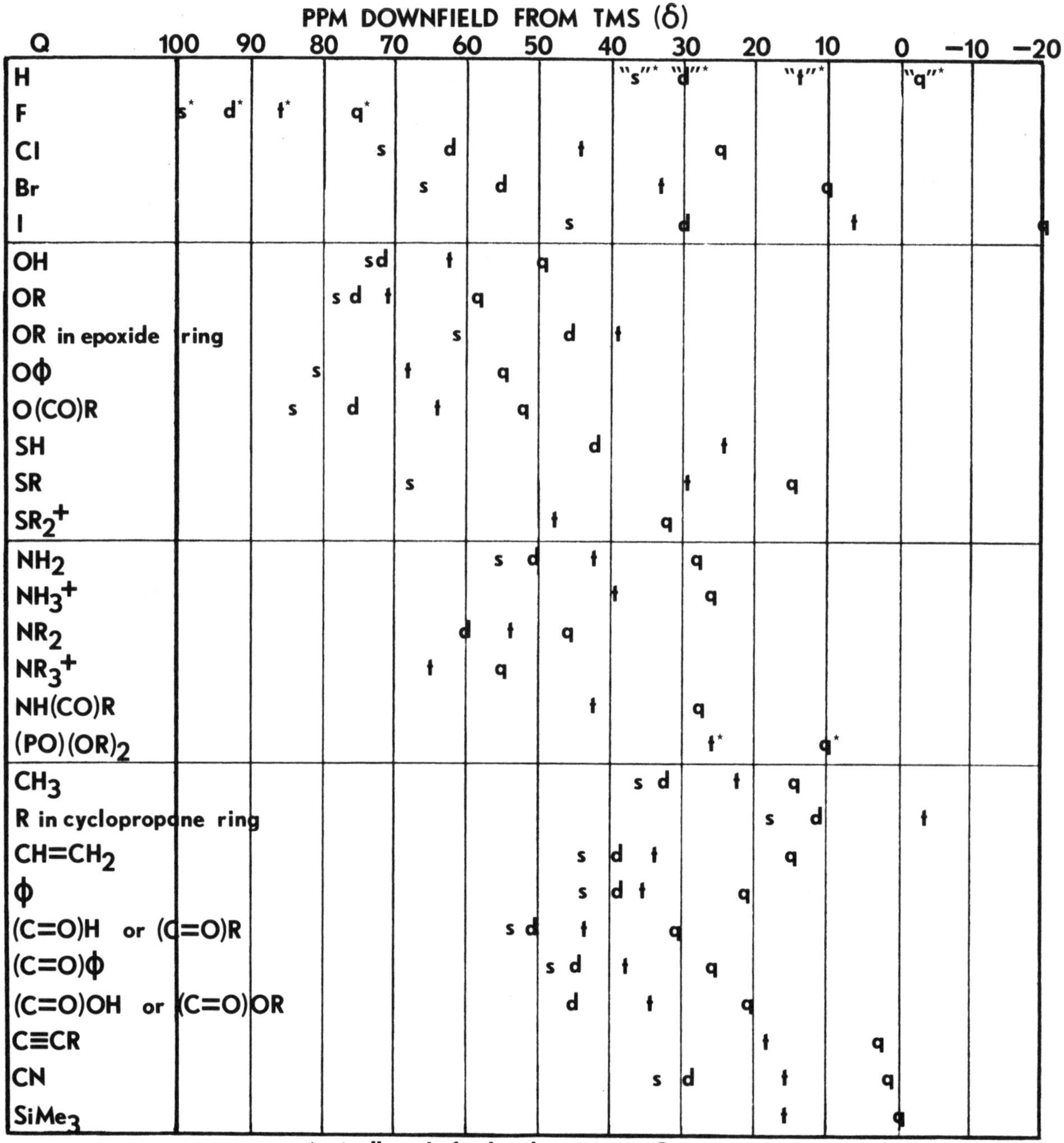

*actually split further by atom in Q